Learn SOLIDWORKS 2025

Gain the skills and knowledge you need to become
a certified SOLIDWORKS Associate or Professional

Tayseer Almattar

Learn SOLIDWORKS 2025

Group Product Manager: Rohit Rajkumar

Publishing Product Manager: Bhavya Rao

Book Project Manager: Srinidhi Ram

Senior Content Development Editor: Feza Shaikh

Senior Editor: Anuradha Joglekar

Technical Editor: Simran Ali

Copy Editor: Safis Editing

Proofreader: Feza Shaikh and Anuradha Joglekar

Indexer: Tejal Soni

Production Designer: Gokul Raj S.T

DevRel Marketing Coordinator: Nivedita Pandey

First published: January 2020

Second edition: January 2022

Third edition: January 2025

Production reference: 1141125

Published by Packt Publishing Ltd.

Grosvenor House

11 St Paul's Square

Birmingham

B3 1RB, UK

ISBN 978-1-83546-308-6

www.packtpub.com

To my parents and siblings for their unbounded love, support, and care.

– Tayseer Almattar

Contributors

About the author

Tayseer Almattar began his career in engineering before transitioning to design. He holds a **Master of Design** (**MDes**) degree in international design and business management and is a strong advocate for the role of design in fostering sustainable business innovation. As the founder of TforDesign, Tayseer has dedicated himself to advancing design education and practice.

With nearly two decades of experience using SOLIDWORKS for 3D design, Tayseer has created numerous online learning programs that have attracted thousands of learners worldwide. In this book, he combines his extensive design and training expertise to offer a unique and practical SOLIDWORKS learning experience. Tayseer is also a SOLIDWORKS Champion, and the founding leader of the Hong Kong SOLIDWORKS user group.

Beyond SOLIDWORKS, Tayseer Almattar works as a design and innovation consultant. He leads workshops on various design innovation topics, including design thinking and systems thinking. Additionally, he has served as a visiting lecturer at the Hong Kong Polytechnic School of Design and has conducted design innovation workshops across Asia and the Middle East.

I want to thank the people who made this book possible, the amazing reviewers, and the Packt team.

About the reviewers

Ritish Hegde is a certified SOLIDWORKS professional and has been a part of the global SOLIDWORKS Champions Program since 2022. He also holds a bachelor's degree in mechanical engineering. He currently works as a mechanical design engineer, with a seasoned amount of experience and expertise in the field of CAD, he has performed a large number of projects across multiple software, including SOLIDWORKS, Catia, NX, Inventor, and Fusion 360. He is passionate about transforming ideas into reality, with a background in engineering design and 3D modeling, he thrives on tackling complex projects and bringing them to life through advanced software tools. He is constantly keen to learn new techniques and keeps up to date on industry trends.

I'd like to thank my family for always being constant pillars of support.

Iqra Bibi has a profound passion for the intricate world of engineering. Iqra received her bachelor's degree in aerospace engineering from the Hong Kong University of Science and Technology. She has experience in various engineering disciplines from **Computer Aided Design (CAD)** to **Computational Fluid Dynamics (CFD)**. Iqra has also actively participated in the research and development of ocular medical devices during her undergraduate years, and she aspires to be part of engineering projects that drive substantial and impactful change.

Table of Contents

Part 3: Basic Mechanical Core Features – Associate Level   103

5

Basic Primary One-Sketch Features   105

6

Basic Secondary Multi-Sketch Features   157

Part 4: Basic Evaluation and Assemblies – Associate Level   203

7

Materials and Mass Properties   205

8

Standard Assembly Mates   225

Part 5: 2D Engineering Drawings Foundation 261

9

Project 1

10

11

Bills of Materials   347

Part 6: Advanced Mechanical Core Features – Professional Level   379

12

Advanced SOLIDWORKS Mechanical Core Features   381

13

Equations, Configurations, and Design Tables   441

Part 7: Advanced Assemblies – Professional Level   473

14

SOLIDWORKS Assemblies Advanced Mates   475

17

Preface

SOLIDWORKS stands as a leading piece of software in the realm of 3D engineering and product design, widely utilized across industries such as aviation, automotive, consumer products, and more. This book adopts a practical methodology to help you master SOLIDWORKS at a professional level. Beginning with the fundamentals, such as navigating the software interface and opening new files, it gradually advances through various topics. You'll learn everything from sketching and constructing intricate 3D models to creating both dynamic and static assemblies.

Our approach is hands-on, ensuring that each tool in SOLIDWORKS is introduced through practical exercises. These exercises will guide you in creating sketches, 3D part models, assemblies, and drawings. To support your learning, we provide downloadable files that you can use to follow along with the concepts and exercises at your own pace. Additionally, the book features two comprehensive projects that integrate the different sections, offering practical applications of the skills you've acquired.

Whether you're a complete beginner or looking to deepen your expertise, this book is structured to be followed sequentially, like a story, but also allows for flexibility to jump between chapters as needed.

Who this book is for

This book is aimed at individuals eager to begin their journey with SOLIDWORKS and gain confidence in using the software. It caters to aspiring engineers, designers, makers, drafting technicians, and hobbyists alike. Additionally, it is tailored for those aspiring to become **Certified SOLIDWORKS Associates (CSWAs)** or **Certified SOLIDWORKS Professionals (CSWPs)**.

No specific background is required to follow this book, as it starts with the basics of what SOLIDWORKS is and how to use it. However, having a basic theoretical understanding of 3D modeling would be beneficial.

What this book covers

Chapter 1, Introduction to SOLIDWORKS, covers what SOLIDWORKS is and the applications that utilize the software. It also explores the professional certifications that are offered by SOLIDWORKS.

Chapter 2, Interface and Navigation, explores how to navigate the SOLIDWORKS interface.

Chapter 3, SOLIDWORKS 2D Sketching Basics, covers what sketching is in SOLIDWORKS. It also covers how you can sketch basic entities such as lines, circles, rectangles, arcs, and ellipses.

Chapter 4, Special Sketching Commands, covers commands that enable us to sketch more efficiently. These include the fill, mirror, offset, trip, and pattern commands.

Chapter 5, Basic Primary One-Sketch Features, explores the most basic features used for generating 3D models from sketches. Each of these features requires you to have one sketch to apply it. The features include extruded boss and cut, revolved boss and cut, fillets, and chamfers.

Chapter 6, Basic Secondary Multi-Sketch Features, explores another set of basic features that require more than one sketch to apply. They include the swept boss and swept cut and the lofted boss and lofted cut. It also explores reference geometries and how to generate new planes.

Chapter 7, Materials and Mass Properties, explores structural materials for our 3D parts. It also teaches you how to calculate mass properties, such as mass, volume, and the center of gravity.

Chapter 8, Standard Assembly Mates, explores what assemblies are in SOLIDWORKS. You will learn how to generate simple assemblies using the standard mates: coincident, parallel, perpendicular, tangent, concentric, lock and set distance, and angle.

Chapter 9, Introduction to Engineering Drawing, explores what engineering drawings are and how to interpret them as per the commonly recognized international standards.

Project 1, 3D *Modeling a Pair of Glasses*, presents a comprehensive practical exercise linking the topics in *Chapters 2* to *9* to 3D model a pair of glasses.

Chapter 10, Basic SOLIDWORKS Drawing Layout and Annotations, explores how to generate basic engineering drawings using SOLIDWORKS drawing tools.

Chapter 11, Bills of Materials, explores what bills of materials are and how to generate and adjust bills of materials with SOLIDWORKS drawing tools.

Chapter 12, Advanced SOLIDWORKS Mechanical Core Features, explores the advanced features used to generate more complex 3D models. These include the draft feature, shell feature, Hole Wizard, features mirroring, rib feature, and multi-body parts.

Chapter 13, Equations, Configurations, and Design Tables, explains how you can apply equations to link different dimensions within the model. You will also learn how to utilize configurations and design tables to generate multiple variations of a single part within one SOLIDWORKS file.

Chapter 14, SOLIDWORKS Assemblies and Advanced Mates, explores using advanced mates to generate more dynamic assemblies. These include the profile center, symmetric, width, distance and angle range, path, and linear/linear coupler mates.

Chapter 15, Advanced SOLIDWORKS Assembly Competencies, explores additional assembly features to better evaluate and generate more sound and flexible assemblies. These include the Interference and Collision Detection tools, assembly features, configurations, and design tables for assemblies.

Project 2, 3D Modeling an RC Helicopter Model, presents a comprehensive practical exercise covering topics from across the book to 3D model a remote-control helicopter.

Chapter 16, Introduction to SOLIDWORKS Cloud Services, explores SOLIDWORKS cloud services, enhancing design sharing, change management, and design files storage through seamless integration with 3DEXPERIENCE.

To get the most out of this book

You will need access to the SOLIDWORKS software for most chapters. Some chapters will also require Microsoft Excel on the same machine.

To get the most out of this book, you should follow all the steps and examples in your version of SOLIDWORKS as you go through them. This book is designed to provide hands-on practical experience.

There are no prerequisite knowledge or skills required to follow this book. However, having a basic theoretical understanding of 3D modeling and interpreting engineering drawings would be helpful.

Download the project files

You can download the project files for this book from GitHub at `https://github.com/PacktPublishing/Learn-SOLIDWORKS-2025-Third-Edition` If there's an update to the code, it will be updated in the GitHub repository.

We also have other code bundles from our rich catalog of books and videos available at `https://github.com/PacktPublishing/`. Check them out!

Code in Action

The Code in Action videos for this book can be viewed at `https://packt.link/ci6I9`

Download the color images

We also provide a PDF file that has color images of the screenshots and diagrams used in this book. You can download it from `https://packt.link/8jkQD`

Conventions used

There are a number of text conventions used throughout this book.

`Code in text`: Indicates code words in text, database table names, folder names, filenames, file extensions, pathnames, dummy URLs, user input, and Twitter handles. Here is an example: "Now, we can use the distance range advanced mate to limit the movement of the drawer support from `0.00mm` to `30.00mm`."

Bold: Indicates a new term, an important word, or words that you see onscreen. For instance, words in menus or dialog boxes appear in **bold**. Here is an example: "Navigate to the **Sketching** mode using the **Top Plane**."

> **Tips or important notes**
> Appear like this.

Get in touch

Feedback from our readers is always welcome.

General feedback: If you have questions about any aspect of this book, email us at `customercare@packtpub.com` and mention the book title in the subject of your message.

Errata: Although we have taken every care to ensure the accuracy of our content, mistakes do happen. If you have found a mistake in this book, we would be grateful if you would report this to us. Please visit `www.packtpub.com/support/errata` and fill in the form.

Piracy: If you come across any illegal copies of our works in any form on the internet, we would be grateful if you would provide us with the location address or website name. Please contact us at `copyright@packt.com` with a link to the material.

If you are interested in becoming an author: If there is a topic that you have expertise in and you are interested in either writing or contributing to a book, please visit `authors.packtpub.com`.

Share Your Thoughts

Once you've read *Learn SOLIDWORKS 2025*, we'd love to hear your thoughts! Scan the QR code below to go straight to the Amazon review page for this book and share your feedback.

https://packt.link/r/1-835-46308-8

Your review is important to us and the tech community and will help us make sure we're delivering excellent quality content.

Free Benefits with Your Book

This book comes with free benefits to support your learning. Activate them now for instant access (see the "*How to Unlock*" section for instructions).

Here's a quick overview of what you can instantly unlock with your purchase:

PDF and ePub Copies

Next-Gen Web-Based Reader

Access a DRM-free PDF copy of this book to read anywhere, on any device.

Use a DRM-free ePub version with your favorite e-reader.

Multi-device progress sync: Pick up where you left off, on any device.

Highlighting and notetaking: Capture ideas and turn reading into lasting knowledge.

Bookmarking: Save and revisit key sections whenever you need them.

Dark mode: Reduce eye strain by switching to dark or sepia themes

How to Unlock

Scan the QR code (or go to `packtpub.com/unlock`). Search for this book by name, confirm the edition, and then follow the steps on the page.

UNLOCK NOW

Note: Keep your invoice handy. Purchases made directly from Packt don't require an invoice.

Part 1:
Getting Started

This part provides an essential introduction to SOLIDWORKS and its applications, including the professional certifications available. It covers the basics of navigating the SOLIDWORKS interface, starting new files, and understanding the main components. Additionally, it covers an introduction to parametric modeling and how to adjust the measurement system for open documents, setting a solid foundation for your journey with SOLIDWORKS.

This part has the following chapters:

- *Chapter 1, Introduction to SOLIDWORKS*

- *Chapter 2, Interface and Navigation*

1

Introduction to SOLIDWORKS

SOLIDWORKS is a **three-dimensional (3D) computer-aided design (CAD)** software that runs on Windows computer systems. It was launched in 1995 and has grown to be one of the most common pieces of software used globally regarding engineering design. Currently, SOLIDWORKS is owned by Dassault Système.

This book covers the fundamental skills for using SOLIDWORKS. It will take you from knowing nothing about the software to acquiring all the basic skills expected of a **Certified SOLIDWORKS Professional (CSWP)**. En route, we will also cover all the skills needed for the more basic **Certified SOLIDWORKS Associate (CSWA)** level. In addition to knowing what the tools are, you will also need to develop software fluency, which you will gain gradually as you practice using the software for different applications. Both the tools and fluency are essential to acquiring any official SOLIDWORKS certifications. If you are new to SOLIDWORKS, we recommend that you follow the book like a story, from this chapter onward. If you are already familiar with SOLIDWORKS, feel free to jump between chapters.

This chapter will provide you with a brief introduction to what SOLIDWORKS is and the fields it can support. We will learn about all the features and capabilities of SOLIDWORKS and will gain a clearer idea of what types of certifications or fields you can strive for. Learning about applicable certifications will enable you to plan your personal SOLIDWORKS development.

The chapter will also explain the governing principle with which SOLIDWORKS functions: parametric modeling. Equipped with a knowledge of SOLIDWORKS' operating principles, we will be able to deepen our understanding of how the software works and what to expect from it. Understanding the software's operating principles will help us to manage the different software commands that are used when building 3D models.

The following topics will be covered in this chapter:

- Introducing SOLIDWORKS
- Exploring the 3DEXPERIENCE platform
- Understanding parametric modeling
- Exploring SOLIDWORKS certifications

For this chapter, there is no Code in Action (CiA) video available.

> **Free Benefits with Your Book**
>
> Your purchase includes a free PDF copy of this book along with other exclusive benefits. Check the *Free Benefits with Your Book* section in the Preface to unlock them instantly and maximize your learning experience.

Introducing SOLIDWORKS

SOLIDWORKS is a 3D design software that's officially capitalized to SOLIDWORKS. It is one of the leading pieces of engineering 3D design software globally. Today, more than 2 million organizations use SOLIDWORKS to develop products and innovations, which represent a large proportion of over 6 million SOLIDWORKS users in total. In this section, we will explore the different applications that SOLIDWORKS supports.

SOLIDWORKS applications

SOLIDWORKS mainly targets engineers and product designers. It is used in a variety of applications and industries. Some of these industries are as follows:

- Consumer products
- Aerospace
- Construction
- High-tech electronics
- Medicine
- Oil and gas
- Packaging
- Machinery
- Engineering services
- Furniture design

- Energy

- Automobiles

Each of these industries utilizes SOLIDWORKS for its design applications to some extent. Within SOLIDWORKS, several disciplines correspond to different design and analysis approaches. They are as follows:

- **Core mechanical design**: This involves the creation of parametric 3D models for parts and assemblies in various machines and systems

- **Two-dimensional (2D) drawings**: This encompasses the generation of detailed and accurate engineering schematics necessary for manufacturing, inspection, and documentation

- **Surface design**: This utilizes advanced surfacing techniques to create complex and visually appealing designs, particularly vital in industries such as automotive and consumer products

- **Sheet metal**: This covers the design and manufacturing process for sheet metal parts and assemblies, incorporating features such as bends, flanges, and cuts

- **Sustainability**: This uses integrated tools to ensure environmentally friendly design practices, considering factors such as material selection and energy usage to conduct **life cycle assessments (LCAs)**

- **Motion analysis**: This employs kinematics and dynamics to provide insights into the performance of mechanical systems, focusing on elements such as movement and force

- **Weldments**: This facilitates the design of welded structures, the production of cut lists, and the creation of accurate weld bead documentation

- **Simulations**: This involves conducting stress, thermal, and vibration analyses to aid engineers in validating and optimizing their designs for improved performance and durability

- **Mold making**: This entails the use of precise tools to streamline the creation of injection molded and cast parts, which covers aspects such as mold base assemblies

- **Electrical**: This integrates electrical design approaches to better control aspects such as wire routing and component placement, easing collaboration between mechanical and electrical fields

Even though the preceding list highlights some possible domains where SOLIDWORKS can be applied, it is not necessary for a single individual to master them all. However, they do demonstrate the capabilities enabled by the software and the fields it can serve. This book will focus on addressing applications within the core mechanical design disciplines.

Core mechanical design

Core mechanical design skills are the basic tools every SOLIDWORKS user needs. These skills include the main 3D modeling features used to create mechanical parts. This book will help you learn these important skills, allowing you to design complex parts and assemblies. With these skills, you can create things like engines, furniture, and everyday items like phones and laptops

We will cover all the knowledge and skills needed to achieve the two major SOLIDWORKS certifications, CSWA and CSWP, under the core mechanical design discipline. Also, mastering core mechanical design concepts can be considered a prerequisite to learning most other specialized modeling disciplines, such as sheet metal and mold making. Because of that, we will only cover a common foundation for core mechanical design in this book. We will discuss all the certifications and levels in more detail in the *Exploring SOLIDWORKS certifications* section.

Sample SOLIDWORKS 3D models

As SOLIDWORKS caters to a variety of fields, it is possible to create 3D models with varying complexity using the software. *Figure 1.1* and *Figure 1.2* show samples of 3D models from different fields that have been made using SOLIDWORKS:

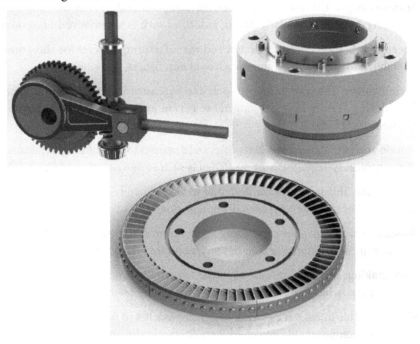

Figure 1.1 – 3D models of a gears assembly, turbine rotor, and mechanical seal

Figure 1.2 – 3D models of Gallon, bookshelf design, and headset design

These models are selections from different fields that can show flexibility and the range of possible applications. In reality, SOLIDWORKS is a tool, and it will remain up to you as to what you will use it for. At this point, we have an understanding of SOLIDWORKS as a leading 3D design software used by various industries, with varied applications such as mechanical design and simulations. Next, we will explore what the 3DEXPERIENCE platform is and how it relates to SOLIDWORKS.

Exploring the 3DEXPERIENCE platform

If you are working with SOLIDWORKS, you will most likely also come across or engage in conversations about the 3DEXPERIENCE platform. In case you have not heard about it yet, here is a small introduction.

The **3DEXPERIENCE** platform is developed by **Dassault Systèmes**, SOLIDWORKS's parent company. You can think of the 3DEXPERIENCE platform as a full operating system tailored for product design and development. This entails a unified digital workspace that includes a selection of integrated apps designed for tasks such as CAD modeling, simulation, project management, and data analytics. The intention of the platform is to alter the methodologies of designing, testing, and manufacturing products.

A user's access to these apps and the rules that govern them depends on the individual user's role. The platform aims to house these various apps in one system to enhance workflow. It strives to eliminate isolated work, promote collaboration among global teams, and stimulate business innovation.

Properties of the 3DEXPERIENCE platform

The platform aims to enhance digital workflows through aspects such as communication improvements, faster design cycles, and predictive analytics capabilities.

Some of the capabilities in offers include the following:

- **Global connection**: Real-time collaboration with peers regardless of their geographical location is made possible.

- **All-in-one access**: The platform houses various design, engineering, and manufacturing applications together with cloud tools in a single place to ease design tracking and collaboration.

- **Efficient monitoring**: Management of projects and data is facilitated through dashboards and reporting features.

- **Informed decision-making**: Analytics and simulation capabilities allow scenario modeling to guide strategic decision-making from a position of informed prediction.

Why should you know about the 3DEXPERIENCE platform?

In this book, our focus will be on the SOLIDWORKS desktop app to build your capabilities in building robust 3D models. However, the SOLIDWORKS desktop app can also connect with the 3DEXPERIENCE platform, which can be used more excessively by other members of your team for project management, simulations, 3D modeling, or other functions.

At this point, all you should know is that the 3DEXPERIENCE platform exists and that it would likely impact you in the future as a SOLIDWORKS user.

We will include a link to learn more about the 3DEXPERIENCE platform in the *Further reading* section at the end of this chapter.

Let us summarize what we have learned about the 3DEXPERIENCE platform. It is a product of Dassault Systèmes, which serves as a comprehensive operating system for product design and development. Aiming to enhance digital workflows, the platform promotes real-time collaboration, centralized access to tools, efficient project management, and data-driven decision-making. Next, we will learn the principle under which parametric 3D software such as SOLIDWORKS operates: parametric modeling.

Understanding parametric modeling

Parametric modeling is the core principle that SOLIDWORKS operates on. It governs how SOLIDWORKS constructs 3D models and how a user should think when dealing with SOLIDWORKS.

In parametric modeling, the model is created based on relationships and a set of logical arrangements that are set by the designer or draftsman. In the SOLIDWORKS software environment, they are represented by dimensions, geometric relationships, and features that link different parts of a model to each other. Each of these logical features is called a **parameter**. A collection of parameters in a 3D model forms our **design intent**, which guides how a 3D model behaves when subjected to modifications. We will address design intent more as we progress through the book.

For example, a simple cube with a side length of 1 mm would contain the following parameters:

- **Four lines in one plane** with the following relationships listed and noted in *Figure 1.3*:

  - All two-line endpoints are merged at the same point. This is presented with the **merged** parameter in *Figure 1.3*.

  - Two opposite angles are right angles (90 degrees).

  - Two adjacent lines are equal to each other in length.

  - The length of one line is **1 mm**, as follows:

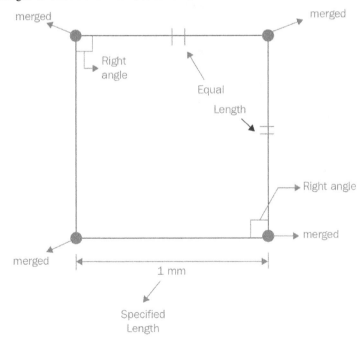

Figure 1.3 – Four lines in one plane

- A **vertical extrusion** that is perpendicular to the square defined in the first set of parameters. This extrusion is by an amount equal to the length of the square's side (1 mm). This vertical extrusion will result in the shape shown in *Figure 1.4*:

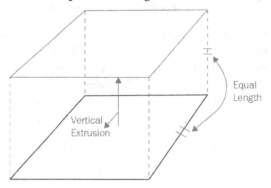

Figure 1.4 – Extruding four base lines upward to make a cube

The parameters shown in *Figure 1.3* and *Figure 1.4* show how software such as SOLIDWORKS interprets and constructs 3D models. The user of the software should specify all those parameters to create a cube or any other 3D model.

Creating 3D models based on parameters/design settings has many notable advantages. One major advantage is the ease of applying design updates. Let's go back to our cube to see how this works.

Notice that up to the point we reached the cube shown in Figure 1.4 we have only specified the length of one side of the base square, as shown in *Figure 1.3*. The other specifications are relationships that ensure the model is a cube, with equal, parallel, and perpendicular sides.. Those parameters make all the parts of our cube interconnected based on what we decide is important. Thus, updating the length of the side of the cube will not sabotage the cube's structure. Rather, the whole cube will be updated while keeping the parameters intact. In other words, our design intent was to build a cube.

To clarify this, we can revisit the cube we just made to update it. In the same model, let's change the dimension we identified earlier from 1 mm to 5 mm, as highlighted in *Figure 1.5*:

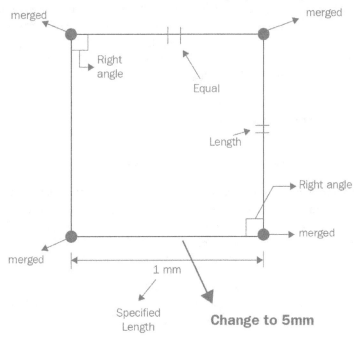

Figure 1.5 – Adjusting the elements in a parametric design propagated to the different parts

With that single step, the cube is fully modified, with all the sides changing to 5 mm in length. Again, this is because our cube parameters must have equal perpendicular and parallel sides. Given that we have defined our intended parameters/design settings for the software, all of those will be retained, resulting in the whole cube model being updated with one single adjustment.

This can be contrasted with pure direct modeling methods. In pure direct modeling, the user creates the cube more abstractly by drawing each line separately and constructing a cube of a certain size. Even though creating the initial cube might be faster, updating it would require updating all the elements separately as they don't relate to each other with any intent or logical features. This would result in considerably more time and effort being invested in creating variations, which is an essential requirement for industrial application.

Other advantages of parametric modeling are as follows:

- The ease of modifying and adjusting models throughout the design and production cycles.
- The ease of creating families of parts that have similar parameters.
- The ease of communicating the design to manufacturing establishments for manufacturing.

All the advantages of parametric modeling make it a popular modeling method for technical applications relating to engineering or product design. On the other hand, direct modeling can perform better in more abstract applications, such as modeling more artistic objects used in gaming or architecture. Understanding parametric modeling will enable us to use the software more easily as we are aware of its limitations, as well as how the software interprets the commands we apply. As we progress through this book, we will expand our understanding of parametric modeling as we tackle more advanced functions, such as design tables and other features.

Now that we know more about SOLIDWORKS and parametric modeling, we will discuss the certifications offered by SOLIDWORKS.

Exploring SOLIDWORKS certifications

SOLIDWORKS provides certifications that cover different aspects of its functionality. As a user, you don't need to gain any of those certifications to use the software; however, they can prove your SOLIDWORKS skills. SOLIDWORKS certifications are a good way of showing employers or clients that you have mastery over a certain aspect of the software that would be required for a specific project.

Certifications can be classified under four levels: **associate**, **professional**, **professional advanced**, and **expert**. Associate certifications represent the entry level, expert certifications represent the highest level, and professional and professional advanced represent the middle levels. The following subsections list the certification levels provided by SOLIDWORKS. Note that SOLIDWORKS may add or remove certifications over time.

You can check the SOLIDWORKS certification program for more information. The link to the program is provided in the *Further reading* section.

Associate certifications

Associate certifications are the most basic ones offered by SOLIDWORKS. Some of those certifications require hands-on testing, while others require the student to have theoretical knowledge related to the certification topic. Brief details pertaining to each certification are as follows:

- **CSWA**: This is the most popular SOLIDWORKS certification. It covers the basic modeling principles involved in using the software. This certification allows the user to prove their familiarity with the basic 3D modeling environment in SOLIDWORKS. It touches on creating parts, assemblies, and drawings. The test for this certification is hands-on, so the student will need to have SOLIDWORKS installed before attempting the test.

- **Certified SOLIDWORKS Associate – Electrical (CSWA-E)**: This covers the general basics of electrical theory, as well as aspects of the electrical functionality of SOLIDWORKS. This certification test does not involve practical work, so the student will not need to have SOLIDWORKS installed.

- **Certified SOLIDWORKS Associate – Sustainability (CSWA-Sustainability)**: This covers theoretical principles of sustainable product design. To take this certification, SOLIDWORKS software is not required.

- **Certified SOLIDWORKS Associate – Simulation (CSWA-Simulation)**: This covers basic simulation principles based on the **finite elements method (FEM)**. This mainly includes stress analysis and the effect of different materials and forces on solid bodies. This is a hands-on test, so the student is required to have SOLIDWORKS installed.

- **Certified SOLIDWORKS Associate – Additive Manufacturing (CSWA-AM)**: This is one of the newer certifications offered by SOLIDWORKS, due to the emergence of the common use of additive manufacturing methods such as 3D printing. This certification covers basic knowledge regarding the 3D printing market. This is not a hands-on test, so the student does not need to have the SOLIDWORKS software installed.

Professional certifications

Professional certifications demonstrate a higher mastery of SOLIDWORKS functions beyond the basic knowledge of a certified associate. All the certifications in this category involve hands-on demonstrations. Thus, the student is required to have access to SOLIDWORKS before attempting any of the tests. Brief details regarding each certification are as follows:

- **CSWP**: This level is a direct sequence of the CSWA level. It demonstrates the user's mastery over advanced SOLIDWORKS 3D modeling functions. This level upgrade focuses more on modeling more complex parts and assemblies.

- **Certified SOLIDWORKS Professional – Model-Based Definition (CSWP-MBD)**: MBD is one of the newer SOLIDWORKS functionalities. This certification demonstrates the user's mastery of MBD functions, which enable the communication of models in a 3D environment rather than in a 2D drawing.

- **Certified PDM Professional Administrator (CPPA)**: PDM stands for **product data management**. This certification focuses on managing projects with a wide variety of files and configurations. Also, it facilitates collaboration in teams working on the same design project.

- **Certified SOLIDWORKS Professional – Simulation (CSWP-Simulation)**: This is an advanced sequence of the CSWA-Simulation certificate. It demonstrates a more advanced mastery of the simulation tools provided by SOLIDWORKS, as well as the ability to evaluate and interpret more diverse simulation scenarios.

- **Certified SOLIDWORKS Professional – Flow Simulation (CSWP-Flow)**: This is another advanced sequence of the CSWA-Simulation certificate. However, it focuses on the ability to set up and run different fluid flow simulation scenarios.

- **Certified SOLIDWORKS Professional API (CSWP-API): API** stands for **application programming interface**. This certificate addresses the user's skill in programming and automating functions within the SOLIDWORKS software.

- **Certified SOLIDWORKS Professional CAM (CSWP-CAM): CAM** stands for **computer-aided manufacturing**. SOLIDWORKS provides a suite of CAM tools that can facilitate the manufacturing of parts by enabling the user to simulate and plan different manufacturing processes. The CSWP-CAM certificate assesses your ability to use those tools in SOLIDWORKS.

Professional advanced certifications

Professional advanced certifications address very specific functions within SOLIDWORKS. Often, these certifications apply to more specific industries compared to the CSWP certificate. All these certificates are advanced specializations of the CSWP certificate.

The advanced certificates offered by SOLIDWORKS are as follows:

- **Certified SOLIDWORKS Professional Advanced – Sheet Metal (CSWPA-SM):** This focuses on applications related to sheet metal. This covers the design, analysis, and manufacturing aspects of sheet metal components.

- **Certified SOLIDWORKS Professional Advanced – Weldments (CSWPA-WD):** This focuses on applications related to welding. This includes welding both sheet metals and different formations such as frames.

- **Certified SOLIDWORKS Professional Advanced – Surfacing (CSWPA-SU):** This focuses on modeling surfaces of irregular shapes, such as car bodies and computer mice.

- **Certified SOLIDWORKS Professional Advanced – Mold Making (CSWPA- MM):** This focuses on making molds for production. This includes molds for both metal and plastic parts.

- **Certified SOLIDWORKS Professional Advanced – Advanced Drawing Tools (CSWPA-DT):** This focuses more on generating 2D engineering drawings to help communicate models to different parties. These can include internal quality teams or external manufacturers.

Expert certifications

Expert certifications are the highest level of certification offered by SOLIDWORKS. Obtaining an expert certificate indicates your mastery of a large array of functions in the software. Also, expert certificates are the only ones with required prerequisites. There are two offered expert certificates, as follows:

- **Certified SOLIDWORKS Expert (CSWE):** This demonstrates mastery over all SOLIDWORKS modeling and design functions. To qualify for this exam, the user must have the CSWP certificate, in addition to four CSWPA certificates.

- **Certified SOLIDWORKS Expert in Simulation (CSWE-S)**: This demonstrates mastery over all the areas of the SOLIDWORKS Simulation software. To qualify for this exam, the user must have the CSWP, CSW-Simulation, and CSWP-Simulation certificates.

A SOLIDWORKS user doesn't need to obtain all these certifications. It is rare to find one person with all these certificates. This is because each certification level can address very different needs and serve different industries and/or positions. Also, some certification levels are more in demand than others as they are more essential and, hence, used in more industries. Sequentially, the certifications can be viewed as in *Figure 1.6*:

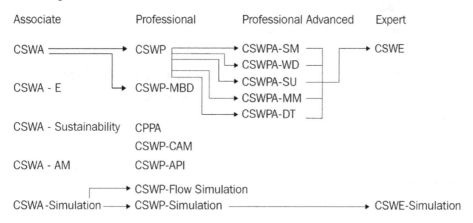

Figure 1.6 – A map of the different SOLIDWORKS certifications

All the certifications mentioned in *Figure 1.6* and the others that we have discussed in this section are related to the SOLIDWORKS desktop application, which is the focus of this book. Other than those, additional certifications are available concerning the 3DEXPERIENCE platform as well. You can learn more about those using the links in the *Further reading* section.

This book covers the two most essential, sequential certification levels: CSWA and CSWP. These two certifications cover the common usage scenarios within SOLIDWORKS.

Summary

In this chapter, we learned what SOLIDWORKS is, how parametric modeling works, and the different certifications offered by SOLIDWORKS. This will help us set our expectations and create our future development roadmap concerning SOLIDWORKS. It will also help us to understand the capabilities of the software and its vast scope. We also learned about the 3DEXPERIENCE platform, a comprehensive suite from Dassault Systèmes that enhances collaboration and streamlines workflow within the product development cycle, which is an integral part of the modern SOLIDWORKS ecosystem. In the next chapter, we will cover the SOLIDWORKS interface and its navigation. This will enable us to navigate the software and identify the different components that exist in its interface.

Questions

Answer the following questions to test your knowledge of this chapter:

1. What is SOLIDWORKS?

2. Name some industries that utilize SOLIDWORKS.

3. What is the 3DEXPERIENCE platform and how does it complement SOLIDWORKS?

4. Describe at least two benefits of using the 3DEXPERIENCE platform for a team working on a product design project.

5. How is parametric modeling defined?

6. What are the major advantages of parametric modeling?

7. What is the difference between parametric modeling and direct modeling?

8. What are the SOLIDWORKS certifications and why are they important?

9. What are the main categories of certification levels offered by SOLIDWORKS?

> **Important note**
> The answers to the preceding questions can be found at the end of this book.

Further reading

More information about the 3DEXPERIENCE platform can be found here: `https://www.3ds.com/3dexperience`

More information about the certifications offered by SOLIDWORKS can be found here: `https://www.solidworks.com/solidworks-certification-program`

Get This Book's PDF Version and Exclusive Extras

UNLOCK NOW

Scan the QR code (or go to `packtpub.com/unlock`). Search for this book by name, confirm the edition, and then follow the steps on the page.

Note: Keep your invoice handy. Purchases made directly from Packt don't require an invoice.

Interface and Navigation

In this chapter, we will look at SOLIDWORKS and its software interface, as well as its main components. In addition, we will cover how to navigate through the software interface so that you will be able to easily find your way around the software in the upcoming chapters. We will also talk about the document's measurement system in terms of the different standard units it uses globally, such as feet, inches, centimeters, and millimeters for measurements of length. Interacting and setting up an interface with the software and setting up our measurement system will be the first two actions we will perform in any new project.

The following topics will be covered in this chapter:

- Starting a new part, assembly, or drawing file

- Main components of the SOLIDWORKS interface

- The document's measurement system

Technical requirements

In this chapter, you will need to have access to SOLIDWORKS.

The project files for this chapter are available at the following GitHub repository: `https://github.com/PacktPublishing/Learn-SOLIDWORKS-2025-Third-Edition`

The CiA video for this chapter can be found at `https://packt.link/Vsh29`

Starting a new part, assembly, or drawing file

This section addresses the three types of SOLIDWORKS files: parts, assemblies, and drawings. Here, we'll briefly cover what each file is for and how we can use each of them; however, more about each type of file will be covered throughout this book.

What are parts, assemblies, and drawings?

As we just mentioned, SOLIDWORKS files fall into three distinctive categories: parts, assemblies, and drawings. Each file type corresponds to a certain deliverable when we're making a product. By deliverable, we mean whether we need to deliver a **three-dimensional (3D)** part file, a 3D assembly file, or a **two-dimensional (2D)** engineering drawing. To illustrate these three file types, let's break down the simple cylindrical box shown in the following diagram:

Figure 2.1 – A cylindrical box assembly consisting of two parts

We can identify three distinctive categories from the preceding cylindrical box diagram: **parts**, **assemblies**, and **drawings**. Let's take a look at each of these here:

- **Parts**: Parts are the smallest elements that make up an artifact. They are the first step in building any product in SOLIDWORKS. Since SOLIDWORKS is used to create 3D models, all of its parts are 3D. Also, each part can be assigned to one type of material. Our cylindrical box contains two parts: a main **cylindrical** container and a **cap**, as shown in *Figure 2.2*:

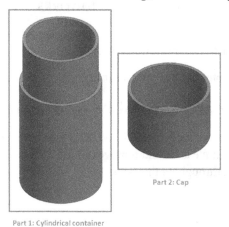

Part 2: Cap

Part 1: Cylindrical container

Figure 2.2 – Separating the parts making the cylindrical box assembly

After creating the two parts separately in two different part files, they can be put together into an assembly file.

- **Assemblies**: SOLIDWORKS assemblies enable you to join multiple parts to create a cohesive assembly, much like the complex objects we use daily, such as cars, phones, water bottles, and tables. These assemblies are crucial for evaluating how separately created parts interact and fit together, ensuring they function correctly as a whole. Additionally, they allow design and engineering teams to assess the overall aesthetic and functionality of the product. Through SOLIDWORKS assemblies, you can also simulate the movements of mechanical products, providing a comprehensive view of the product's performance. For instance, in a cylindrical box assembly, you can visualize how the parts come together, as illustrated in *Figure 2.3*.

Figure 2.3 – A closed cylindrical box assembly

- **Drawings**: SOLIDWORKS drawings allow you to create 2D engineering drawings out of your parts or assemblies. Engineering drawings are the most common way to communicate designs on paper. They often show dimensions, tolerances, materials, costs, parts **identifiers (IDs)**, and so on. Engineering drawings are often required when designs need to be reviewed by certain parties. Also, they are often required if you wish to talk about your designs with clients or manufacturing/prototyping establishments. For our cylindrical box, an engineering drawing might look like this:

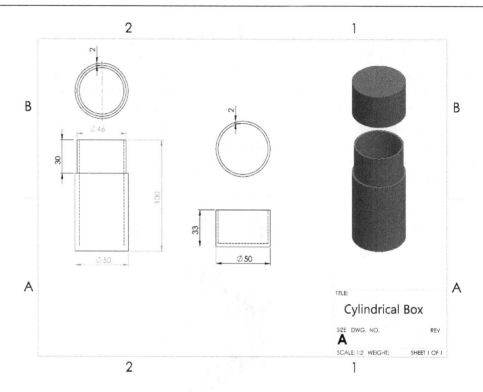

Figure 2.4 – A 2D engineering drawing communicating the design of the cylindrical box

All three types of files—parts, assemblies, and drawings—are essential to SOLIDWORKS users. This is because they are all necessary for the creation of products.

Now that we understand what parts, assemblies, and drawings are, let's look at how we can start using them in SOLIDWORKS.

Opening a part, assembly, or drawing file

Now that we know the difference between parts, assemblies, and drawings, we will explore how to start using each type of file. Once you open SOLIDWORKS 2025, a **Welcome** window will appear, along with some shortcuts. One of those options is opening a new **Part**, **Assembly**, or **Drawing** file. These options are highlighted in the following screenshot. Once you click on any of these options, that type of file will be opened:

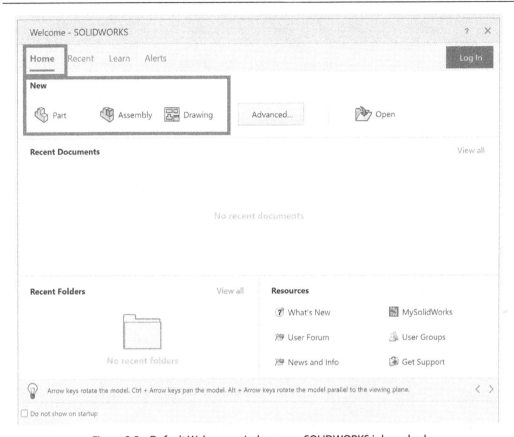

Figure 2.5 – Default Welcome window once SOLIDWORKS is launched

If the **Welcome** window does not appear, there is another way to open a new file, as follows:

1. Click on **File** in the top-left corner of SOLIDWORKS.

2. Select **New...**, as shown in the following screenshot:

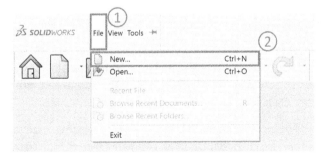

Figure 2.6 – Opening a new file in SOLIDWORKS

3. After selecting **New...**, you will be able to pick one of the three options—that is, to open a new **Part**, **Assembly**, or **Drawing** file, as shown in the following screenshot. You can select the type of file you want to work with and click **OK**. Alternatively, you can double-click on the file type you would like to start with:

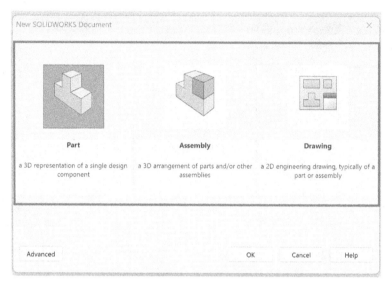

Figure 2.7 – The different options for a new SOLIDWORKS document

In this book, first, we will focus on creating parts, then assemblies, and—finally—drawings. Being able to distinguish between the different types of files is very important, as everything we do afterward will be built on top of the file type we choose. Now that we understand how to open parts, assemblies, and drawings in SOLIDWORKS, let's look at how to use the software's interface further.

Main components of the SOLIDWORKS interface

In this section, we will discuss the main components of the SOLIDWORKS interface. These main components are the Command Bar / CommandManager, the Task Pane, the canvas/graphics area, and the FeatureManager design tree.

Being familiar with these components is essential if we wish to use the software to a good extent. For a practical follow-up, you can download the SOLIDWORKS part linked with this chapter, which will be used to explain the main components of the SOLIDWORKS interface.

In this chapter, we will be focusing on the interface that's used when we need to deal with parts, instead of assemblies and drawings. However, the main components of the interface are the same when we deal with each file type.

When opening a part in SOLIDWORKS, regardless of whether it is new or existing, you will be faced with the view shown in the following screenshot. We will cover the four main sections of this screen: the Command Bar, the FeatureManager design tree, the Task Pane, and the canvas/graphics area. These are the main sections of SOLIDWORKS that we'll be interacting with and referring to throughout this book:

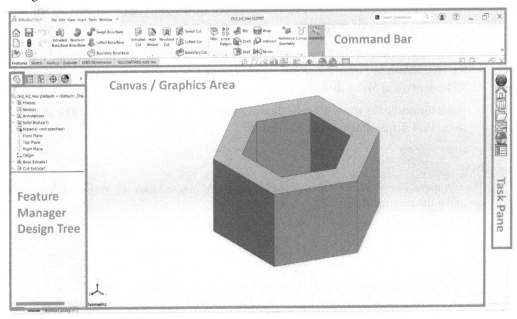

Figure 2.8 – A breakdown of the SOLIDWORKS interface

We will look at the Command Bar, the FeatureManager design tree, the canvas/graphics area, and the Task Pane in more detail in the following sections.

The Command Bar

The Command Bar is located at the top of the screen. It contains all the SOLIDWORKS commands that are used for building models. It contains different categories of commands, and each category contains a set of different commands. A close-up of the Command Bar is shown in the following screenshot:

Figure 2.9 – A breakdown of the Command Bar

Different categories (tabs) of commands correspond to different functions. For example, in the **Sketch** category/tab, you will find all the commands that we will need in the sketching phase. Those will enable us to sketch lines, circles, rectangles, and so on. In the **Features** category/tab, you will find all the commands that we will need to go from the sketching phase and start creating a 3D model. Those will enable us to build cubes, spheres, prisms, and so on. The categories that are shown in *Figure 2.9* are not the only ones SOLIDWORKS provides, but they are the most common ones we will use. To show the hidden Commands categories, we can do the following:

1. Right-click on any of the Commands categories, then expand the **Tabs** menu. You will get the view shown in *Figure 2.9*, which contains more Commands categories, such as **Surfaces**, **Weldments**, and **Mold Tools**.

2. Select the categories you want to be shown. By doing this, these categories will be added to the Command Bar, as illustrated in *Figure 2.10*:

> **Note**
>
> The commend bar is also refered to as the CommandManager. However, in this book, we will use call it the command bar.

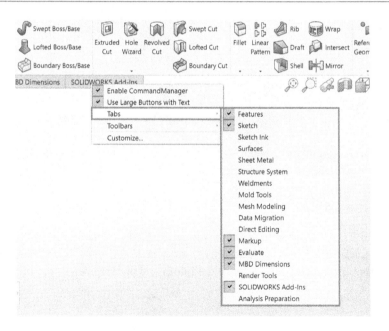

Figure 2.10 – List of command categories that can be added to the Command Bar

This concludes our overview of the Command Bar, which contains the different commands we will use as we build 3D models. Now, we will look at the FeatureManager design tree.

The FeatureManager design tree

The FeatureManager design tree details everything that goes into creating your parts. The following screenshot shows the FeatureManager design tree for the part we explored in this chapter. We can simplify the FeatureManager design tree by splitting it into four parts, as illustrated in the following screenshot:

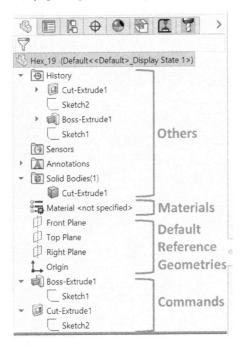

Figure 2.11 – A breakdown of the FeatureManager design tree

The four parts of the FeatureManager design tree are listed here:

- **Commands/features**: These are the commands that are used to build the model. This includes sketches, features, and any other supporting commands that were added during the modeling phase (since we are building a 3D model). In the preceding screenshot, two features were used to create the model, as indicated by **Commands**. The first is **Boss-Extrude1** and the second is **Cut-Extrude1**. Note that these commands are listed in the order of when they were applied.

- **Default reference geometries**: The SOLIDWORKS canvas can be understood as endless space. These default reference geometries are what can fix our model to a specific point or plane. Without these, our model will be floating in an endless space without any fixtures. Throughout this book, we will start our models from these default references. There are three planes (**Front Plane**, **Right Plane**, and **Top Plane**), in addition to the origin.

- **Materials**: All of the artifacts we have around us are made of a certain material. Some examples of materials include plastic, iron, steel, and rubber. SOLIDWORKS allows us to assign which structural material the part will be made of. In the preceding screenshot, the **Material** feature is classed as **<not specified>**.

- **Others**: This section includes other aspects of our model's creation, such as **History**, **Sensors**, **Annotations**, and **Solid Bodies**. We will explore them later in this book.

> **Note**
>
> Through the book, we will use the term *design tree* as a reference to the FeatureManager design tree.

The design tree helps us to easily identify how the model was built and in what sequence. This makes it easier for us to modify existing models. Now, let's look at the canvas.

The canvas/graphics area

The canvas provides a visual representation of the model we have at hand. It contains three main components, as illustrated in the following screenshot:

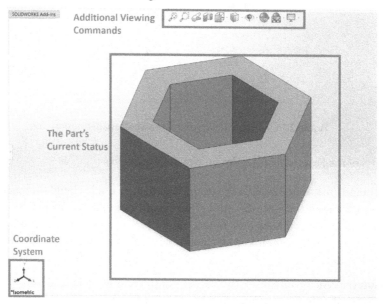

Figure 2.12 – A breakdown of the canvas/graphics area

Let's break down the components, as follows:

- **Coordinate system**: This shows the orientation of the model in relation to the default coordinate system in terms of the x, y, and z axes. They are interactive and can be used to position the viewing angle of the model. By clicking on the various axes, you can arrive at that viewing orientation.

- **The part's current status**: This shows the current status of the part at work. This is updated with every construction command that's used to build the model.

- **Additional viewing commands**: These provide alternative views of the model, such as the wireframe view and section view. They also provide shortcuts that we can use to modify the scene and the appearance of the model and hide/show various properties.

When controlling the model in the canvas, using a mouse with a scroll wheel is recommended due to the functionalities the scroll wheel has. Here are two ways the scroll wheel helps you control the model, which you can try with the provided model:

- When the cursor is within the canvas, rolling the scroll wheel will allow us to zoom in and out of the cursor's location.

- When the cursor is within the canvas, pressing on the scroll wheel and moving the mouse will rotate the model in a certain direction. For example, if we move the mouse to the right, the model will pivot to the right.

> **Note**
> We will use the term *canvas* throughout the book. However, many also use the term *graphics area*.

Now that we have covered the canvas, let's talk about the Task Pane.

The Task Pane

The Task Pane shows to the right of our interface by default. It contains shortcuts for the different tools we will be using in order to enhance the efficiency of our work. This includes access to common online resources and forums, as well as different tools, such as appearance adjustments and the View Palette (mainly for drawing files). In this book, however, we won't be using linked resources while making parts or assemblies. We will use the View Palette in *Chapter 10*.

Now that we know about the major components of the SOLIDWORKS interface, we will learn how to adjust the measurement system of our open document.

The document's measurement system

Since SOLIDWORKS is an engineering software, all of the models are constructed with user-provided (user-input) measurements. To facilitate communication, SOLIDWORKS uses standard systems that are currently used in the industry, including the **International System of Units (SI)**, the **imperial system**, and variations of each.

Different measurement systems

When modeling with SOLIDWORKS, the user must take note of the measurement system that is set in the document. A measurement system is a set of common agreed-upon units that facilitate how we communicate quantities in terms of length, mass, volume, and so on. Some examples of such units are meters and inches, which are measurements of length.

These often correspond with internationally recognized systems such as the SI and the imperial system. The SI system is also commonly known as the **metric system**. Currently, it is used in most countries around the world. Another common system is the imperial system, which is mostly used in the **United States (US)**.

The following table compares the major units that are used in the SI and imperial systems:

	Imperial unit	SI unit
Length	Inches (inch)	Meters (m)
Mass	Pounds (lb.)	Kilograms (kg)
Time	Seconds (s)	Seconds (s)

Figure 2.13 – A comparison between the imperial and SI unit systems

Before we start modeling anything in SOLIDWORKS, we must decide on which system to use. The unit system we use often depends on the standards that have been adopted by the organization we work for or the requirements of our clients.

Adjusting the document's measurement system

Now that you have decided which system to use, you must set it up in the software. You can adjust the unit of measurement by following these steps:

1. Open a new part file.
2. In the bottom-right corner, you will find the current/default measurement system in an abbreviated form. Click on the displayed measurement system, as shown in the following screenshot:

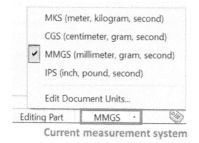

Figure 2.14 – The set and default options for unit systems

3. Choose from the default settings.

You can create your own measurement system by selecting **Edit Document Units...**. This will open the following window, where you can select custom options and customize and implement your own custom units:

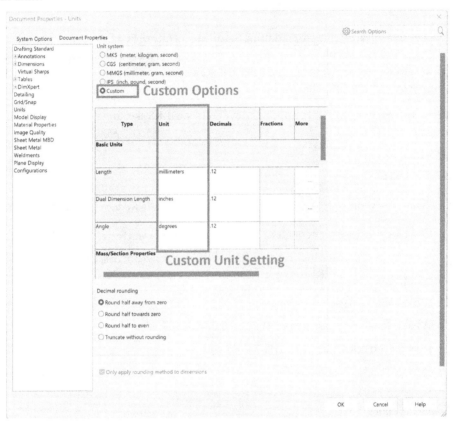

Figure 2.15 – Customizing the unit system

Note that you can change the set units at any time during the modeling process. This will convert all the units that were already set in the file. For example, let's assume that the document's length measurement was set to **IPS (inch, pound, second)** and a line measuring 2 inches was drawn. If we change the measurement system later to **MMGS (millimeter, grams, second)**, the length of the line will be automatically converted into 50.8 millimeters. This is because 1 inch is equal to 25.4 millimeters.

Note

The same procedure for adjusting the measurement system applies to the part, assembly, and drawing file types.

Understanding measurement systems is crucial because our goal is to create a tangible object for production or prototyping. If we don't set the correct parameters from the beginning, our final 3D model may not be usable or accurate.

Summary

In this chapter, we learned how to start using the different types of SOLIDWORKS files—that is, parts, assemblies, and drawings. We also learned about the main components of the SOLIDWORKS interface, as well as the different measurement systems that are available and how to adjust them. These are the first steps we need to follow when we plan to use the software to create a project and its foundations.

In the next chapter, we will start working with SOLIDWORKS sketching. Sketching is foundational to building any 3D model.

Questions

The following questions will help to consolidate the main points we have learned in this chapter:

1. What are the three types of files a SOLIDWORKS user can create?
2. What is the difference between SOLIDWORKS parts, assemblies, and drawings?
3. What is contained in the canvas?
4. What does the design tree show?
5. What is the difference between SI units and imperial units?
6. How do you open a new SOLIDWORKS part file?
7. How do you set the unit for the new SOLIDWORKS file to **MKS (meter, kilogram, second)**?

> **Important note**
> The answers to the preceding questions can be found at the end of this book.

Get This Book's PDF Version and Exclusive Extras

UNLOCK NOW

Scan the QR code (or go to packtpub.com/unlock). Search for this book by name, confirm the edition, and then follow the steps on the page.

Note: Keep your invoice handy. Purchases made directly from Packt don't require an invoice.

Part 2: 2D Sketching

Making 2D sketches in SOLIDWORKS is essential to applying any feature or creating any 3D model. Without a strong foundation in sketching tools, you will not be able to construct 3D models. This part covers all the sketching foundations you would need as a SOLIDWORKS associate and professional.

This part has the following chapters:

- *Chapter 3, SOLIDWORKS 2D Sketching Basics*
- *Chapter 4, Special Sketching Commands*

3
SOLIDWORKS
2D Sketching Basics

The foundation of any 3D SOLIDWORKS model is a 2D sketch. This is because SOLIDWORKS builds 3D features based on the guidance of 2D sketches. This chapter will get you started with SOLIDWORKS 2D sketching. We will cover multiple sketching commands that will allow you to sketch shapes such as rectangles, triangles, circles, and ellipses. You will also learn how to combine those different sketches and create more complex shapes. Then, we will explore the different levels at which SOLIDWORKS defines a sketch. Mastering SOLIDWORKS 2D sketching is essential if we wish to build a 3D model.

The following topics will be covered in this chapter:

- Introducing SOLIDWORKS sketching

- Getting started with SOLIDWORKS sketching

- Sketching lines, rectangles, circles, arcs, and ellipses

- The state of sketches – under-defined, fully defined, and over-defined

Technical requirements

In this chapter, you will need to have access to SOLIDWORKS.

The CiA video for this chapter can be found at `https://packt.link/Jz5AH`

Introducing SOLIDWORKS sketching

In this section, we will discuss what **SOLIDWORKS sketches** are. We will discuss the importance of SOLIDWORKS sketching functions and how to view them when modeling with SOLIDWORKS. SOLIDWORKS sketches are the base of each SOLIDWORKS 3D model. Thus, it is important to master SOLIDWORKS sketching first.

The position of SOLIDWORKS sketches

Sketches are typically viewed as fast drafts of a certain shape. For example, the following diagrams show a hand-drawn sketch of a square and a cube, respectively. The main point of these is to provide a rough idea of an object.

In this hand-drawn sketch, we are communicating the idea of a square, without specifying how big that square is:

Figure 3.1 – A hand-drawn sketch of a square

Similarly, the following hand-drawn sketch communicates the idea of a cube, without specifying how big the cube is:

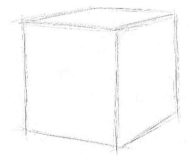

Figure 3.2 – A hand-drawn sketch of a cube

SOLIDWORKS sketches are a bit different. In SOLIDWORKS, a sketch is a fully dimensional and exact shape that's mostly given in two dimensions. SOLIDWORKS also has a 3D sketches function that is more commonly used with surface modeling. In this book, we will only use 2D sketching.

The following diagram shows a SOLIDWORKS sketch of a square with a side dimension of 50 mm. Note that the sketch is different than the one we looked at previously; it is an exact square and not an approximation of a square:

Figure 3.3 – A SOLIDWORKS sketch of a square

SOLIDWORKS sketches are the starting points of any 3D model. They are the basic guiding elements for 3D features. For example, if we want to make a cube, we have to start by drawing the preceding square sketch. After that, we can extrude it to generate a cube, as shown in *Figure 3.4*.

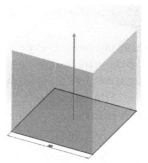

Figure 3.4 – Extruding a cube in SOLIDWORKS

Note that you can see our initial sketch at the bottom of the cube.

When we create a 3D shape in SOLIDWORKS, we often start by creating a 2D sketch and then apply a feature to it. Then, we keep iterating those two steps as the 3D shape becomes more complicated. This is why it is very important that we master SOLIDWORKS sketching before anything else.

The preceding diagram also shows the common sequence of a SOLIDWORKS model, starting with a sketch rather than a feature. The sequence then repeats as the model becomes more complex.

Simple sketches versus complex sketches

SOLIDWORKS has many ready-made commands that we can use to create simple sketch shapes such as lines, squares, circles, ellipses, and arcs. We will learn about these sketching commands in this chapter. However, it is important to understand that all complex sketches are a combination of different simpler sketches.

Let's illustrate this with an example. The following figure shows a sketch of a relatively complex shape:

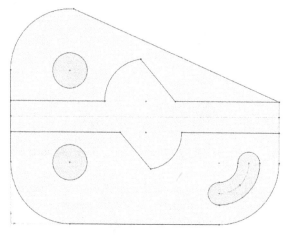

Figure 3.5 – A complex sketch in SOLIDWORKS

Note that the complex shape is made up of a combination of different simple elements. This sketch can be easily made with four different sketching commands: lines, arcs, circles, and slots. *Figure 3.6* shows how we can break down the complex sketch into these four sketching commands. All the unmarked elements of the sketch are repetitions of the marked ones. We can refer to each of these elements (a line or an arc, for example) as a **sketch entity**:

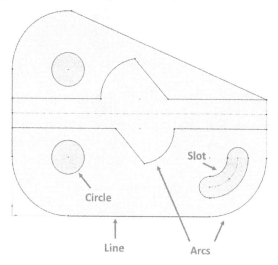

Figure 3.6 – Complex sketches are combinations of simple sketch entities

Simplifying complex sketches so that they're simple elements is a skill that takes time to build. Since we've just started sketching, we may face difficulties with simplifying complex sketches. This book will help you develop this skill, but you will become better at it with experience.

Now that we know what sketches are, we can start defining the elements that are involved in our sketches. We will start with sketch planes.

Sketch planes

Sketch planes are flat surfaces that we can use as bases for our sketches. They are important because they give our sketches a solid location. If we didn't have them, our sketches would float undefined in 3D space.

SOLIDWORKS provides us with default sketch planes when we start a new part. We can find them listed in the design tree. The default sketch planes are the front plane, the top plane, and the right plane. They are listed in the following screenshot:

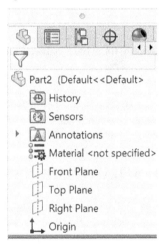

Figure 3.7 – The sketch planes as shown in the software interface

To help you visualize these default planes, imagine a box and the planes being the top, front, and right-hand sides of it. *Figure 3.8* shows the three default planes in the shape of a box:

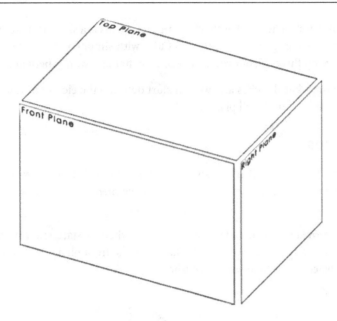

Figure 3.8 – The three base planes reshaped as a box

These three planes are not the only ones that are available to us. In addition to the default planes, we can use any other 2D planar surface as a sketch plane as well. We also have the option of creating our own planes based on different geometrical references. We will explore these possibilities later in the book. Now that we've identified our sketch planes, we can start using those planes to create our sketches.

Getting started with SOLIDWORKS sketching

In this section, we will discuss how to start sketching, what it means to define a sketch, and what the major geometrical relations that exist in SOLIDWORKS sketching are. These topics will be our practical introduction to getting into SOLIDWORKS sketching.

Getting into sketching mode

To start a sketch, we need to have a part file open. Then, we can follow these steps to get into sketching mode:

1. Select one of the default sketch planes: front, top, or right.

2. In the CommandManager, select the **Sketch** option (which is marked as **2** in *Figure 3.9*). This will open up the sketch commands category, which will show all the commands related to sketching.

3. Select the **Sketch** command (which is marked as **3** in *Figure 3.9*). This will allow us to enter sketching mode. When we're in sketching mode, we can apply different sketching commands, such as the category labeled **Simple sketching commands** in *Figure 3.9*:

Figure 3.9 – The Sketch tab and the Sketch command

Now that we are in sketching mode, we can see multiple sketching commands on the command bar. This includes commands for sketching simple shapes such as lines, rectangles, circles, polygons, and arcs. Let's go over these commands in more detail. However, since we are already in sketching mode, take some time to randomly click on these commands and try them out on the canvas.

In addition to the steps we just explored, there are two alternative ways to start sketching mode, as follows:

- Select the **Sketch** command first and then select the sketch plane.

- Click on the **Sketch** shortcut that appears when you click on one of the planes. The following screenshot highlights this shortcut:

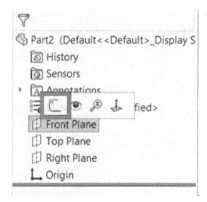

Figure 3.10 – Another way to get into sketching mode is to use the popup after selecting a plane

Before we start applying these sketching commands, we will discuss what it means to define a sketch.

Defining sketches

Now that we know how to start sketching, we will learn what defines a sketch. Constructing a sketch requires two elements; the first is the sketch entities, such as lines, arcs, and circles, while the second is the dimensions and relations that define the sketch entities. Remember that SOLIDWORKS is a form of engineering software that aims to support the design and manufacture of products. Defining sketches ensures the design intent is integrated into the design. Thus, defining shapes is very important.

To define sketches, we can use dimensions and relations. Let's start by defining them:

- **Dimensions**: These represent distances and angles that can be defined with a numerical value. Some examples are as follows:

  - Lengths of lines

  - The diameters and radii of circles and arcs

  - An angle between two lines

- **Relations**: These represent geometric relations between the different parts of a sketch. Some examples are as follows:

  - A line that has a **horizontal** relation to a sketch plane

  - Two lines are **perpendicular** to each other

  - Two circles are **concentric** to each other

  - A line is a **tangent** to a curve

To illustrate this, let's look at a visual comparison between two lines. One is **fully defined** and one is **under defined**:

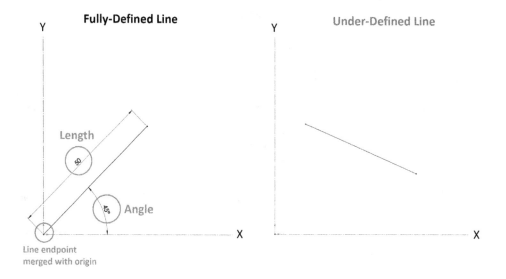

Figure 3.11 – A visual comparison between a fully defined and an under-defined line

Of the two preceding lines, the one on the left is fully defined while the one on the right is under-defined. The line on the left is defined as such because of the following:

- One end coincides with the origin of the coordinate system

- The length of the line was defined as 50 millimeters

- The angle between the line and the x axis is 45 degrees

These dimensions and relations are what make a sketch defined. It may not look like there's much of a difference between them when we look at the two lines, assuming that we drew them on paper. However, when we work with a computer program, we need to let the computer know everything about the line (or any other sketch entity); otherwise, it won't know what we want.

To make it easier to distinguish between fully defined and under-defined sketches, SOLIDWORKS color-codes them. Black parts are fully defined, while blue parts are under defined. In addition to color-coding, SOLIDWORKS also indicates the status of our sketch below the canvas. The following screenshot shows an indication of a **Fully Defined** sketch. Other classifications include **Under Defined** and **Over Defined**:

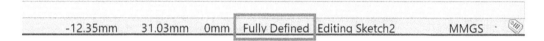

Figure 3.12 – The sketch status classification as found at the bottom of the interface

> **Note**
> You can change the default color code through the settings if needed.

Sketches are defined with measurements and relations, where measurements refer to numbers and relations refer to geometrical relations. We will explore geometrical relations in the next section. Also, in the section titled *Under-defined, fully defined, and over-defined sketches*, we will be discussing those terms in more depth.

Geometrical relations

The following table summarizes most of the geometrical relations we will come across while working with SOLIDWORKS sketching. You don't need to memorize all of these relations for now. Simply read through the following table and use it as a reference as we continue to develop our SOLIDWORKS skills. The relations in the following table have been organized in alphabetical order. As we use the SOLIDWORKS interface throughout this book, we will deal with these icons more and more since they will appear in the sketches themselves, as well as when we choose which relation to apply them to:

Relation's Name	Relation Icon	Relation Function
Coincident		Coincident relations occur between points and lines, arcs, circles, and so on. This would make a point lie in other sketch entities, such as lines.
Colinear		This makes two or more lines lie in one direction.
Concentric		This can make two or more circles or arcs share the same center.
Coradial		This applies to two or more arcs if the different arcs share the same center and radius.
Equal		This makes two or more lines or arcs equal to each other in terms of length.
Equal Curve Length		This relation occurs between an arc or circle and a line. It makes the perimeter equal to the line in terms of length.
Fix		This fixes the selected sketch entity to where it exists at the time of setting the relation.
Horizontal		This makes lines horizontal to the sketch plane. In addition, it can make more than one point lie in a horizontal line.
Intersection		This will position a point at the intersection point of two lines. This includes the extension of the lines, as well as the lines themselves.
Merge		This can merge more than one point together into one point location.
Midpoint		This can position a point so that it's in the middle of a line.
Parallel		This can make more than one line parallel to another.
Perpendicular		This can make two lines perpendicular to each other.
Tangent		This relates to circles, arcs, and other curved entities. It can turn a line tangent into a curved entity. It can also make more than one curved entity tangent to each other.
Vertical		This makes lines vertical to the sketch plane. In addition, it can make more than one point lie in a vertical line or in relation to another point.

Figure 3.13 – The geometrical relations for sketching

These icons and their relations will repeatedly show up as we are creating sketches. Thus, it is important that we know what they mean. We will start using these relations in the next section, that is, when we start sketching different shapes.

Sketching lines, rectangles, circles, arcs, and ellipses

In this section, we will discuss the major sketching functions and how to use them. These include sketching lines, rectangles, circles, arcs, and ellipses. We will address each of these sketching commands separately and find out how to define each one.

The origin

On the canvas, you should be able to see a small dot with arrows, as shown in the following figure. This dot is located exactly where the two arrows meet and represent the origin point of the canvas. You can think of the origin as the only defined and fixed point in our SOLIDWORKS infinite canvas:

Figure 3.14 – The SOLIDWORKS graphical representation of the origin

Because it is the only fixed point, it is very important to always link our sketches to that origin point. Otherwise, our sketch will always be under defined in the infinite canvas.

Sketching lines

To illustrate how to sketch lines, we will sketch the shape highlighted in *Figure 3.15*. Note that the sketch is fully defined, and so SOLIDWORKS will color the lines in black after it's been sketched. Also, take note of the relations and dimensions (in millimeters) shown in the following figure:

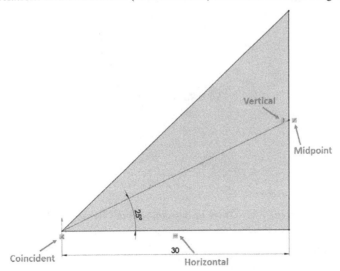

Figure 3.15 – The final output of our sketching exercise

Let's go ahead and start constructing the shown sketch. To do that, we will go through two stages: **outlining** and **defining**.

In the outlining stage, our aim will be to draw a rough outline of the final shape. Here, we will not pay much attention to dimensions or relations. We will follow these steps in the outlining stage:

1. Start a new part file.
2. Set the document's measurement system to **MMGS (millimeter, gram, second)** if it is not already.
3. Navigate to the **Sketching** mode using the **Top Plane**.
4. Select the **Line** sketch command, as indicated in the following figure.

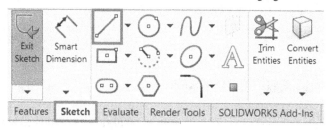

Figure 3.16 – The location of the line command

5. Move the mouse cursor onto the canvas and click on the origin. To start drawing the triangle, move your cursor to the right and, once the line is drawn, left-click on your mouse. Note that the shape of the cursor will change into a pen.
6. Next, move the mouse upward to create the second line, and then left-click on the mouse again.

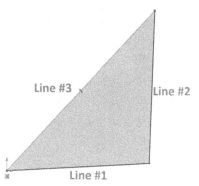

Figure 3.17 – Three under-defined lines forming a triangle

7. Move the cursor again to draw a line linking back to the origin. This concludes the creation of a shape that looks like a triangle, as shown in *Figure 3.17*.

8. Now, we can draw the last line, which links the origin point to any point on the vertical line so that we end up with the following sketch. Note that the sketch on the screen is blue, indicating that it is under defined.

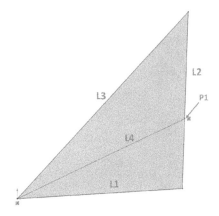

Figure 3.18 – The sketch with all the sketch entities

9. Exit the **Line** sketching command by pressing *Esc* on the keyboard.

In the defining stage, we will work on defining our outline with the necessary dimensions and relations to fully define our sketch. Note that some of the relations are set automatically by SOLIDWORKS, according to how we place our lines. Follow these steps for the defining stage referring to the numbers indicated in *Figure 3.18*:

1. Click on line (**L1**), as indicated by the preceding sketch.
2. A new panel will appear on the left, in place of the design tree. It will be titled **Line Properties**, as shown in the following screenshot. Under **Add Relations**, click on **Horizontal**. This will add a horizontal relation to the line. You will see a small icon appear next to the line, showing that the line is horizontal:

Figure 3.19 – The Line Properties manager showing the applicable geometric relations

> **Note**
>
> SOLIDWORKS can apply relations automatically if they can be inferred by how you sketch an entity. For example, if you try to sketch a horizontal line, SOLIDWORKS will apply the **Horizontal** relation to the line automatically. In this case, you will see the relation listed under **Existing Relations** in **Line Properties**.

3.   Click on line 2 (**L2** in *Figure 3.18*) and add a **Vertical** relation.

> **Note**
>
> To deselect a line, click anywhere else on the canvas.

4.   Press and hold down *Ctrl* on your keyboard, click on the endpoint **P1**, then click on line 2. From the properties that appear on the left, select the **Midpoint** relation.

Tip

We can select multiple sketch entries at the same time by clicking and holding down *Ctrl* on the keyboard and selecting multiple entities.

Note

By following the preceding steps, we will have set up all the relations for our sketch. Note that many parts of the sketch are still blue. Try to click and hold different parts of the sketch, points, or lines and move the mouse around. All the sketch elements will move in a way that preserves all the relations that have been set.

5. On the command bar, select the **Smart Dimension** command, as shown in the following screenshot:

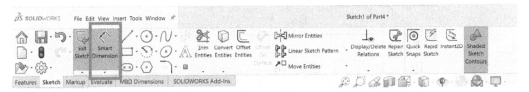

Figure 3.20 – The Smart Dimension command

6. Go back to the canvas and click on line 1. A dimension will appear, displaying the current length of the line. Left-click on an empty space in the canvas once more. You will be prompted to enter a length value for the line. Type in the value 30 and then click on the green checkmark, as shown in the following screenshot. After that, you will notice that the line's length changes to match the new length. Also, note that the line, including its endpoints, will turn black:

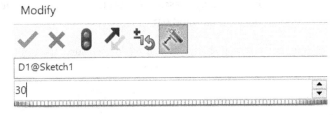

Figure 3.21 – The Smart Dimension interface with space to input the dimension value

7. Click on line 1 again, and then click on line 4. Note that the specified dimension changed to the angle between the two lines, which is the dimension we want to specify. Left-click anywhere on the canvas once more to confirm the dimension's location. In the box, type 25, which indicates the degree, and then click the green checkmark, as highlighted in *Figure 3.22.*

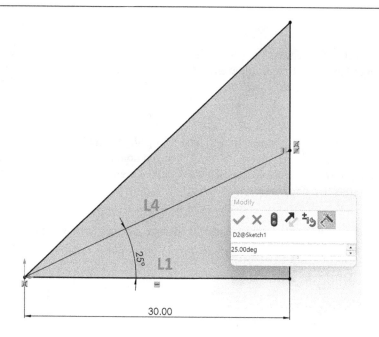

Figure 3.22 – Dimensioning an angle between two lines

> **Note**
>
> To delete a dimension, select it and press *Delete* on the keyboard. Alternatively, we can right-click on the dimension and select **Delete** from the options that are available.

8. Exit the smart dimension mode by pressing *Esc* on the keyboard or clicking on the **Smart Dimension** command on the command bar.

Note the shown sketch in the exercise is shaded indicating an enclosure. You can turn this feature on and off by toggling the **Shaded Sketch Contours** command on the command bar, as shown in *Figure 3.23*.

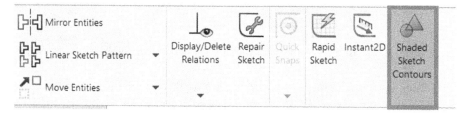

Figure 3.23 – The Shaded Sketch Contours option shades enclosed sketch entities

At this point, we will see that the sketch is fully black, indicating that it is fully defined. This concludes our first sketching exercise. In this simple exercise, we have covered many essential sketching features that we'll keep using when we model with SOLIDWORKS throughout this book, including the following:

- How to start sketching

- How to sketch lines

- How to set up lengths and angles using smart dimensions

- How to set up geometric relations such as vertical, horizontal, and midpoint

These sketching commands are essential to sketching using SOLIDWORKS sketching tools. In the next section, we will use these skills to sketch rectangles and squares.

Sketching rectangles and squares

In this section, we will learn how to sketch rectangles and squares. To illustrate this, we will sketch the diagram shown in *Figure 3.24*. We have already covered most of the concepts we'll need to complete the sketch, including how to get into sketching mode and how to define a sketch. In this example, the dimensions are in **inches, pounds, seconds (IPS)**, where the length is in inches. Note that we labeled the sketch with **R1** and **R2**, indicating rectangles 1 and 2, and **S1**, indicating square 1:

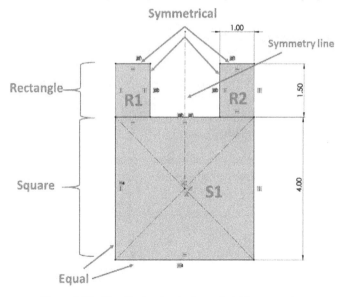

Figure 3.24 – The final output of our sketching exercise

To create this sketch, we will go through the same two stages that we went through previously: outlining and defining.

For outlining, we will draw an arbitrary outline of the final shape we want to draw. Let's get started:

1. Start a new part and make sure that the document measurement system is set to **IPS (inch, pound, second)**. Note that we are using a different measurement system for practice purposes here.

2. Navigate to the **Sketch** mode using any of the sketch planes (for example, **Front Plane**).

3. On the command bar, select the drop-down menu next to the rectangle shape and click on the **Center Rectangle** command. In this exercise, we will use the two rectangle commands, both of which are shown in *Figure 3.25*: **Center Rectangle** and **Corner Rectangle**.

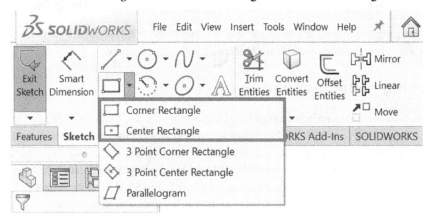

Figure 3.25 – The Corner Rectangle and Center Rectangle sketching commands used in the exercise

Both commands create rectangles; the difference is in how those rectangles are created. A center rectangle is created with two clicks: one indicating the center and the other indicating a corner. A corner rectangle is created with two clicks, indicating the opposing corners. Note that the small figures also show us how to draw that particular type of rectangle by showing us the sequence of clicks that are needed.

4. After selecting **Center Rectangle**, click on the origin, move the mouse to the side, and click away from the origin to form an approximate shape of a square. We are doing this to create the square labeled **S1** in *Figure 3.24*.

> **Tip**
> You can delete any part of the sketch by highlighting or selecting that part and pressing *Delete* on the keyboard. Alternatively, you can right-click and select **Delete**.

5. Select the **Corner Rectangle** command and draw the two rectangles, that is, **R1** and **R2**. For **R1**, click on the top-left corner of square **S1**. Then, move the mouse up and away from the first click, then click again to form the rectangle. Do the same for **R2**. The result should look like the following sketch.

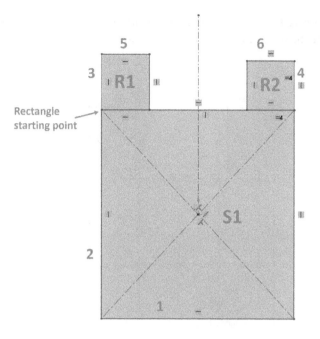

Figure 3.26 – The final result after the outlining stage

6. The last step is to add the symmetry line. The symmetry line is a **centerline** that we will be able to use to create symmetrical relations between parts on opposite sides of that line. To create a centerline, click on the drop-down menu next to the line command and select **Centerline**. The centerline goes through the middle of the sketch, starting at the origin and going upward.

> **Tip**
>
> Since we are creating the centerline, note that we can move the line endpoint slowly at an angle until the vertical relation appears, at which point, you can lock it on that. This is one way in which SOLIDWORKS interprets the relations we want to apply and applies them to us. We can use this approach to make sketching faster.

For our defining stage, we will define the outlined shape by applying relations and dimensions. Follow these steps to do so:

1. Select lines 1 and 2, and apply the **Equal** relation. This condition is what will make a normal rectangle into a square.

2. Select lines 3 and 4, as well as the symmetry line, and apply the **Symmetric** relation. In this case, SOLIDWORKS will automatically interpret the centerline as the symmetry line since it is located in the middle. You can also do the same with lines 5 and 6.

3. Using the smart dimension, set the given lengths so that they match what's shown in *Figure 3.27*. We can use the smart dimension in the same way as we did when we sketched a line:

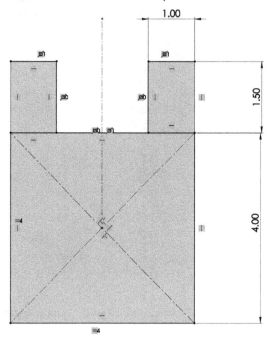

Figure 3.27 – The final sketch being fully defined

At this point, we should have a fully black shape that is fully defined. Also, take note of the status of the sketch, which is shown at the bottom of the SOLIDWORKS screen. It is indicated as **Fully Defined**, as shown in the following screenshot:

Figure 3.28 – The sketch referenced as fully defined

This concludes our sketching exercise. In this simple exercise, we have covered many essential sketching features that we will utilize throughout our SOLIDWORKS interaction, including the following:

*   How to sketch squares and rectangles

*   How to set up equal-length lines

*   How to draw a centerline and set up symmetrical sketch entities

At this point, we already have many of the sketching basics under our belt. Now that we're advancing, you won't need as much guidance as all the commands will start becoming second nature to you. Before moving on, take the time to experiment with creating other types of rectangles, such as **3 Point**

Corner Rectangle and **3 Point Center Rectangle**. In addition, take some time to experiment with creating a parallelogram. We can find all of these shapes in the **Rectangle** command drop-down menu.

Now, we know how to sketch lines, rectangles, and squares. Next, we will develop our skills by addressing circles and arcs.

Sketching circles and arcs

In this section, we will sketch circles and arcs. First, let's break down what a circle and an arc are. The following figure shows a circle. Note that a circle is defined by its center (a point) and its diameter:

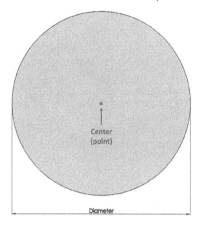

Figure 3.29 – A breakout of a circle

Figure 3.30 shows an arc, as well as the elements that define it. An arc can be defined by its center (a point), as well as other points, which indicate the endpoints of the arc. This is in addition to its radius and various distances:

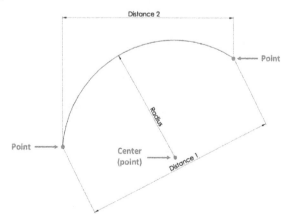

Figure 3.30 – A breakout of an arc

Each of these elements can be controlled with dimensions or relations. Each point can be understood as a standalone entity that we can use for relations or dimensions.

To illustrate these two commands, we will sketch the following shape:

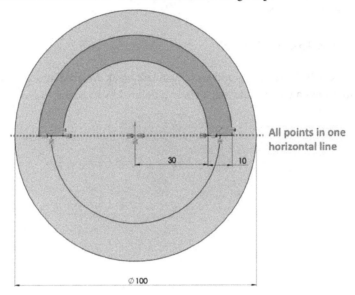

Figure 3.31 – The final result of this sketching exercise

Similar to what we did previously, we will do outlining and then defining. We will start with the outlining stage, as follows:

1. Open a new part file and make sure that the measurement system is set to **MMGS (millimeter, gram, second)**. Note that we have selected **MMGS** for practice purposes only.

2. Navigate to the **Sketch** mode using any of the default planes (for example, **Top Plane**).

3. On the command bar, select the **Circle** command. Click on the origin, and then move the mouse further to form a circle. Click again to finish drawing the circle:

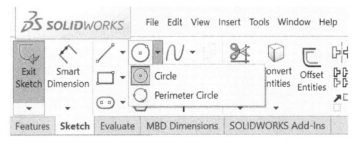

Figure 3.32 – The Circle sketching command

4. Click on the **Centerpoint Arc** command and follow the instructions shown in the small command icon, creating two arcs (labeled **Arc 1** and **Arc 2** in *Figure 3.34*).

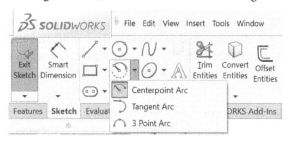

Figure 3.33: The Centerpoint Arc sketching command

5. Connect the endpoints of the arcs using the **Line** sketch command. The result will be similar to *Figure 3.34*. Note that we indicated the different points of the sketch with **P1-P4**, which stand for point 1, point 2, and so on:

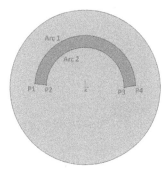

Figure 3.34 – The arc endpoints can be connected using lines

6. Use the **Centerpoint Arc** to create the last lower arc, which links the two lines we created in *Step 5*, to create the following sketch:

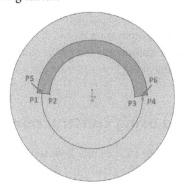

Figure 3.35 – All the sketch entities required for our shape

Now that we've finished outlining, we can start defining the sketch elements in the defining stage. Let's get started:

1. Note that **P1**, **P2**, **P3**, **P4**, and the origin are all in a horizontal line, as highlighted in *Figure 3.31*. To do this, select all the points and the origin and select the **Horizontal** relation. Alternatively, we can do the same in more steps by selecting **P1** and the origin and setting the relation to **Horizontal**. We can then do the same with **P2** and the origin, **P3** and the origin, and **P4** and the origin.

2. Set a **Midpoint** relation between **P5** and line **P1-P2**. Do this by selecting the point, **P5**, and the line it is on, and then select the **Midpoint** relation. Do the same for **P6** and the line **P3-P4**.

3. Set the dimensions shown in *Figure 3.36* using the **Smart Dimension** function:

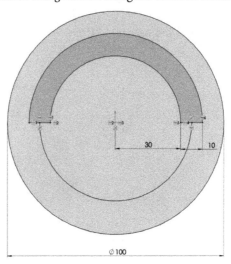

Figure 3.36 – The final sketch fully defined

This concludes this exercise of using circles and arcs. At this point, our sketch is fully defined. Before moving on, take some time to individually experiment with creating a perimeter circle, a tangent arc, and a three-point arc. These are some other ways we can create circles and arcs that we did not explore in this exercise. However, all these commands follow the same principles when it comes to making circles and arcs. Now that we've mastered how to create circles and arcs, we will address ellipses and construction lines.

Sketching ellipses and using construction lines

In this section, we will discuss what ellipses are, how to define them, and how to make them in SOLIDWORKS. We will also touch on the concept of construction lines. We can look at an ellipse as a combination of two axes and five points, as shown in *Figure 3.37*:

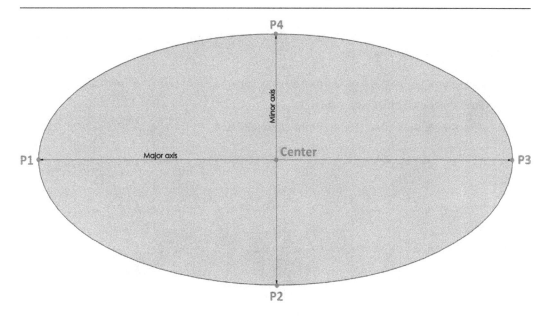

Figure 3.37 – A breakout of what makes an ellipse

When we define an ellipse in SOLIDWORKS, we can use the four points and the center as our defining factors. We can also define an ellipse with the help of construction lines, which we can use to define the size and the location of the ellipse. To illustrate this, we will sketch the following ellipse:

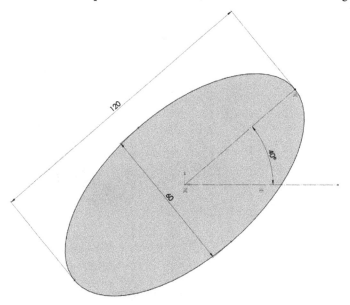

Figure 3.38 – The final result of this sketching exercise

Similar to all our other exercises, we will sketch the ellipse in two stages: outlining and then defining. Let's start with our outlining stage:

1. Start a new part file and set the measurement system to **MMGS (millimeter, gram, second)**.

2. Start the Sketch mode using the right plane (or any other plane).

3. Select the **Ellipse** command from the command bar.

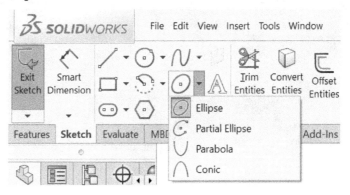

Figure 3.39 – The Ellipse sketching command

4. We will need to click three times to create an ellipse. First, click on the origin. Then, move the mouse to create the major axis and then left-click to confirm it. After that, move the mouse once more to create the minor axis and left-click again to confirm it. Make sure that the ellipse is tilted a bit to avoid unnecessary automated relations being made by SOLIDWORKS. We should have a shape similar to the one shown in *Figure 3.40*:

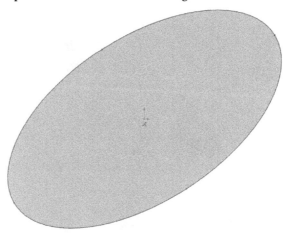

Figure 3.40 – An arbitrary starting sketch for an ellipse

As usual, now that we've finished our shape outline, we can start our definition stage:

1. First, use a smart dimension to define the lengths of the major and minor axes. We will set them to 60 mm for the minor axis and 120 mm for the major axis. We can define the lengths of the axes by defining the distance between the points on the perimeter of the ellipse.

2. To set up the angle of the ellipse, we can use construction lines. Construction lines are dotted lines that are used to support the definition of our sketches. However, they are not accounted for by SOLIDWORKS when building features. To set up construction lines, select the normal line command and then check the **For construction** option in the **Options** panel on the left, as shown in *Figure 3.41*. After that, all the lines we sketch will be construction lines:

Figure 3.41 – The construction line setup from the line PropertyManager

Alternatively, we can sketch normal lines and turn them into construction lines. Do that by clicking on the normal line or any other sketch entity and checking **For construction** from the **Options** tab that appears on the left.

3. Draw the two construction lines that are shown in *Figure 3.42*:

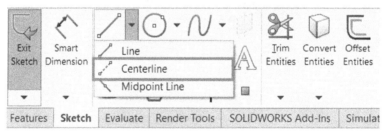

Figure 3.42 – Construction lines can play a role in defining our sketches

4. Using the smart dimension, set the angle between the two construction lines to 40 degrees.

This concludes this exercise on creating an ellipse. In this exercise, we covered how to draw and define an ellipse, as well as what construction lines are and how to utilize them to define our sketches.

In the exercise, we defined construction lines using **Line Properties**. However, you can also sketch one directly by using the drop-down menu for the line command and selecting **Centerline**, as shown in *Figure 3.43*.

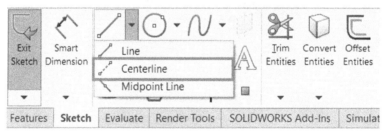

Figure 3.43 – The Centerline command can directly sketch construction lines

Both ellipses and construction lines will be very useful as we advance our SOLIDWORKS skills. Now, we will look at the fillets and chamfers commands so that we can improve our sketching skillset further.

Fillets and chamfers

In this section, we will discuss making fillets and chamfers for our sketches. Fillets and chamfers can be applied between two sketch entities, usually between two lines. They are defined as follows:

- **Fillets**: Fillets can be viewed as a type of arc. Thus, they are defined in the same way, that is, with a center and a radius.

- **Chamfers**: Chamfers can be defined in different ways. These include two equal distances, two different distances, or a distance and an angle.

Figure 3.44 illustrates the shapes of fillets and chamfers, as well as how they are defined:

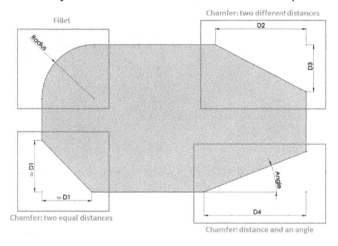

Figure 3.44 – The sketch fillets and the different types of chamfers

To illustrate how to create fillets and chamfers, we will sketch what's shown in *Figure 3.45*:

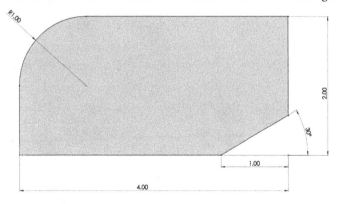

Figure 3.45 – The final result of this sketching exercise

Fillets and chamfers are different than other sketching commands in that we apply and define them at the same time. Thus, they don't follow our typical procedure of outlining and then defining. To create the sketch, we will use the IPS measurement system. To achieve the preceding sketch, we will need to sketch a 4 x 2-inch rectangle, which will be our starting point. Note that we marked the different lines in the following figure with the letter **L** and marked the different vertices with the letter **V** for ease of reference:

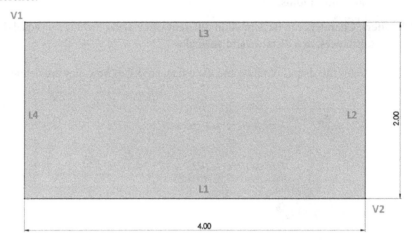

Figure 3.46 – A fully defined rectangle as a starting point to apply the fillet and chamfer to

Now, we can start creating the fillet, as follows:

1. A sketch is positioned based on a vertex or the two lines around a vertex. Select lines 3 and 4 (**L3** and **L4**) and then select the **Sketch Fillet** command from the command bar, as follows. Alternatively, we can select vertex 1 (**V1**) and then select the **Sketch Fillet** command:

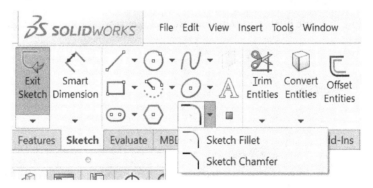

Figure 3.47 – The Sketch Fillet command

2. This will bring up a preview of the fillet on the canvas. You can find more options on the left-hand side, in place of the design tree, as shown in *Figure 3.48*. Under **Fillet Parameters**, fill in the radius of the fillet as 1 inch. Then, click on the green checkmark. This concludes making the fillet. Now that we've made the fillet, we can press *Esc* on the keyboard to exit the fillet sketching mode:

Figure 3.48 – The Sketch Fillet PropertyManager showing the fillet parameters

> **Tip**
> Press *Ctrl* + *Z* to undo the fillet you've applied.

Now that we've sketched the fillet, we will move on to sketching the chamfer:

1. The chamfer that we had in our final sketch is defined by an angle and a distance. Note that the angle displayed indicates the angle measurement between **L2** and the chamfer itself. Thus, hold down *Ctrl*, select **L1**, and then select **L2**. After that, select the **Sketch Chamfer** command. Similar to the **Fillet** command, we will get a preview of the chamfer. Also, on the left-hand side, we will find more options that we can use to define our chamfer.

2. Select **Angle-distance**. Then, set the angle to 30 degrees and the distance to 1 inch.

3.   Click on the green checkmark afterward:

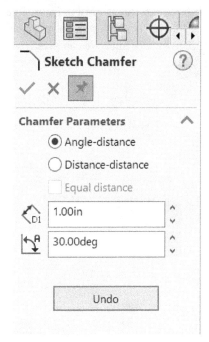

Figure 3.49 – The Sketch Chamfer PropertyManager showing the chamfer parameters

> **Tip**
> If you apply the wrong chamfer, you can click on the **Undo** button. Alternatively, you can press *Ctrl + Z* on your keyboard.

This concludes how to create fillets and chamfers. In this exercise, we learned what fillets and chamfers are, what defines them, and how to create fillets and chamfers in SOLIDWORKS sketching mode.

At this point, we know how to use all the major basic sketching commands. You will use these commands over and over again when working with the software. Now, we will dig deeper into what the different types of definition statuses mean, that is, under defined, fully defined, and over defined.

Under-defined, fully defined, and over-defined sketches

SOLIDWORKS sketches can fall under three status categories according to how they are defined. They can be under defined, fully defined, or over defined. These terms have already been mentioned briefly, but in this section, we will explore what those statuses are, as well as some ways to deal with them. We will explore these different statuses by drawing and defining a triangle and forcing it to go over the different statuses.

Under-defined sketches

Usually, the starting point of a sketch is under defined. Under-defined sketches have parts of them that are loose or lack proper definition; for example, a line without a specific length. To find out more about under-defined sketches, we'll examine the following sketch. We will use the MMGS measurement system for this exercise. We have indicated the lines and points with the letters **L** and **P** for reference:

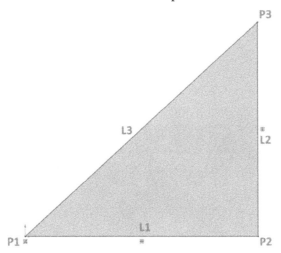

Figure 3.50 – An under-defined sketch consisting of three unrestrained lines

Take some time to sketch the preceding sketch and examine it. From this sketch, we can see the following:

- Note that, in your SOLIDWORKS canvas, **P1** and **L1** are black. Black colors indicate that those parts are fully defined. To test this further, click and hold **P1** or **L1** and try to move the mouse. You will notice that the sketch entity doesn't move. This is an indication that that part of the sketch is fully defined.

- The other parts of the sketch (**P2**, **P3**, **L2**, and **L3**) are blue. This indicates that those parts haven't been defined yet. To test this, click and hold any of those parts and move the mouse. You will notice that those sketch parts move around the canvas. If any part of the sketch is blue, the sketch will be labeled as **Under Defined** at the bottom of the SOLIDWORKS interface, as shown in *Figure 3.51*:

Figure 3.51 – The sketch indicated as under defined

We will always try to make our sketches fully defined. To fully define a sketch, we can simply add more dimensions and/or relations to turn the blue parts black.

To find out which parts of the sketch need definition, we can click and hold on to any of the blue parts and move them. The resulting movement tells us which parts need definition. For example, if we hold and move **P2** left and right, we will notice that **L1** changes in length. This indicates that we can define **P2** by setting a dimension for **L1** (or between **P1** and **P2**), as shown in *Figure 3.52*. After defining the length, we will notice more lines turned black. Now, if we click on **P2** and try to move it, it will be fixed:

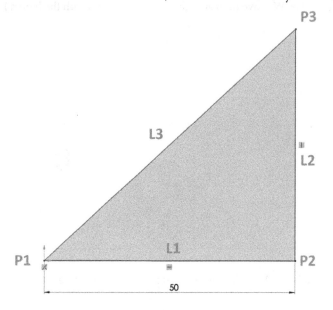

Figure 3.52 – A sketch that is partially defined, with the interface showing blue and black lines

We can do the same movement test for **L3** and **P3** and decide what elements of the sketch we can define further.

Fully defined sketches

Fully defined sketches are where all of the parts of the sketch are fully fixed. In other words, no part of the sketch can be moved from its current position. To illustrate this, we will take another look at the sketch we started and fully define it. We will do this by adding an angle of **45** between **L1** and **L3**. The fully defined result is as follows:

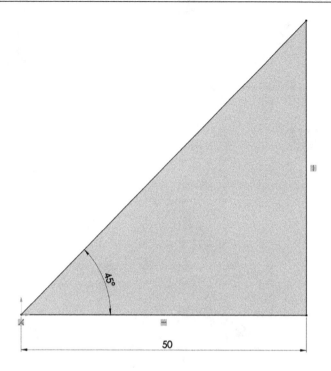

Figure 3.53 – A fully defined triangle

Note that the sketch is now fully black on your screen. Also, if you hold and try to move any of the sketch parts, they will not move because they are fully restrained. When the sketch is **Fully Defined**, SOLIDWORKS will take note of this at the bottom of the interface. Remember that we fully defined the sketch by adding relations and dimensions until all the sketch elements were fixed:

Figure 3.54 – The sketch indicated as Fully Defined

When sketching with SOLIDWORKS, we will mostly try to make our sketches fully defined in order to capture our full design intent. Now, let's look at over-defined sketches.

Over-defined sketches

Over-defined sketches are those with more relations and dimensions than are needed for the sketch elements to be fully fixed. This is not a recommended status to have for a sketch. Over-defined sketches occur when we apply contradicting relations to define a particular part of a sketch. To illustrate this, we can add an extra dimension to **L2**. Once we do that, we will get the following message:

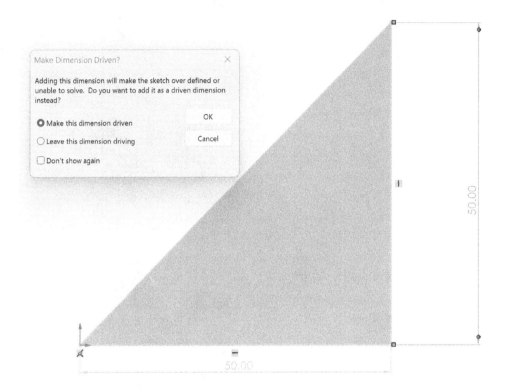

Figure 3.55 – A warning for over-defined sketches

Let's examine these two options and their results:

- **Make this dimension driven**: This will make the dimension so that it's just for show. This option will only tell us the length of the line. If the lengths of the other lines change, the displayed length here will change accordingly. Choosing this option will not cause any issues.

- **Leave this dimension driving**: This will give the dimension driving power, just like the other dimensions. However, since the sketch is fully defined already, selecting this option will leave our sketch with conflicting relations and a warning message that the sketch is **Over Defined**, as shown in the following screenshot. Once we are in an over-defined sketch situation, we will need to delete some existing relations or dimensions to get rid of the over-defined status.

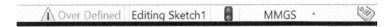

Figure 3.56 – A sketch indicated as Over Defined

Keep in mind that the best practice is to have our sketches fully defined to ensure that we are fully capturing the design intent of our sketch.

All the definition statuses are related to how many relations and dimensions we add to our sketch. A fully defined sketch has all the sketch elements fully fixed with the proper number of relations and dimensions to fully capture the design's intent. Under-defined sketches have fewer relations than they require, and over-defined sketches have more relations than needed.

Summary

In this chapter, we learned about the different aspects of SOLIDWORKS sketching that form our sketching foundations. We learned what sketching is and how to sketch different sketching elements, including lines, rectangles, circles, arcs, ellipses, fillets, and chamfers. We also covered using dimensions and relations to define sketches, as well as the meaning of the different sketch definition statuses, that is, under defined, fully defined, and over defined.

All of this information is part of our sketching foundation, which we will use every time we build a 3D model with SOLIDWORKS. In the next chapter, we will address additional sketching commands that can greatly enhance our sketching performance and speed, such as patterns and mirrors.

Questions

1. What is the position of SOLIDWORKS sketching in modeling?
2. What are the two stages we commonly follow when sketching?
3. Sketch the following shape using the MMGS measurement system:

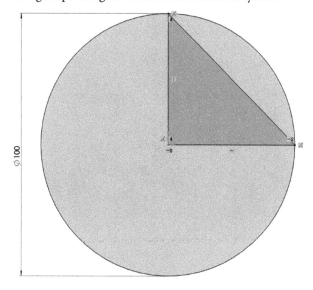

Figure 3.57 – The final output of question 3

4.   Sketch the following shape using the IPS measurement system:

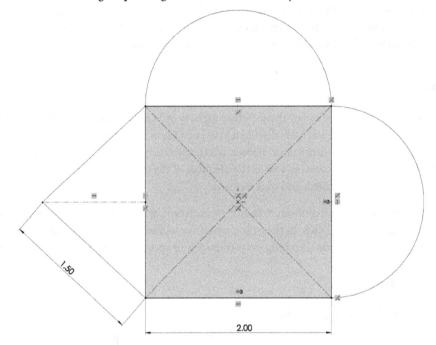

Figure 3.58 – The final output of question 4

5.   Sketch the following shape using the CGS measurement system:

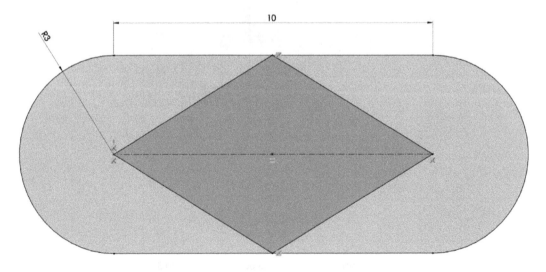

Figure 3.59 – The final output for question 5

6. Sketch the following shape using the MMGS measurement system:

Figure 3.60 – The final output for question 6

7. What are under-defined, fully defined, and over-defined sketches?

> **Important note**
> The answers to the preceding questions can be found at the end of this book.

Get This Book's PDF Version and Exclusive Extras

Special Sketching Commands

Gaining the ability to sketch rectangles and ellipses is not enough to master SOLIDWORKS sketching. We also need some special commands that can make sketching complex shapes faster and simpler.

In this chapter, we will introduce sketching commands such as mirroring, offsets, patterns, and trimming. We will also cover examples where we will use multiple shapes and commands to create relatively complex sketches. Even though we can continue without using these commands, they will greatly enhance the efficiency of our sketching creation process.

In this chapter, we will cover the following topics:

- Mirroring and offsetting sketches
- Creating sketch patterns
- Trimming in SOLIDWORKS sketching

By the end of this chapter, you will be able to use the previously mentioned sketching commands to both optimize and speed up your sketching process.

Technical requirements

In this chapter, you will require access to the SOLIDWORKS software.

The CiA video for this chapter can be found at `https://packt.link/wZbaB`

Mirroring and offsetting sketches

Some of the sketching commands in SOLIDWORKS allow us to easily create more sketch entities based on ones we already have, including circles, rectangles, lines, or any combination of sketch entities. Examples of such sketching commands are mirroring and offsetting. Using these commands will help us avoid creating similar sketch entities more than once. Here, we will start by exploring the mirroring and offsetting sketching commands. We will learn about what these commands do and how we can use them.

Mirroring a sketch

As the name suggests, mirroring a sketch means to reflect one or more sketch entities around a mirroring line. It is very similar to reflecting an image in a mirror. *Figure 4.1* illustrates the components of mirroring in SOLIDWORKS:

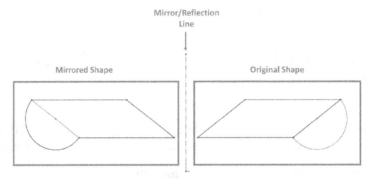

Figure 4.1 – Mirroring a sketch creates a reflection of it around a mirror line

In the preceding diagram, we can see that there are two parts that we need in order to use mirroring:

- A sketch entity to mirror
- A mirroring or reflection line

Since the two shapes are mirror sketch entities of each other, any changes that happen to one shape will automatically happen to the other.

To highlight how we can use mirroring, we will create the following shape. Note that the shape is a right-angle triangle with a mirrored reflection and that we will use the **millimeter, gram, second (MMGS)** measurement system for this exercise:

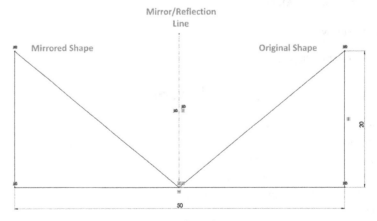

Figure 4.2 – The end result of the sketch mirroring exercise

To create this shape, we need to complete the outlining and defining stages. Here, we will create the base/original sketch and then mirror it. Then, we will define the resulting sketch.

Let's start by outlining our general shape and applying the **Mirror Entities** command. Follow these steps:

1. Sketch the outline of the first triangle and the mirroring line, using a centerline or construction line for the latter:

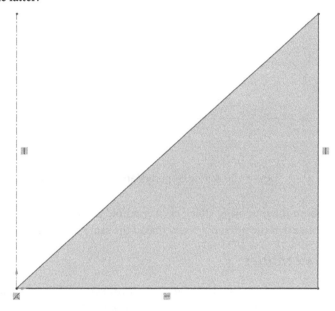

Figure 4.3 – An underdefined outline of a right-angle triangle

2. Select the **Mirror Entities** command, as shown in *Figure 4.4*:

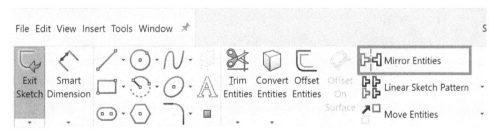

Figure 4.4 – The Mirror Entities command

3. On the left-hand side, we will see the available mirroring options. For **Entities to mirror**, select the three lines that make up the right-angle triangle (**Line2**, **Line3**, and **Line4**). For **Mirror about**, select the centerline (**Line1**). Make sure that the checkbox for **Copy** is also checked.

Now, we will see a preview of the triangular mirrored shape on the left. Click on the green checkmark to approve the mirroring:

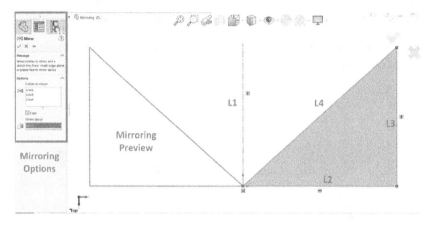

Figure 4.5 – Mirror PropertyManager and preview

To find out more about the mirroring effect, click and hold any of the blue parts of the sketch and move it. You will notice that the mirrored entity makes the same movement as the original shape.

Defining mirrored entities

Now that we have our shapes outlined, we can start defining our whole sketch. We can define the sketch by adding the dimensions and relations shown in *Figure 4.6*. Note that, when defining one line, the mirrored line will also be defined in the same way, as shown in the **20** mm dimension. Also, note that we can add dimensions between the mirrored entities, as shown in the **50** mm dimension. After adding those dimensions, we will notice that the sketch becomes fully defined:

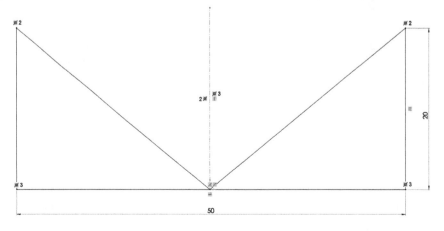

Figure 4.6 – The resulting sketch after fully defining it

In this exercise, we defined the whole sketch after applying the **Mirror Entities** command. An alternative approach is to fully define the first triangle before using the **Mirror Entities** command. In this scenario, the mirrored triangle will also be fully defined, directly after mirroring.

The **Copy** option in mirroring is used if we wish to keep the original shape. If the **Copy** option is checked, we will keep the original shape and create a mirrored shape. If it is unchecked, the original shape will be deleted, and we'll be left with the mirrored shape only. In addition, whatever dimensions and relations are on the original shape will be removed.

> **Tip**
> Centerlines/construction lines are not explicitly required to mirror sketch entities. An alternative to using construction lines is to use an existing line in the sketch as a reflection line for mirroring.

This concludes the section on using the **Mirror Entities** sketching command. Note that, in practice, the order in which we outline and define our sketches can vary. As you gain more experience with sketching, you will develop your own approach to doing things. Now, we can start learning about the **Offset Entities** sketch command.

Offsetting a sketch

Offsetting a sketch allows us to generate sketch entities that are similar to existing ones by oving the original sketch entities a certain distance away while maintaining all their features. The following figure shows an example of a sketch and its offset. Note that the original sketch is the one we sketch first; after that, the offset sketch is created by applying the **Offset Entities** sketch command. The offset sketch is defined by inputting an offset distance as shown in *Figure 4.7*

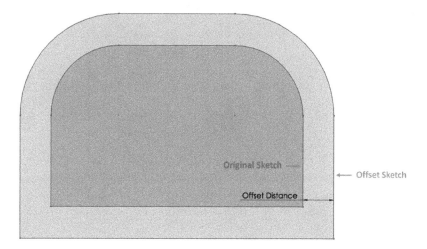

Figure 4.7 – A sketch and its offset

To illustrate how we can create an offset distance, we will create the sketch shown in *Figure 4.8*:

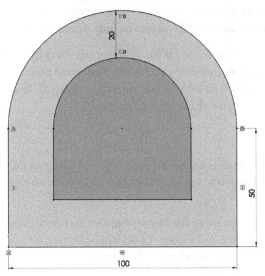

Figure 4.8 – The end result after running the Offset Entities command

We will use the MMGS measurement system for this exercise. To sketch the given shape, we will create and fully define the original sketch. After that, we will create the offset sketch using the **Offset Entities** command. Follow these steps:

1. Sketch and define the original sketch using the sketching commands we covered in *Chapter 3*. The result is shown in *Figure 4.9*:

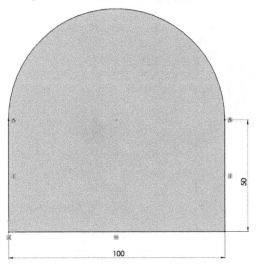

Figure 4.9 – The original sketch before applying an offset

2. Select the **Offset Entities** sketching command, as highlighted in the following screenshot:

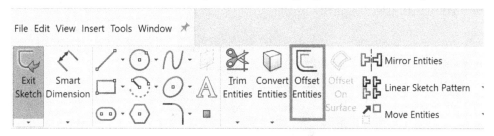

Figure 4.10 – The Offset Entities command

3. An options panel for the command will appear on the left-hand side of the screen. From there, we can customize our way of defining our sketch. Set the options that are shown in *Figure 4.11*, which will give us our desired result:

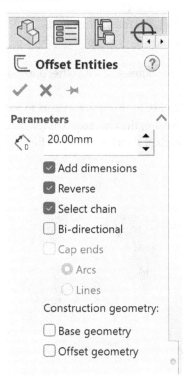

Figure 4.11 – The Offset Entities PropertyManager options

4. Select any of the lines in the original sketch we created. A preview will then be shown on the sketch canvas, illustrating the result of the offset. The preview will be shown in yellow, as

highlighted in *Figure 4.12*. Any changes that are made to the options will be reflected in the preview as well:

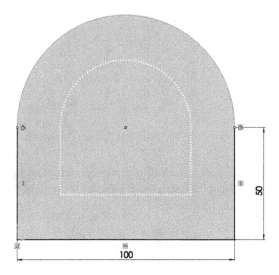

Figure 4.12 – An offset preview

5. Click on the green checkmark in the options panel to apply the offset. This will result in the following figure, which matches the final required sketch. Note that the offset sketch is already fully defined since our original sketch was fully defined as well:

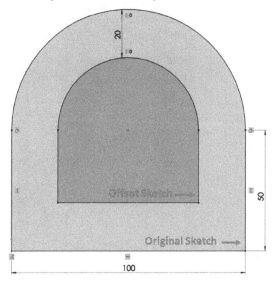

Figure 4.13 – The final result of the offset sketch exercise

In addition to using the **Offset Entities** sketching command, we need to learn how to delete an offset and learn about the customization options associated with offsets. Let's explore those.

Deleting an offset

To delete an offset entity, we can select that entity and press *Delete* on the keyboard. Alternatively, we can right-click on the entity and select **Delete**. Deleting an offset is just like deleting a normal sketch entity.

Customization options

- The following are some of the different options that accompany the **Offset Entities** command. Those options allow us to customize the command to best meet our needs and are displayed in the **Offset Entities** PropertyManager, as shown in *Figure 4.11*:

- **Add dimensions**: This option makes the offset fully defined upon implementation, as in our previous example. Unchecking this option will create the offset; however, it will not add the offset dimension, thus making it undefined.

- **Reverse**: This changes the direction of the offset. For example, the default offset direction of an enclosed circle is outward, forming a bigger circle. Checking this option changes the offset direction to inward.

- **Select chain**: This command helps us select all the sketch entities that are linked together. For example, with this option, selecting one line of a rectangle will automatically select the whole rectangle, since the four lines that make up the rectangle are connected. We should uncheck this option if we want to offset only a part of the shape; for example, only one line of a rectangle.

- **Bi-directional**: This will apply two offsets in two different directions, outward and inward.

- **Cap ends**: This command is applied when the offset is not an enclosed shape, such as an enclosed rectangle or a circle. The **Cap ends** option allows us to easily enclose open loops between the original and the offset sketch entities.

- **Construction geometry**: This allows us to make the original or the offset sketch entities construction entities (for example, construction lines, arcs, and more). Checking **Base geometry** will change the original sketch entities into construction entities. Checking **Offset geometry** will do the same for the offset entities.

This concludes the section explaining how we can use the **Offset Entities** sketching command. We have learned what the **Offset Entities** command does and how to use it. We have also covered how to use the **Mirror Entities** sketching command.

To recap, the **Mirror Entities** command allows us to quickly reflect a copy of a sketch around a reflection line; this means both the original and the mirrored entities will keep imitating each other. The **Offset Entities** command allows us to create a copy of a sketch entity by offsetting it by a certain distance.

Now, we can start learning about how to create patterns.

Creating sketch patterns

Sketch patterns allow us to easily copy a sketch entity multiple times in a pattern formation. Such sketch patterns can be created in a **linear** or **circular** formation. In this section, we will cover creating patterns in both formations.

Defining patterns

Patterns are repeated formations that are commonly found in consumer products, architecture, fabrics, and more. In patterns, we often have a base shape, sometimes called a **base cell** or **patterned entity**, which is created from scratch. Then, this basic shape is duplicated multiple times to form a bigger pattern. There are two common types of patterns: **linear patterns** and **rotational/circular patterns**. Examples of both types of patterns will be shown in this section.

Linear patterns are ones in which you have a base shape (patterned entity) that is repeated linearly in different directions. Linear patterns are commonly found in curtains, carpets, building tiles, floors, and architecture. The following diagram is an example of a linear pattern. The patterned entity shape is highlighted with a red square:

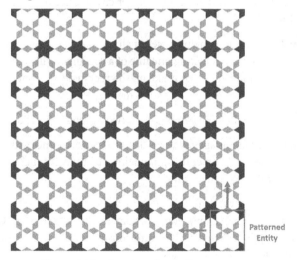

Figure 4.14 – An example of a linear pattern

Rotational patterns are ones in which we have a base shape that is repeated as we rotate it. A common application for the rotational pattern is in car rims, as highlighted in the following figure. Within the SOLIDWORKS interface, rotational patterns are known as circular patterns:

Figure 4.15 – An example of a rotational/circular pattern

SOLIDWORKS' pattern tools make it easier for us to create similar linear and circular patterns when sketching. Now that we know what these two types of patterns are, we can start exploring them within SOLIDWORKS, starting with linear patterns.

Linear sketch patterns

Linear sketch patterns allow us to pattern sketch entities in a linear direction. The following sketch shows us how we can define linear patterns in SOLIDWORKS sketching:

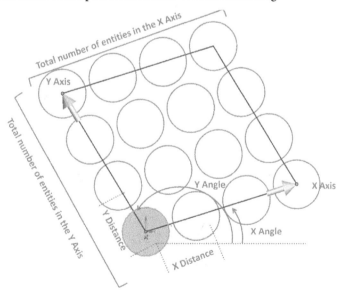

Figure 4.16 – The elements needed to define a linear pattern in SOLIDWORKS

In the preceding sketch shown in *Figure 4.16*, the shaded circle is the base circle, while the other ones are additions to be made with the pattern command. The annotations in writing are the parameters that we need in order to define a pattern in SOLIDWORKS sketching. Each of the annotations is repeated twice: once for the *X* axis and once for the *Y* axis. They are as follows:

- **Axes**: The *X* axis and *Y* axis represent the direction in which our pattern is implemented.

- **Total number of entities**: The number of times we want the entities to be sketched, including the base entity. In the preceding sketch, the number of entities is four for the *X* and *Y* directions.

- **Distance**: This specifies the distance between two entities in a certain direction. In the preceding sketch shown in *Figure 4.16*, these distances are highlighted as **X Distance** and **Y Distance**.

- **Angle**: This specifies how tilted our axes should be, since the axes determine the direction of the patterns. Therefore, the whole pattern will shift as we change the direction of an axis. In the preceding sketch, the angles are highlighted as **X Angle** and **Y Angle**. Note that the *X* and *Y* angles start from the same baseline.

Now that we know what elements define a linear pattern, we can start creating one in SOLIDWORKS. To highlight this, we will sketch the following figure. We will use the MMGS measurement system for this exercise:

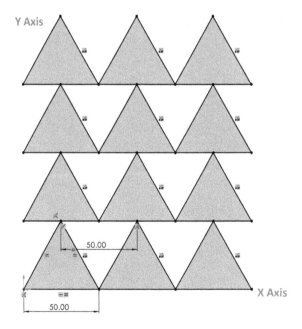

Figure 4.17 – The final result of the linear pattern exercise

To sketch the preceding diagram, follow these steps:

1. Sketch and define the base equilateral triangle, as in *Figure 4.18*:

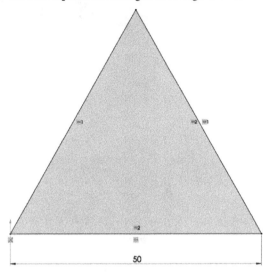

Figure 4.18 – The base triangle used for the pattern

2. Select the **Linear Sketch Pattern** command as shown in *Figure 4.19*:

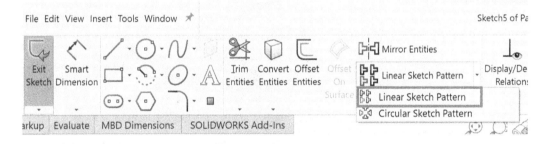

Figure 4.19 – The Linear Sketch Pattern command

3. Select the three lines that form our base triangle. Now, we can set up our linear sketch pattern using the options that appear on the left-hand side of SOLIDWORKS. Set the following options:

Figure 4.20 – The options for setting up a linear sketch pattern

As we are adjusting those options, a preview of the final shape will appear on the canvas. From top to bottom, the options we are using are as follows:

- **Direction**: For **Direction 1**, we can see the direction of the X axis. If we click on the two black and gray arrows next to the X axis, the direction of the pattern will shift by 180 degrees. For **Direction 2**, the direction of the pattern will go toward the Y axis.

- **Distance**: This determines the distance between every two entities that are adjacent to each other.

- **Dimension X spacing**: This makes the dimension set a fixed driving dimension. Unchecking this option will not add a dimension to the pattern; instead, the listed dimension in **Distance** will be the starting point. The **Dimension Y spacing** option does the same in the second direction.

- **Number of instances**: This indicates how many times we want an entity to be drawn, including the base sketch.

- **Display instance count**: Checking this option will show the number of instances on the drawing canvas.

- **Angle**: This sets the direction of the X axis and Y axis, which governs the pattern's direction.

- **Fix X-axis direction**: Checking this option will make the angle a driving dimension. Unchecking this option will not add the angle direction to the pattern; therefore, we need to identify it separately.

- **Dimension angle between axes**: Checking this option will make the angle between the X and Y axes a driving dimension. This option is used to define the Y-axis direction in our canvas.

- **Entities to Pattern**: This lists all the entities that will be patterned with the command. We can delete entities by deleting them from the list. We can add entities by selecting them in the sketch canvas.

Our preview may look something like the following *Figure 4.21*:

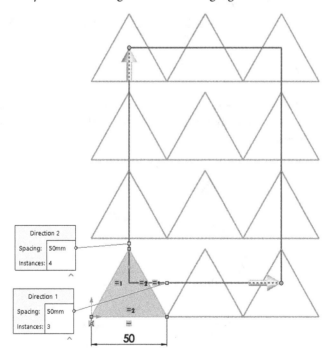

Figure 4.21 – The resulting preview of the pattern

4. After adjusting these options, we can click on the green checkmark. We will see the shape shown in *Figure 4.22*. Note that the first row of triangles in the x direction is fully defined, while the

other triangles extending in the *y* direction are not. To understand how such patterns work, we can click and drag the blue parts around our screen. We will see that all the patterned shapes move together:

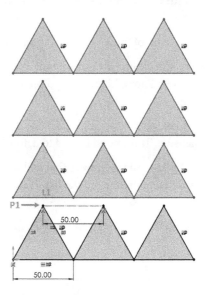

Figure 4.22 – The pattern might not be fully defined after application

5. To fully define the pattern, we can define the shapes in the *y* direction. We will need to add a midpoint relation between **L1** and **P1** from *Figure 4.22*. This will make the sketch fully defined, as shown in *Figure 4.23*.

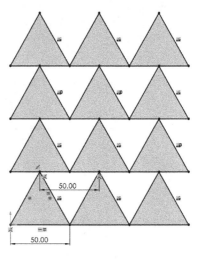

Figure 4.23 – The final result of the linear sketch pattern exercise

> **Tip**
>
> We can also define a dimension for the y-direction pattern using the **Linear Pattern** command PropertyManager, as shown in *Figure 4.20*, by checking the **Dimension Y spacing** option. However, in this exercise, we used a relation to define the Y-axis pattern.

Let's have a look at some related commands:

- **Instances to Skip**: At the bottom of the **Linear Pattern** PropertyManager options, we will find the **Instances to Skip** option. We can use this to skip instances of the pattern. For example, in the preceding exercise, we could have removed the two middle triangles from the pattern by adding them to **Instances to Skip**:

Figure 4.24 – The Instances to Skip option in sketch patterns

- **Edit Linear Pattern**: To edit an existing linear pattern, we can right-click on any of the patterned instances and select the **Edit Linear Pattern** option, as shown in the following *Figure 4.25*:

Figure 4.25 – The Edit Linear Pattern command

- **Delete Linear Pattern**: To delete the pattern (or parts of it), we can simply select or highlight the relevant entities and press *Delete* on the keyboard.

This concludes this subsection on using linear patterns. In this subsection, we have covered the following topics:

- How to set up and define linear patterns
- The different options that we can use to define a linear pattern
- How the linear pattern entities interact as under-defined entities

Now that we have mastered linear patterns, let's move on and look at circular sketch patterns.

Circular sketch patterns

Circular sketch patterns allow us to pattern sketch entities in a circular direction. The following figure highlights how we can define a circular pattern in SOLIDWORKS sketching:

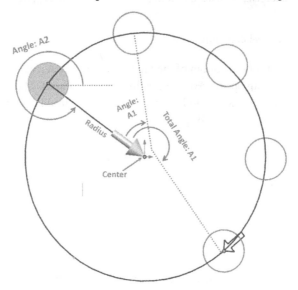

Figure 4.26 – The elements needed to define a circular pattern

In the preceding sketch, the shaded circle is the base circle, while the others are additions to be made with the pattern command. The annotations are the parameters that we need in order to communicate a pattern in SOLIDWORKS sketching. The following is a short description of the different annotations in the preceding figure:

- **Center**: This represents the center of rotation for the circular pattern. This can be determined with specific x and y coordinates or by relating it to another point.
- **Radius**: This is the distance between the original entity and the center of the pattern.
- **Angle: A1**: This is the angle between two adjacent patterned entities.
- **Total Angle: A1**: This is the angle between the original and the last patterned entity.
- **Angle: A2**: This is the angle between the original patterned entity and the **center**.
- **Number of patterned entities**: This shows the total number of patterned entities, including the base sketch.

To illustrate how to use circular sketch patterns, we will create the following sketch, using the **inch, pound, second (IPS)** measurement system in this exercise:

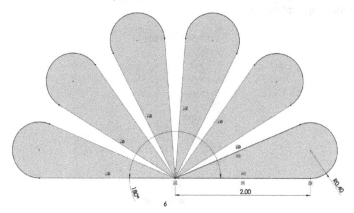

Figure 4.27 – The final result of the circular pattern exercise

To sketch the preceding diagram, follow these steps:

1. Sketch and fully define the base entity, as shown in the following figure:

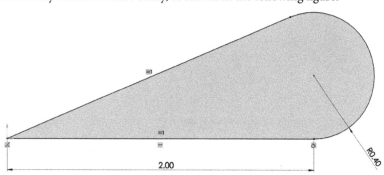

Figure 4.28 – The base sketch for the circular pattern

2. Select the **Circular Sketch Pattern** command from the **Sketch** command bar:

Figure 4.29 – The Circular Sketch Pattern command

3. Select the three lines that form our base sketch. Now, we can set up our circular pattern using the options that appear on the left-hand side of SOLIDWORKS. Set the options that are shown in the following screenshot:

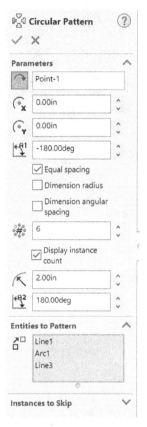

Figure 4.30 – The Circular Pattern options relating to this exercise

As we are adjusting those options, a preview of the final shape will appear on the canvas. From top to bottom, the options we will be using are as follows:

- **Center**: This field is populated with **Point-1** by default which represents the center of the pattern. By default, the center of the pattern will be the origin; however, we can change it by selecting other points. **Rotational direction**, which is to the left of the **Point-1** selection, is a button that will flip the direction of the rotation.

- **X and Y center locations**: The two fields marked with **X** and **Y** represent the location of the center of our circular pattern. Since our center is the origin, the location is marked as 0.00 inches for the *x* and *y* directions. We can use these fields to set up an exact center in the coordinate system.

- **A1 angle**: This defines the angle that will govern the locations of the circular pattern. In our example, it is set as -180 since the pattern goes counterclockwise by 180 degrees. Note that this does not define the dimension; instead, it helps us approximate the location and the look of the pattern. We can fully define the pattern after its implementation.

- **Equal spacing**: This will ensure that all the patterns are equally distributed in the angle range.

- **Dimension radius**: This will add a driving dimension to the radius of the pattern. Note that this is not needed if we merge the center with a fixed point, such as the origin.

- **Dimension angular spacing**: Checking this option will allow us to dimension the angle between the adjacent patterned instances, instead of the angle between the base sketch and the last patterned entity.

- **Number of patterned instances**: This indicates how many times we want the entity to be drawn or patterned, including the base sketch.

- **Display instance count**: Checking this option will show the number of instances on the drawing canvas.

- **Radius**: This allows us to linearly increase or decrease the radius between the patterned entities and the center.

- **A2 angle**: This will shift the center of the pattern so that it's at a certain angle.

4. After adjusting these options, we can click on the green checkmark. This will give us the shape highlighted in *Figure 4.31*:

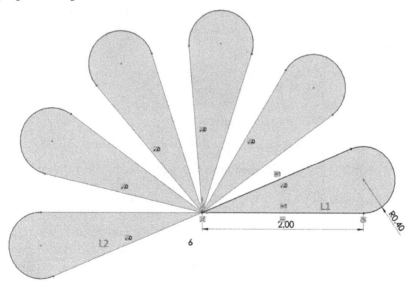

Figure 4.31 – The resulting sketch after applying the pattern might not be fully defined

> **Important note**
> The patterned entities are not fully defined. To understand how circular patterns work, we can click and drag the blue parts of the sketch around. We will see that the patterned shapes move together.

Set the angle between **L1** and **L2** so that it's equal to 180 degrees. Doing this will fully define the sketch, as shown in the following figure:

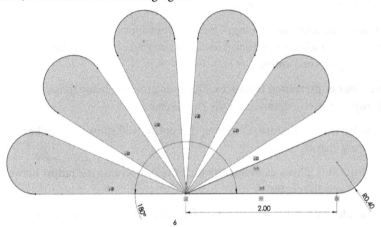

Figure 4.32 – The final resulting sketch from the exercise

Similar to linear patterns, we have the option of skipping instances, editing the circular patterns, and deleting instances. To do these things, we can follow the same procedure that we followed for linear patterns.

This concludes this subsection on using circular patterns. In this subsection, we have covered the following topics:

- How to set up and define circular patterns
- The different options that we can use to define a circular pattern
- How the circular pattern entities interact with each other as under-defined entities

In this section, we talked about linear and circular sketch sketching patterns. Both allow us to quickly repeat a selection of sketch entities multiple times, which saves a lot of time and energy. A linear pattern enables us to pattern entities linearly while a circular pattern enables us to pattern entities rotationally. We can also use both linear and circular patterns to build one sketch. However, we have to apply them separately. Now, we can start looking at another special sketching command – trimming.

Trimming in SOLIDWORKS sketching

Trimming in SOLIDWORKS allows us to easily remove unwanted sketch entities or unwanted parts of sketch entities. This makes it easier for us to create complex sketches. In this section, we will cover what trimming is, why we use trimming, and how to use trimming within SOLIDWORKS.

Understanding trimming

Trimming allows us to delete parts of sketches that are unwanted. This makes it easier to create complex sketches that go beyond what we can create with the standard sketch commands. This is because it makes it easier to sketch irregular shapes using different elements from various standard sketch commands. To explore this further, let's examine the sketch in *Figure 4.33* and how the trimming command can help create it:

Figure 4.33 – Trimming allows us an easy or alternative way of sketching this

We can start this simple sketch by sketching two circles, as shown in the following figure. After that, we can trim/remove the interfering parts to get our desired shape:

Figure 4.34 – Trimming unwanted sketch entities can get us new shapes

Also, depending on the sketch we want, trimming could make it easier for us to define the sketch according to our specific design intent. Now that we know what trimming is, we can start using the command.

Using power trimming

To show you how we can use the trimming tool in SOLIDWORKS, we will create the following sketch, which consists of a circle and a rectangle. We will use the MMGS measurement system for this exercise:

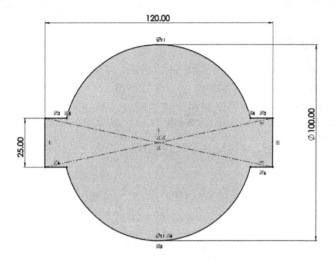

Figure 4.35 – The end result of the power trim tool exercise

To create the given sketch, follow these steps:

1.   Sketch and fully define the base shapes of a circle and a rectangle, as in the following *Figure 4.36*:

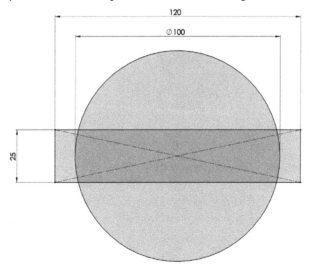

Figure 4.36 – Original sketches with different intersecting enclosures

2. Select the **Trim Entities** command from the command bar, as shown in the following *Figure 4.37*:

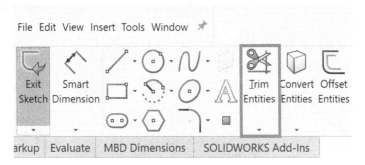

Figure 4.37 – The Trim Entities command

After selecting this command, the PropertyManager for it will appear on the left of the interface. This will show different types of trimming tools. Here is a brief description of the different trimming tools that are available:

- **Power trim**: This is the easiest method for trimming entities. **Power trim** is a multipurpose trimming tool that allows us to cut entities by hoering over them in the canvas. Power trim is what we will use in this section.

- **Corner**, **Trim away inside**, **Trim away outside**, and **Trim to closest**: These are different and more specific ways we can trim. A **Corner** trim makes it easier to trim two entities till they intersect at a projected corner. **Trim away inside** allows us to trim an entity that lies inside bounding entities, while **Trim away outside** allows us to trim entities that lie outside bounding entities. **Trip to closest** trims an entity to its closest two boundaries. We will utilize these specific trim options in this book.

- **Keep trimmed entities as construction geometry**: When checked, trimming will not remove entities from the canvas. Instead, entities will be converted to construction lines.

- **Ignore trimming of construction geometry**: When checked, trimming will not function with construction lines.

As shown in *Figure 4.38*, we have picked **Power trim** and, as we didn't want to revoke the construction lines that define our rectangle, we checked the **Ignore trimming of construction geometry** option:

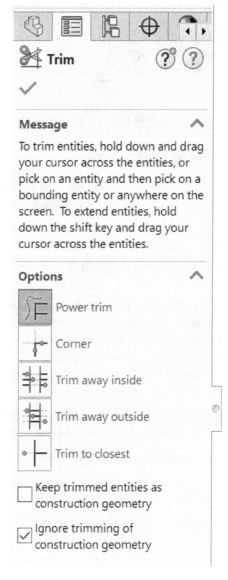

Figure 4.38 – The different options in the Trim Entities PropertyManager

3. Go back to the canvas and start trimming. To trim unwanted parts, we can click, hold, and move the mouse, as illustrated by the red lines in the following *Figure 4.39*:

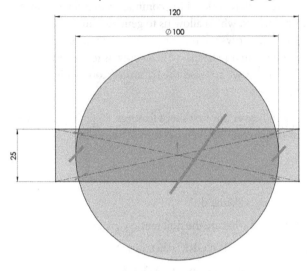

Figure 4.39 – The lines indicating the power trim movement for trimming unwanted entities

Inspect the final sketch after trimming. In our case, our sketch should look like *Figure 4.35*. As we cross each of these lines, we will notice that the line will disappear. By doing this, we will only remove the lines that are not desirable. If we trim the wrong line, we can simply undo this by pressing *Ctrl + Z*. After doing this, we will end up with the final shape, which was shown at the beginning of this section.

Tip

After trimming an entity using the **Power trim** command, a small red square will appear after each trim. Moving the mouse back to that red rectangle will undo that trim.

This concludes this section on trimming entities using SOLIDWORKS. In this section, we have covered the following topics:

*   How to set up the **Trim Entities** command
*   How to use trimming to get our desired result

The trimming tool is a powerful enhancement feature in SOLIDWORKS that allows us to quickly create complex sketches, significantly improving our sketching efficiency

Summary

In this chapter, we covered different sketching commands that can enhance our sketching capabilities. These included sketch mirrors, which allow us to generate mirrored sketch entities around a mirror line, and offsetting, which allows us to generate duplicated sketch entities at an offset. Then, we learned about linear and circular patterns, which allow us to create many instances of a sketch entity in a specified pattern. Finally, we covered the trimming tool, which allows us to remove unwanted parts of our sketches.

In the next chapter, we will cover our first set of features. So far, we've only learned about 2D sketches. Features will allow us to turn our 2D sketches into 3D models.

Questions

1. What does mirroring a sketch do?

2. What are patterns, and what are the different types of patterns we can create in SOLIDWORKS?

3. What is trimming in SOLIDWORKS sketching?

4. Sketch the following diagram using the MMGS measurement system:

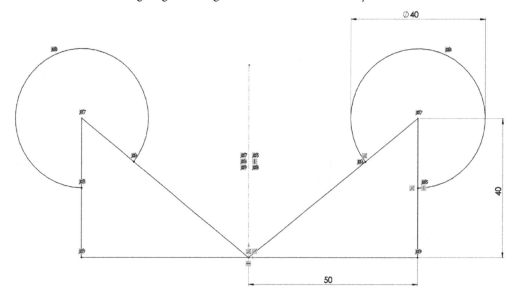

Figure 4.40 – The resulting sketch from question 4

5. Sketch the following diagram using the IPS measurement system:

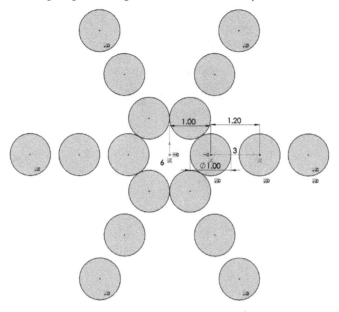

Figure 4.41 – The resulting sketch from question 5

6. Sketch the following diagram using the IPS measurement system (tip—use ellipses):

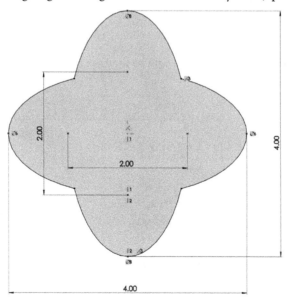

Figure 4.42 – The resulting sketch from question 6

7. Sketch the following diagram using the MMGS measurement system:

Figure 4.43 – The resulting sketch from question 7

Important note

The answers to the preceding questions can be found at the end of this book.

Get This Book's PDF Version and Exclusive Extras

Scan the QR code (or go to packtpub.com/unlock). Search for this book by name, confirm the edition, and then follow the steps on the page.

Note: Keep your invoice handy. Purchases made directly from Packt don't require an invoice.

Part 3:
Basic Mechanical Core Features – Associate Level

The Associate level is the first level of proficiency for SOLIDWORKS users. This part will cover all the 3D modeling features expected of that level. Those include extruded boss and cut, fillets, chamfers, revolved boss and cut, swept boss and cut, loft boss and cut, and reference geometries.

This part has the following chapters:

Basic Primary One-Sketch Features

In SOLIDWORKS, features are what can turn a 2D sketch into a 3D model. In this chapter, we will move on from 2D sketches and start creating 3D models. We will explore the most basic features, such as extruded boss and extruded cut, fillets, chamfers, revolved boss, and revolved cut. We will study how to apply, modify, and delete features. We will also start creating more complex models by applying multiple features. Each feature that's covered in this chapter requires only one sketch to apply or no sketch at all.

In this chapter, we will cover the following topics:

- Understanding features in SOLIDWORKS
- Understanding and applying extruded boss and cut
- Understanding and applying fillets and chamfers
- Understanding and applying revolved boss and revolved cut

By the end of this chapter, we will be able to create 3D models using the most common features in SOLIDWORKS. Even though the features covered in this chapter are simple, they will enable us to create complex 3D models.

Technical requirements

In this chapter, you will need to have access to SOLIDWORKS.

The CiA video for this chapter can be found at `https://packt.link/yXmKr`

Understanding features in SOLIDWORKS

SOLIDWORKS features are our way of moving from 2D to 3D. Similar to sketches, SOLIDWORKS provides many features that can help us create simple shapes. For more complex shapes, we will have to use more features that are applied on top of each other. In this section, we will discuss SOLIDWORKS features, simple versus complex models, and additional sketch planes.

Understanding SOLIDWORKS features and their role in 3D modeling

Features is the term we use to refer to the tools that allow us to construct 3D models based on sketches. We usually use features directly after sketching to go from two-dimensional sketches to 3D models.

For example, if we were to model a cube, we would follow these steps:

1. First, we would create a sketch of a square.

2. Then, we would apply a feature in order to make the square into a cube. We do this by extruding it.

Figure 5.1 illustrates these two steps:

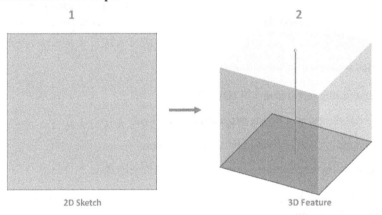

Figure 5.1 – A 2D sketch is used to build a 3D feature

Now that we know what features are, as well as their purpose, we can address how features differ in terms of simple and complex 3D models.

Simple models versus complex models

SOLIDWORKS offers a wide variety of features that can help us easily create different shapes. Most of these features are for simple shapes such as simple hexahedrons (cubes), rotational shapes such as spheres and tubes, and much more. Thus, we will apply more features that build on top of each other to create more complex models.

When we were sketching, we applied and mixed multiple sketch commands to create more complex sketches. This is similar to what we do with features. The more complex the model is, the more features it may require.

To highlight this, take a look at the models shown in *Figure 5.2*. The one on the left is a simple model of a cube. We only used one feature to create this cube. The model on the right is a turbine rotor. It is a more complex model, and we had to use 11 features to build it:

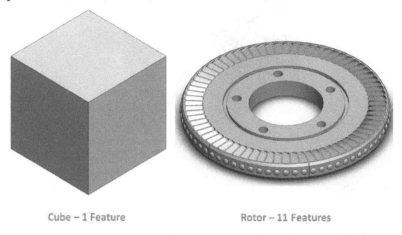

Cube – 1 Feature   Rotor – 11 Features

Figure 5.2 – More complex models can require more features to build

As we continue using SOLIDWORKS, we will be able to create more complex models. Now, let's learn about one fundamental aspect of all features – **planes**.

Sketching planes for features

By default, SOLIDWORKS provides three default planes: the front plane, the top plane, and the right plane. We will use one of these planes to create our first sketch and feature. As we start applying features, these three basic planes may not fulfill our needs for further sketches and features.

Thus, by creating more features, the resulting planner surfaces can also be used as sketching planes. We can use these to create even more sketches and features.

For example, for a new file, we will only have three sketch planes – the default base ones. If we create a cube, each face of the cube will also be a possible sketch plane. Thus, after creating the cube, we will be adding five potential sketch planes for the five new faces of the cube. *Figure 5.3* shows the three base planes, as well as some of the new planes that were created with the new cube. Note that some of the sketch planes may coincide with each other:

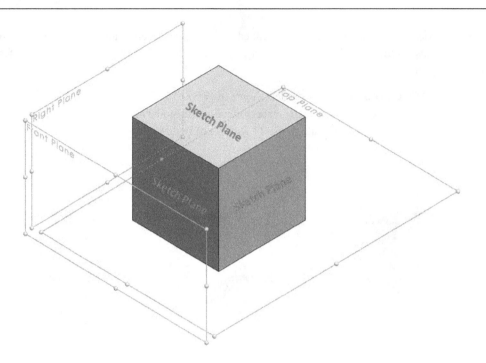

Figure 5.3 – A visual representation of the base planes and a cube's surfaces

> **Tip**
> Any planner surface can also be used as a sketch plane.

We have just learned what features are, which features are used in complex and simple models, and how sketch planes relate to features. Now, let's explore our first set of features – extruded boss and cut.

Understanding and applying extruded boss and cut

Extruded boss and extruded cut are the most basic and easiest features to apply. They are direct extensions of a sketch that push it into the third dimension. In this section, we will cover what extruded boss and extruded cut are, how to apply them, how to edit them, and how to delete them.

What are extruded boss and extruded cut?

Extruded boss and extruded cut are two of the most basic features we'll use when modeling with SOLIDWORKS. Let's look at them in more detail:

- **Extruded boss**: This is a direct extension of a sketch that pushes it into the third dimension, resulting in adding materials.

- **Extruded cut**: This is a direct extension of a sketch that pushes it into the third dimension, resulting in removing or subtracting the materials of an already existing solid model.

From these definitions, you can see that extruded boss and extruded cut are quite similar, but they have opposite effects. The extruded boss adds materials, while the extruded cut removes material. *Figure 5.4* illustrates the effect of the extruded boss feature. Note that we were able to go from a 2D sketch to adding materials to form a cube:

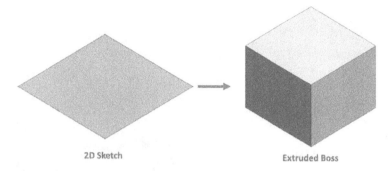

2D Sketch   Extruded Boss

Figure 5.4 – A representation of what the extruded boss feature does

Figure 5.5 illustrates the effect of an extruded cut. Note that we were able to use a sketch to remove materials:

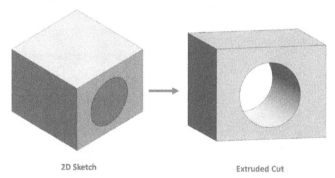

2D Sketch   Extruded Cut

Figure 5.5 – A representation of what the extruded cut feature does

Let's learn how to apply the extruded boss feature.

Applying extruded boss

In this section, we will discuss how to apply the extruded boss feature. To show this, we will create the model shown in *Figure 5.6*. We have added annotations for each view, including information about the view's type and its dimensions.

> **Important note**
> We will keep building with the same model as we explore extruded boss and extruded cut. Thus, keep saving the model as we go along.

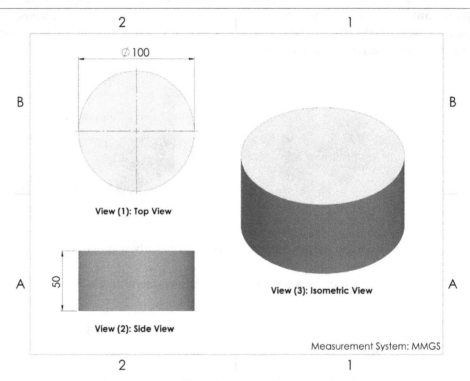

Figure 5.6 – The end shape of the extruded boss exercise

When applying the extruded boss and extruded cut features, we will always start with a 2D sketch and then apply that feature based on that 2D sketch. Thus, for this exercise, we will split each feature application into two stages – the sketching stage and the applying-the-feature stage.

One important aspect to keep in mind is that as we continue modeling, we will need to plan a strategy when it comes to how to model the targeted object. There is no right or wrong way to create a model. Thus, different people will have different plans for making the same model. It is always good to plan ahead when it comes to creating a model. We can do that by either sketching or writing down our ideas. Since we are taking our first steps toward 3D modeling, we will need to have a brief written plan before we start modeling:

1. **Planning**: We start by creating a circle and then extruding that into a cylinder.

2. **Sketching**: Next, we sketch and fully define a circle with a diameter of 100 mm. We can see this in *Figure 5.6*, where it says **View (1): Top View**. The circle will look as follows:

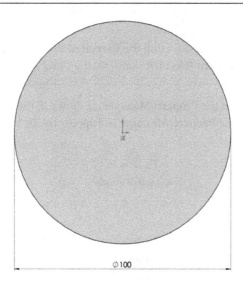

Figure 5.7 – A circular sketch used as a foundation for extruded boss

3. **Applying the feature**: We will apply the extruded boss feature.

To apply the features, follow these steps:

1. Click on the **Features** tab and select the **Extruded Boss/Base** command, as shown in the following screenshot. We don't need to exit sketch mode. As soon as we select this command, SOLIDWORKS will understand that we want to apply this feature to the active sketch:

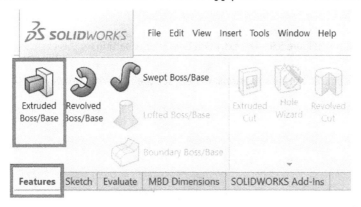

Figure 5.8 – The location of the Extruded Boss/Base feature

2. Once we click on the **Extruded Boss/Base** command, we will notice that an options panel appears on the left-hand side. The extrusion preview will also appear on the canvas.

> **Important note**
>
> If we exit sketch mode before selecting the **Extruded Boss/Base** command, we can simply select the command and then select the sketch on the canvas.

3. Fill out the options in the **PropertyManager**, as shown in the following screenshot. Fill in the height as 50 mm. The **PropertyManager** will appear on the left-hand side of the interface:

Figure 5.9 – The extruded boss feature setting for this exercise

4. Once we've filled in these options, we will see the following preview:

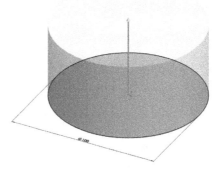

Figure 5.10 – A preview of the extruded boss application

5. After adjusting the options for our extrusion, we can click on the green check mark at the top of the **PropertyManager** panel to apply the extrusion, as highlighted in *Figure 5.11*.

Figure 5.11 – The green check mark on top of the PropertyManager to apply the feature

The result will be the 3D model shown in *Figure 5.12*.

Figure 5.12 – The resulting 3D model after the extrusion

Before we finish looking at the extruded boss feature, let's take a look at options in the **PropertyManager**. We will look at them based on their listing order, that is, from top to bottom, as shown in *Figure 5.13*. We will start with the options we used in this exercise and then move on and look at the options we didn't use:

Figure 5.13 – The different options available for the extruded boss feature

The following options are available for the extruded boss feature:

- **From:** This determines where the extrusion features should start. Since we are still beginners, we will mostly use **Sketch Plane**. This means that the extrusion will start from the sketch that was used to create it. Other options include starting from **Surface**, **Vertex**, and **Offset**. The first two can't be used in this case because our model doesn't have multiple surfaces and vertices. The last option, **Offset**, can be used to offset the whole extrusion by a certain distance. You can see the **From** option on top of the **PropertyManager** feature, as shown in *Figure 5.13*.

- **Direction 1:** This is active by default. Under this heading, we can customize the previewed extrusion that's shown on the canvas. We can hover the mouse over the options to see their names. The options under **Direction 1** are as follows:

  - **End Condition:** This determines how the extrusion stops. In this exercise, we will only be using the **Blind** option, which is selected by default. This means that the extrusion will be extended by the dimensions that we indicate. We will explore other end conditions later in the book.

  - **Reverse Direction:** This is the two-arrows symbol to the left of **End Condition**. This can easily reverse the direction of the extrusion from up to down and vice versa.

  - **Depth (D1):** This determines the depth of the extrusion. In our case, we wanted the extrusion to be 50 mm deep, so we inputted 50 mm.

  - **Draft:** The icon below **Depth** is used to draft the extrusion. We can activate drafting by clicking on the icon. We will cover this option at a more advanced level later in this book.

- **Direction 2:** This is very similar to **Direction 1**; however, it applies the extrusion in the opposite direction. We can use this if we want to have different length extrusions in two directions on the sketch. We can simply check this box if we require the second direction. This wasn't needed in our example.

- **Thin Feature:** This applies an extrusion based on the thin borders of the sketch rather than the enclosed shape. It can be activated by checking the box next to it. If we apply **Thin Feature** to our circle, we will get a result similar to the one shown in *Figure 5.14*. Take some time to draw another circle and experiment with **Thin Feature**:

Figure 5.14 – A sketch of one circle can translate to a ring using Thin Feature

> **Note**
>
> To apply a thin feature, the sketch does not have to be enclosed, as would be the case with a closed extruded boss. This will allow us to extrude a line, an arc, or other open shapes.

- **Selected Contours**: This can be used if we have more than one enclosed area. Then, we can select which ones we want to apply the extrusion to. *Figure 5.15* shows a sketch with three enclosed areas and the extrusion result using the selected contours option to only extrude enclosure **2**:

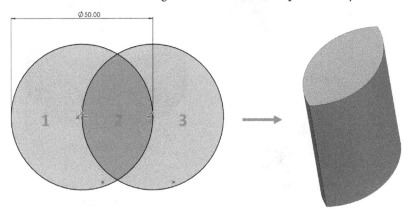

Figure 5.15 – A sketch with three enclosed areas

This concludes the section on creating the requested cylinder using the extruded boss feature. We learned about the following topics:

- The sequence we follow to create a model using features, that is, planning, sketching, and applying features
- How to use a sketch to apply the extruded boss feature
- How to set up the extruded boss feature and apply the extrusion

Now that we know about the extruded boss feature, we will look at the extruded cut feature.

Applying extruded cut and building on existing features

In this section, we will discuss how to apply the extruded cut feature. The extruded cut feature is very similar to the extruded boss feature in terms of the options that are available to us. Due to this, we will explain this feature in less detail. We will build upon the model we created previously with the extruded boss so that we can learn how to build upon existing features. To demonstrate this, we will create the following model highlighted in *Figure 5.16*.

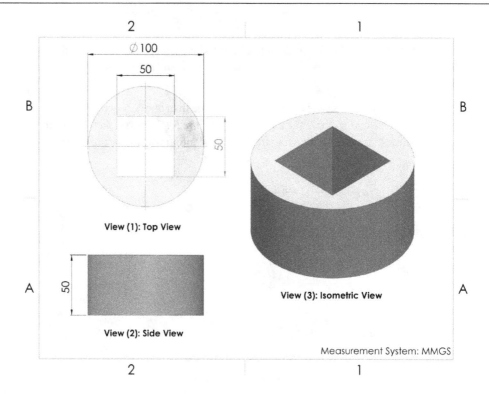

Figure 5.16 – The end model of the extruded cut exercise

Note that, in this model, we are only applying an extra cut over the cylinder that we created with the extruded boss feature. Thus, we will start from the cylinder we created earlier and create an extra extruded cut. We will go through the following phases to do so:

1. **Planning**: We will draw a square on top of the cylinder and apply it using the extruded cut feature.

2. **Sketching**: The cutoff shape is a square, so we will sketch a square on the top surface of the cylinder. Note that the top surface is not a default sketch plane. However, it is a planner surface, which means we can use it as a sketch plane.

3. **Applying the feature**: When we have our sketch, we can apply our feature; in this case, this is an extruded cut.

Follow these steps to create the sketch:

1. Select the top surface of the cylinder and click on the **Sketch** command, as shown in the following screenshot. We can also do this the other way around, that is, we can select the **Sketch** command first and then the surface we want to sketch on, as shown in *Figure 5.17*.

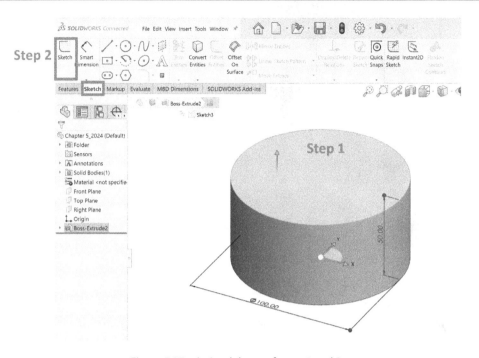

Figure 5.17 – A visual demo of steps 1 and 2

Now we need to start sketching. However, we may have a tilted view of our new sketch surface, which will make it harder for us to sketch. We can adjust our view so that it's normal to the sketch surface to make it easier to sketch. To do that, we can *right-click* on the new sketch at the bottom of the design tree and select **Normal To**, as shown in *Figure 5.18.*:

Figure 5.18 – The Normal To command changes our view to facing the sketch plane

This will change our view of the canvas so that it's facing the sketch surface. If we click on **Normal To** again, the model will flip 180 degrees.

2. Sketch and fully define the required square, as shown in *Figure 5.19*. The side of the square is 50 mm in length, as shown by **View (1): Top View** in *Figure 5.16* at the beginning of this section:

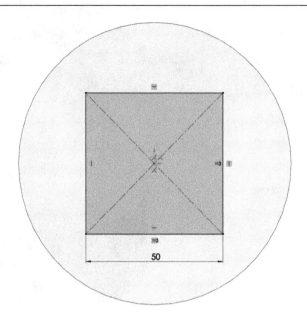

Figure 5.19 – The fully defined sketch for the extruded cut feature

Now that we have our sketch, we can apply our extruded cut feature.

3. Select the **Features** tab and select the **Extruded Cut** command, as shown in *Figure 5.20*. (As with the extruded boss feature, we don't need to exit sketch mode.) As soon as we select the command, SOLIDWORKS will interpret that we want to apply the feature to the active sketch:

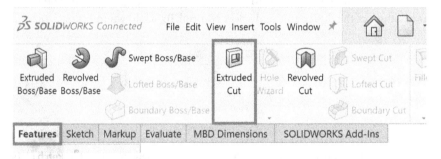

Figure 5.20 – The location of the extruded cut feature

4. Set the options in the **PropertyManager** as shown in *Figure 5.21*.

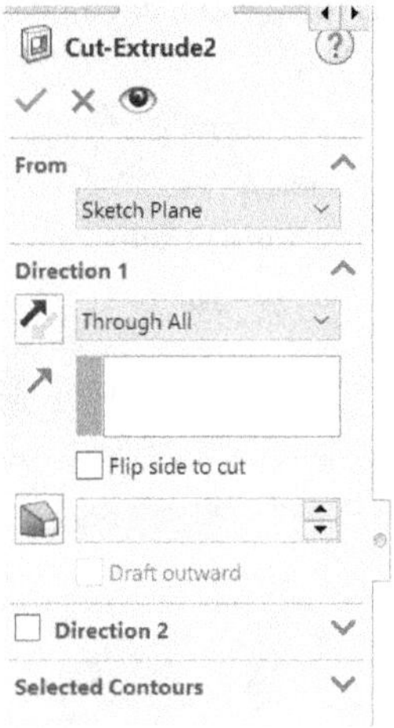

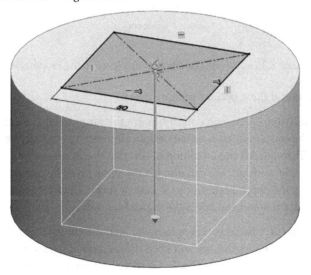

Figure 5.21 – The PropertyManager setting for the extruded cut feature

5. As with the extruded boss feature, we will also see a preview of the cut appear on the canvas. This preview is shown in *Figure 5.22*.

Figure 5.22 – A preview of the extruded cut

Note that the options that are shown in the PropertyManager for the extruded cut feature are almost the same as the options for the extruded boss feature. We will only elaborate on those that are highlighted in *Figure 5.23*.

Figure 5.23 – The end condition as displayed in the PropertyManager

Let's take a look at these options in more detail:

- **End Condition**: Many end conditions are the same as for the extruded boss feature. By default, the end condition is usually **Blind**. However, in our case, we changed it to **Through All**. The **Through All** end condition means that the cut will extend to the end of the model, which is what we want. We can also get the same result by using the **Blind** end condition and setting the depth of the cut to 50 mm or more. To change the end condition, click on the drop-down menu and select the desired condition.

- **Flip side to cut**: Checking this option will turn the cut part around. In our example, if we have this option checked, we will keep the contained square and delete everything else. Having this option unchecked will delete the contained square and keep everything else. Experiment with this option to understand what it does.

6. Click the green check mark at the top of the PropertyManager to apply the extruded cut feature. We will end up with the following model highlighted in *Figure 5.24*.

Figure 5.24 – The final model after applying the extruded cut feature

Recall that to make sketching easier, we changed the view to **Normal To** after entering a sketch. You can also set the software to automatically switch to the **Normal To** view when you create or edit a sketch. To do this, go to the system options and enable the auto-rotate feature. You can find the option by going to **Tools** and then **Options**. Under **System Options**, click on **Sketch**, and then find the **Auto-rotate view normal to sketch plane on sketch creation and sketch edit** option, as shown in *Figure 5.25*.

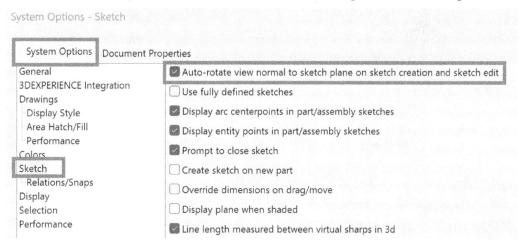

Figure 5.25 – Auto-rotate to a normal view

This concludes using the extruded cut feature. In this section, we covered the following topics:

- How to sketch over existing planner surfaces
- How to apply the extruded cut feature

Now that we know how to apply the extruded boss and cut features, we will learn how to modify and delete them.

Modifying and deleting extruded boss and extruded cut

Often, we apply a feature and then need to edit it or delete it. In this section, we will address how to edit and delete features. To illustrate this, we will apply the modifications shown in *Figure 5.26* to the model we created earlier:

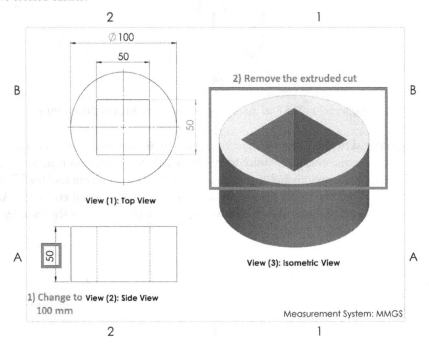

Figure 5.26 – The changes to the model to be applied in this exercise

As shown in *Figure 5.26*, the changes that we are going to make are as follows:

- Changing the height of the cylinder from 50 mm to 100 mm
- Removing the extruded cut that goes through the cylinder

Let's go ahead and apply these changes.

Editing a feature – changing the height of the cylinder from 50 mm to 100 mm

To edit an implemented feature, we can *right-click* (or *left-click*) on it on the design tree and select the **Edit Feature** option. In this case, we want to edit the boss extrude feature we applied. Thus, we can *right-click* on the first **Boss-Extrude** feature and select **Edit Feature**, as shown in *Figure 5.27*.

Figure 5.27 – The Edit Feature command as shown in the SOLIDWORKS interface

Once we select **Edit Feature**, the extruded boss features will be shown on the left, while a preview of the extruded boss feature will be shown on the right. Note that the preview doesn't show the extruded cut feature because it is located lower in the design tree. From the available options, we can change **D1** from 50 to 100, as shown in *Figure 5.28*. After making this change, we can click on the green check mark to implement it:

Figure 5.28 – A feature can be edited by directly editing its PropertyManager

Now our cylinder will become longer, as shown in *Figure 5.29*. If the shape of the model failed to update, try clicking on the traffic light (**Rebuild**) icon on top of the SOLIDWORKS interface or using the *Ctrl + B* shortcut:

Figure 5.29 – The shape of our model after editing the extruded boss feature

This concludes how to edit a feature. Now, let's look at deleting a feature.

Deleting a feature – removing the extruded cut feature that goes through the cylinder

To delete a feature, *right-click* on a feature in the design tree and select the **Delete…** option. In this case, we want to delete the extruded cut feature. Thus, we can *right-click* on it and select **Delete…**, as shown in *Figure 5.30*.

Figure 5.30 – Features can be deleted with the Delete… command

Once we've selected **Delete…**, we will get the message highlighted in *Figure 5.31*, asking us to confirm that we want to delete the feature. The message will specify the item to be deleted, which is **Cut-Extrude1 (Feature)** in our case. It will also specify any dependent items that will be deleted with the feature; there are none in this case. We will cover dependent items at a more advanced level later in this book. We can click **Yes** to confirm that we want to delete the feature:

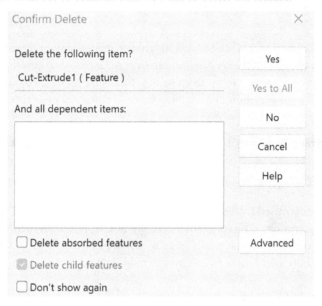

Figure 5.31 – A confirmation box appears when deleting a feature

Now, we will end up with the following model. Note that we only deleted the feature, so the sketch will still remain in the model for us to use for any other purpose, as shown in *Figure 5.32*.

Figure 5.32 – The sketch still remains even though its feature was deleted

If we want to delete both the feature and its sketch, we can check the following **Delete absorbed features** option, as shown in *Figure 5.31*, when deleting the feature.

> **Tip**
> We can delete the feature by directly selecting it on the design tree and pressing *Delete* on the keyboard.

This concludes the sections on editing and deleting features. Every feature can be edited and deleted in the same way.

As designers and 3D modelers, we will be faced with many situations where we receive models from other individuals and are asked to edit them. Also, we ourselves will modify our models as part of improvement cycles. Thus, it is very important for us to know how to modify models. As our SOLIDWORKS skills grow, we will pay special attention to modifying models, especially pre-existing ones.

We have just learned about our first set of features, that is, extruded boss and extruded cut. We learned how to apply and modify them. Now, we will move on to another set of features – fillets and chamfers.

Understanding and applying fillets and chamfers

Fillets and chamfers are used to modify edges and vertices on our models by making them less sharp. If we look at everyday objects around us, such as phones, laptops, and furniture, we will notice the common use of small fillets and chamfers on the edges. In this section, we will discuss what fillets and chamfers are, how to apply them, and how to modify them.

Understanding fillets and chamfers

Fillets and chamfers are modifications that are made to the edges and vertices of our models. A fillet is a curved surface defined by a radius, while a chamfer is a transitional planner surface defined by lengths and angles. They help remove sharp edges and turn them into softer ones in order to provide a safer product or a better user experience. They are similar to fillets and chamfers in sketching. *Figure 5.33* illustrates the effect of the fillet feature:

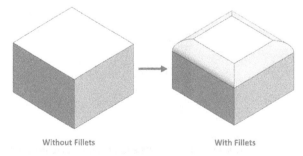

Without Fillets   With Fillets

Figure 5.33 – Edges with and without fillets

Figure 5.34 illustrates the effect of the chamfer feature:

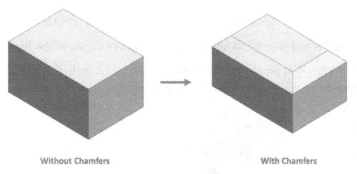

Figure 5.34 – Edges with and without chamfers

Note that fillets and chamfers can only be applied to existing features. In *Figure 5.34*, we applied the extruded boss feature first, and then we were able to apply fillets or chamfers. Also, to apply fillets and chamfers, we don't need to start with a 2D sketch.

Now that we know what fillets and chamfers are, we can start using them in SOLIDWORKS.

Applying fillets

In this section, we will discuss how to apply the fillet feature. To show this, we will create the model shown in *Figure 5.35*:

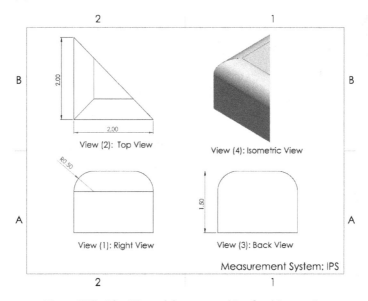

Figure 5.35 – The 3D model we are making for this exercise

As usual, we will start with planning, sketching, and then applying a feature:

- **Planning**: Here, we will create a triangular prism and then apply fillets to the top edges.

- **Sketching**: We will sketch and fully define a triangle, as shown by the **Top View** in *Figure 5.35*.

- **Features**: We will apply the extruded boss feature by 1.5 inches, as shown by the **Back View** in the preceding figure. This will result in a triangular prism that looks as follows *Figure 5.36*:

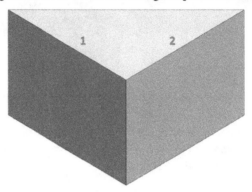

Figure 5.36 – The base triangular prism we will use to apply the fillet

To apply the fillet feature, select the **Fillet** command, as shown in *Figure 5.37*:

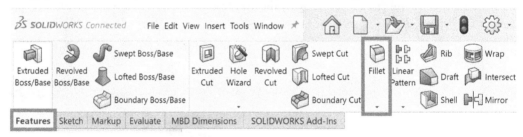

Figure 5.37 – The location of the Fillet command

Select **Edge<1>** and **Edge<2>**, which have fillets applied to them. Adjust the **Fillet** options, as shown in *Figure 5.38*:

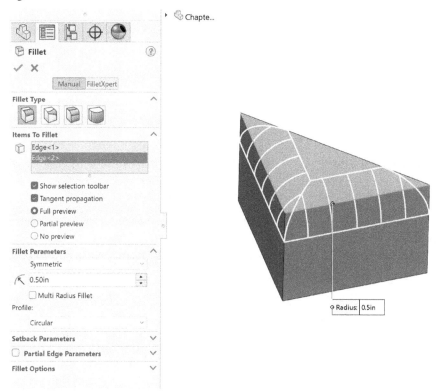

Figure 5.38 – The fillet feature PropertyManager and a preview

The following is a brief explanation of the options that we used in this exercise:

- **Fillet Type**: This specifies the type of fillet we're using. The first type is the constant-size fillet, which we are using in this exercise. As we are still beginners, we will only use this type.

- **Items to Fillet**: This highlights where the fillet will be applied. In our example, we're applying it to **Edge <1>** and **Edge <2>**. We can use this window to remove edges from the selection if we've added one by mistake. For the selection, we can select individual edges or faces. Selecting a face will apply the fillet to all the edges related to that face.

- **Show selection toolbar**: If this option is checked, then we can select an edge; a small selection toolbar will appear with some shortcuts that we can use to select where we want to apply the fillet. The toolbar looks as in *Figure 5.39*. Since the selection toolbar only contains selection shortcuts, we can disregard it for now and start using it later, once we are comfortable with the different selections it provides.

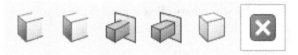

Figure 5.39 – The selection toolbar is an easy way to select the edges to apply the fillet to

- **Tangent propagation**: This only applies if there's more than one edge and they are tangent to each other, for example, two curved edges that have a tangent relationship with each other. In the case of our triangular prism, none of the edges are tangent to each other. Thus, selecting or deselecting this option will not make any difference.

- **Full preview**, **Partial preview**, and **No preview**: These options decide the type of fillet preview we can see when we are making the shape. Selecting **Full preview** will show us what the fillet looks like in its entirety, whereas selecting **No preview** won't show it at all.

- **Fillet Parameters**: Here, we have two options: **Symmetric** and **Asymmetric**. A **Symmetric** fillet is a uniform fillet in which we only need to define one radius. An **Asymmetric** fillet, on the other hand, has different curvatures on each side and requires us to define more than one dimension.

- **Radius**: Here, we can input a numerical value for the fillet radius. In this exercise, it is 0 . 5 inches, as shown by **View (1): Right View** in *Figure 5.35* at the beginning of this section.

- **Multi Radius Fillet**: Clicking this option will allow us to determine a different radius for different edges. For example, if we want the fillet radius for **Edge <1>** to be different from the fillet radius for **Edge <2>**, we can use this option.

- **Profile**: This determines the profile of the fillet. The **Circular** profile option projects the fillet profile as a quarter of a circle. Other options include **Conic Rho**, **Conic Radius**, and **Curvature continuous**.

We won't be using the **Setback Parameter**, **Partial Edge Parameter**, and **Fillet Options** just yet. While we are here, take some time to play with some of the preceding options and look at the result in the canvas preview. Once done, click on the green check mark at the top of the **Options** tab to apply the fillet. We should get the following shape highlighted in *Figure 5.40*:

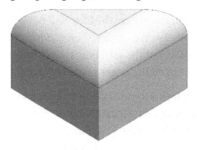

Figure 5.40 – The final 3D model after applying the fillet

This concludes how to apply fillets. In this section, we covered the following topics:

- How to apply fillets in SOLIDWORKS
- The different options we can use to define fillets in SOLIDWORKS

Now that we know how to apply fillets, let's learn how to apply chamfers.

Applying chamfers

In this section, we will discuss how to apply the chamfer feature. To do this, we will create the model highlighted in *Figure 5.41*:

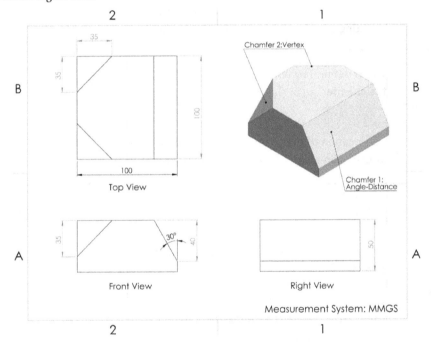

Figure 5.41 – The 3D model we are making with this exercise

Figure 5.41 specifies two different types of chamfers, each with different specifications. In SOLIDWORKS, there are five different types of chamfers, based on how they are defined.

The following table shown in *Figure 5.42* illustrates the angle-distance, distance-distance, and vertex chamfer types:

Chamfer type	Description	Illustration
Angle-distance	This is defined by specifying an angle and the distance that makes up the chamfer. The location of the chamfer is identified by selecting an edge, a face, or multiples of both.	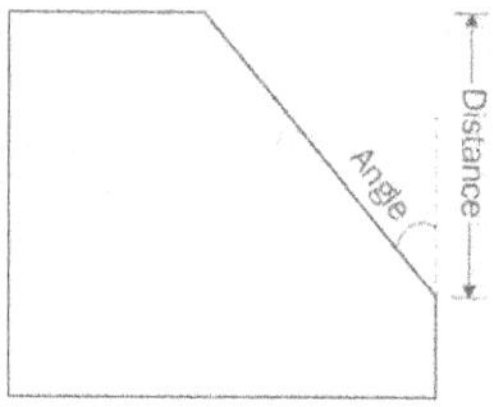
Distance – distance	This is defined by two distances that make up the chamfer. The location of the chamfer is identified by selecting an edge, a face, or multiples of both.	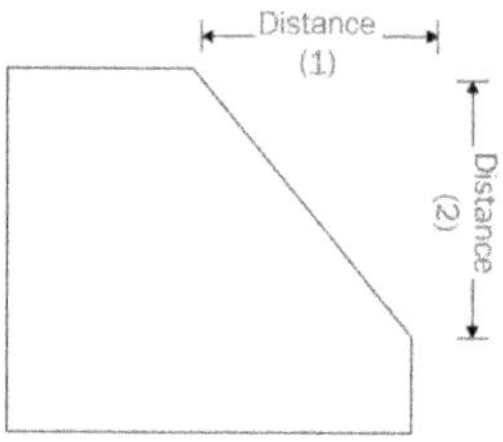
Vertex	This is defined by three distances that move away from a vertex. The location of the chamfer is identified by selecting a vertex.	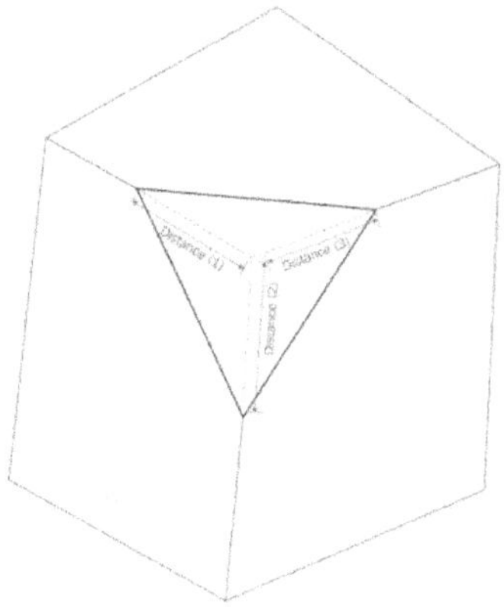

Figure 5.42 – Illustrations of the angle-distance, distance-distance, and vertex chamfer types

The following table illustrates the offset face and face-face chamfer types. Each chamfer type is defined differently giving us different options to help maintain our required design intent:

Chamfer type	Description	Illustration
Offset face	This is defined by an offset distance for the two faces that are adjacent to the chamfer's edge. This gives the chamfer a special definition if the two faces are not perpendicular to each other. The location of the chamfer is identified by selecting one or more edges.	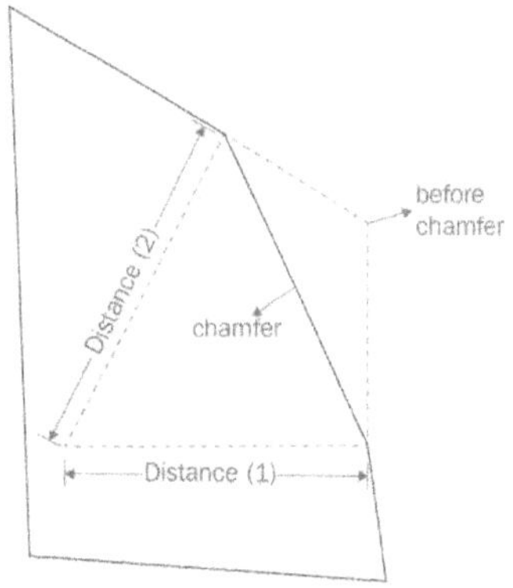
Face – face	This is defined by the distance between two or more faces. The location of the chamfer is defined by selecting the faces that surround the chamfer.	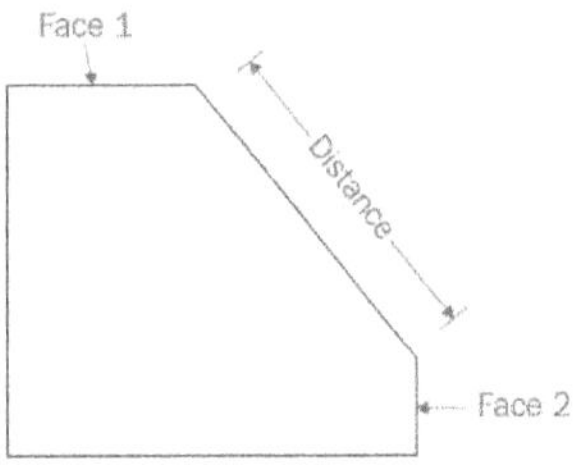

Figure 5.43 – Illustrations of the offset face and face-face chamfer types

The model we are going to create uses two types of chamfers: angle-distance and vertex. To create the chamfer, we will follow our usual procedure, where we start by planning, then creating sketches, and then applying a feature:

1. **Planning**: Here, we will create a rectangular prism and then create an angle-distance chamfer, followed by two vertex chamfers.

2. **Sketch**: We will sketch and fully define a 100 x 100 mm square, as shown in the top view of the diagram at the beginning of this section.

3. **Feature**: We will be applying the extruded boss feature and extruding the square by 50 mm, as shown in the right view in *Figure 5.41* at the beginning of this section. At this point, we will have the rectangular prism shown in *Figure 5.44*:

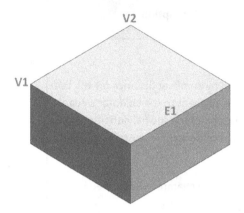

Figure 5.44 – The base rectangular prism we will use to apply the chamfers to

To apply angle-distance chamfers, follow these steps:

1. To apply the chamfer feature, select the **Chamfer** command, as shown in *Figure 5.45*. The command can also be accessed from the drop-down menu, under the **Fillet** command:

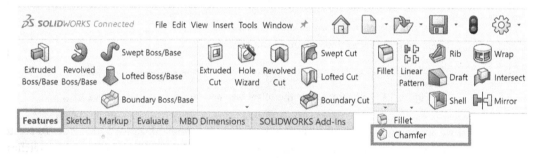

Figure 5.45 – The location of the Chamfer command

2. Set up the chamfer's options in the PropertyManager, as shown in *Figure 5.46*. Some options that are available for the chamfer feature are the same as for the fillet feature, so we won't go over them in as much detail here:

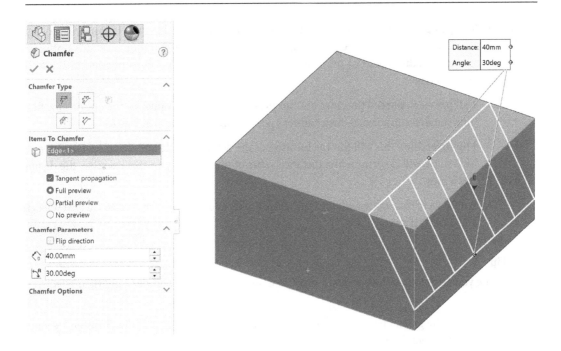

Figure 5.46 – Chamfer's PropertyManager and a preview of the chamfer

The following is a brief description of the options that we used in this exercise:

- **Chamfer Type**: From here, we can select one of the five types of chamfers, as specified in the preceding tables. We can hover over each one to see their names. For our first chamfer, we will use the **Angle-Distance** chamfer.

- **Items To Chamfer**: This specifies the location of the chamfer. We can apply the angle-distance chamfer by selecting edge 1 (**E1**), as highlighted in the rectangular prism highlighted in *Figure 5.47*. If we select another edge or face by mistake, we can remove it by deleting it from the list under **Items to Chamfer** in the PropertyManager:

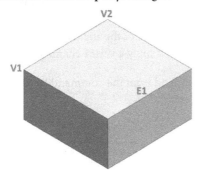

Figure 5.47 – Select edge 1 (E1) to apply the angle-distance chamfer to

- **Tangent propagation**: This works the same as for fillets. Checking this option will propagate the chamfer across any edges that are tangent to each other. In our case, this option will not make a difference, as there are no other edges that have a tangent relationship to our selected edge.

- **Full preview**, **partial preview**, and **no preview**: These are similar to the options for fillets. This can determine what our chamfer's preview looks like on the canvas.

- **Flip direction**: This will flip the location of the defining angle and distance. **Distance (D):** This value will determine the distance value of our chamfer. In this exercise, the distance is 40 mm.

- **Angle (A)**: This value will determine the angle value of our chamfer. In this exercise, the angle is 30 degrees.

> **Tip**
> While you are here, play around and mix these different options to understand their effects.

3. Click on the green check mark to apply the chamfer. After applying the chamfer, we should have the model shown in *Figure 5.48*:

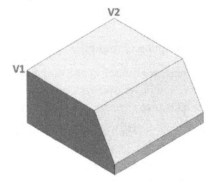

Figure 5.48 – The shape of the 3D model after applying the angle-distance chamfer to E1

At this point, our model has one chamfer of the angle-distance type. Now, we will apply the other two chamfers, which are vertex types. To apply a vertex chamfer, follow these steps:

1. Select the **Chamfer** command from the command bar to apply a second chamfer.

2.   Set up the chamfer's **PropertyManager** options as shown in *Figure 5.49*:

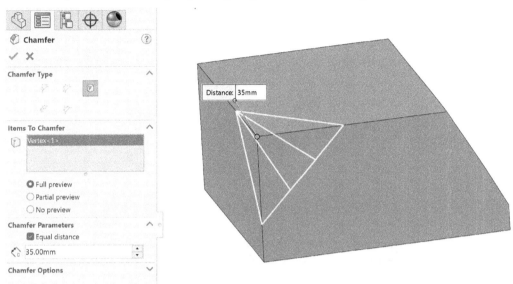

Figure 5.49 – The chamfer PropertyManager for the vertex chamfer type and a preview of it

Here is a brief description of some of the special features we can apply to the vertex chamfer feature:

- **Chamfer Type**: From here, we select the type of chamfer we want to create; in this case, we want to create a vertex chamfer.

- **Items To Chamfer**: Select **V1** in the canvas to locate the chamfer. Note that, when using the vertex chamfer, we can only apply chamfers to one vertex at a time. In other words, we can only select one vertex in **Items To Chamfer**.

- **Equal distance**: The vertex chamfer can be defined with three distances that extend away from the vertex into the edges it consists of (refer to the preceding table regarding chamfer types). Checking this option will make all three distances equal, so we only need to input one dimension. If we uncheck this option, we will need to enter three different distances.

- **Distance (D)**: This value will determine the distance of the chamfer. In this exercise, the distance is 35 mm in all three directions.

3. Click on the green check mark to apply the chamfer. After applying the chamfer, we should have the model shown in *Figure 5.50*:

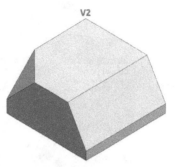

Figure 5.50 – The 3D model after applying the vertex chamfer

4. Now, we can repeat *steps 1 – 3* for **V2** to apply the second vertex chamfer. After repeating those steps, we will have the final model. This will look as in *Figure 5.51*:

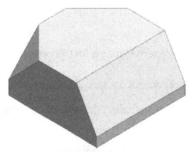

Figure 5.51 – The final model after applying the two vertex chamfers

This concludes how to apply chamfers. In this section, we discussed the following topics:

- The five different types of chamfers, as well as how are they defined

- How to apply different types of chamfers in SOLIDWORKS

- The different options we can use to define chamfers in SOLIDWORKS

At this point, we know how to apply fillets and chamfers. Now, let's learn how to modify them.

Modifying fillets and chamfers

To edit or delete a fillet or chamfer, we can follow the same procedure that we followed to edit and delete the extruded boss and cut features. In fact, every feature can be modified in the same way. Here, we *right-click* on a feature in the design tree and then select **Edit** to edit it or **Delete** to delete it.

One special aspect when it comes to modifying chamfers is that there are some limitations when switching from one chamfer type to another. For example, if we modify an **Angle-Distance** chamfer, we won't have the option to change the type to an **Offset Face** chamfer. If we are faced with such a limitation, we can simply delete the chamfer and start again with a new one.

Applying partial fillets and chamfers

So far, we have talked about making fillets and chamfers that extend to a full edge. However, we can also apply partial fillets and chamfers that are only applied to a specific section of an edge. Let's talk about those here. To do this, we are going to create the 3D model highlighted in *Figure 5.52*:

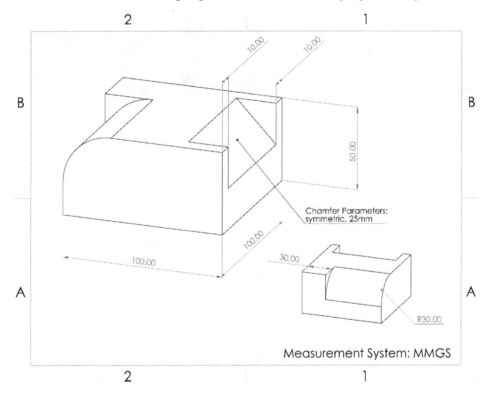

Figure 5.52 – The 3D model we are making with this exercise

We are going to create it in three phases, as follows:

1.   Create the rectangular prism.
2.   Apply the partial fillet.
3.   Apply the partial chamfer.

The first stage is simple; you can start by sketching a **100** mm square, then extrude it by **50** mm using the **Extruded Boss** feature. We are already familiar with applying extruded boss, as shown in *Figure 5.53*:

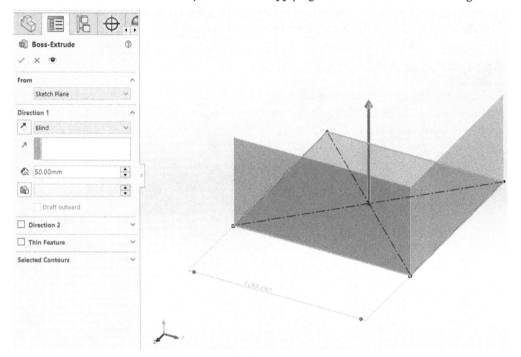

Figure 5.53 – The first step is to extrude the base square by 50 mm

Now that we have our base rectangular prism, we can start applying the partial fillet and chamfers. We will start by applying the fillet.

Applying a partial fillet

To apply a partial fillet, we can follow these steps:

1. Select the **Fillet** command. Then, under **Fillet Type**, select **Constant Size Fillet**, select an edge to apply the fillet to, and set the parameters to **Symmetric** with the radius being 3 0 mm, as shown in *Figure 5.54*:

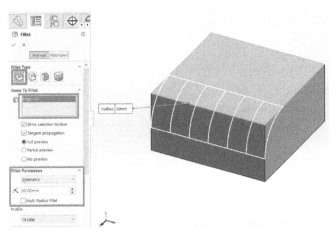

Figure 5.54 – The fillet preview and the PropertyManager applying a full fillet

> **Important note**
> Partial fillets can only be applied to the **Contact Size Fillet** type.

2. Scroll to the bottom of the fillet **PropertyManager**, and check the **Partial Edge Parameters** option.

3. Set the start condition to **Distance Offset** and input 3 0 mm, as shown in *Figure 5.55*:

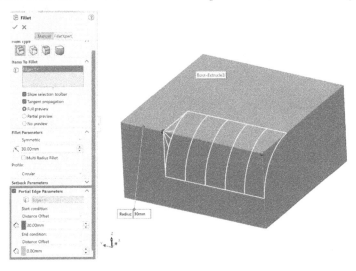

Figure 5.55 – The preview and the PropertyManager applying the fillet to a partial edge

4. Click on the green checkmark on top of the **PropertyManager** to apply the partial fillet.

Apart from the Distance Offset start and end conditions, there are another two conditions that can be used. Let's explore them:

- **Percentage offset**: This allows you to offset the fillet by a certain percentage in relation to the edge.

- **Reference offset**: This allows you to offset the fillet in relation to an added reference. For example, you can sketch a point on the edge and then use it as a reference to the start or end of the partial fillet.

> **Tip**
> While you are here, take some time to experiment with the different types of start and end conditions.

Now that we know how to apply a partial fillet, let's move to apply a partial chamfer to our model.

Applying a partial chamfer

With a few exceptions, applying a partial chamfer follows a similar procedure to applying a partial fillet. We can follow these steps:

1. Select the **Chamfer** command.

2. Under **Chamfer Type**, select the **Offset Face** type, select an edge to apply the chamfer to, and set the parameters to **Symmetric** with the distance being 25 mm, as shown in *Figure 5.56*:

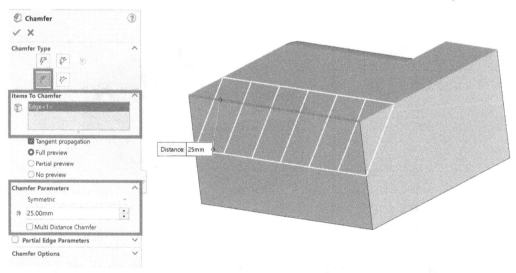

Figure 5.56 – The full chamfer preview and its PropertyManager

> **Important note**
>
> Partial chamfers can only be applied to the **Offset Face** type.

3. Check the **Partial Edge Parameters** option found at the bottom of the PropertyManager.

4. Set the start condition and end condition to **Distance Offset** and input 10 mm for both, as shown in *Figure 5.57*:

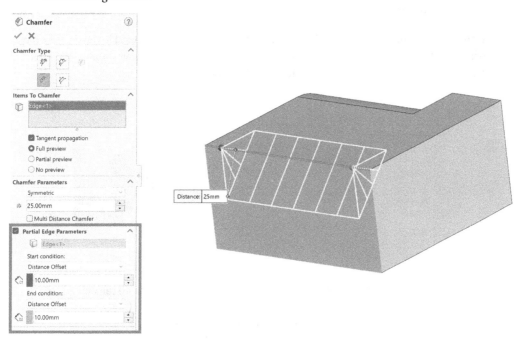

Figure 5.57 – The partial chamfer preview and its PropertyManager

5. Click on the green checkmark to apply the partial chamfer.

This concludes this section on fillets and chamfers. In this section, we have learned how to apply and modify them. Now, we can start learning about another feature set – revolved boss and revolved cut.

Understanding and applying revolved boss and revolved cut

Revolved boss and revolved cut are two of the most common features in SOLIDWORKS and are also easy to apply. They capitalize on rotational movements to add or remove materials. In this section, we will discuss what revolved boss and revolved cut are, how to apply them, and how to modify them.

What are revolved boss and revolved cut?

Revolved boss and revolved cut are among the most basic features in SOLIDWORKS. Let's explain them in more detail:

- **Revolved boss**: This adds materials by rotating a sketched shape around an axis
- **Revolved cut**: This removes materials of an existing solid body by rotating a sketched shape around an axis

From these definitions, we can see that revolved boss and cut are similar. However, they have the opposite effect. Revolved boss adds materials, while revolved cut removes materials. *Figure 5.58* illustrates the elements and effects of the **Revolved Boss** feature:

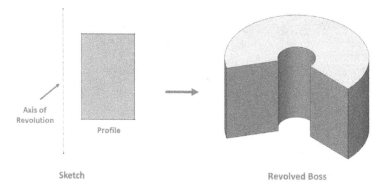

Figure 5.58 – An illustration of what the revolved boss feature can do

Figure 5.59 illustrates the elements and effect of the **Revolved Cut** feature:

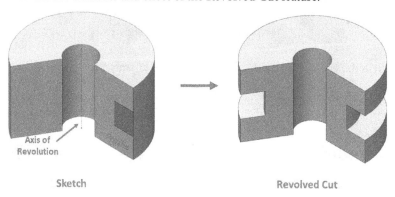

Figure 5.59 – An illustration of what the revolved cut feature can do

As we can see, two elements are required if we want to apply a revolved boss and revolved cut – a profile and an axis of revolution. We have to sketch both of these in sketch mode before we can apply our features.

> **Important note**
> The axis of revolution can also be part of the profile.

Now that we know what the revolved boss and revolved cut features are, we will learn how to apply them.

Applying revolved boss

The revolved boss feature adds materials by revolving a sketch around an axis of revolution. To show you how to apply the revolved boss feature, we will create the model highlighted in *Figure 5.60*. Now that we are using more and more features, we will start to notice that we use the same options repeatedly. For example, when applying the revolved boss, most options are the same ones that we use while applying extruded boss. Therefore, we won't explain these again here:

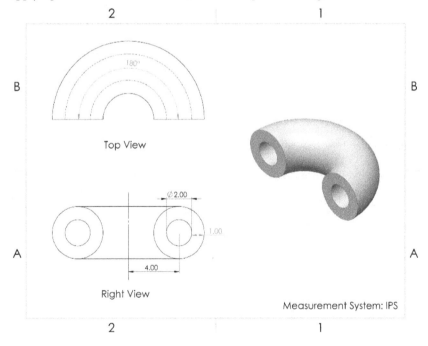

Figure 5.60 – The 3D model we are building in this exercise

As usual, we will follow our standard procedure of planning, sketching, and applying features:

1. **Planning**: Here, we'll draw the profile shown in the right view of the preceding drawing, and then we'll rotate that by **180** degrees.

2. **Sketching**: Here, we're going to sketch the profile highlighted in **Right View** using the right plane. This includes using the axis of revolution. Our sketch should look as highlighted in *Figure 5.61*:

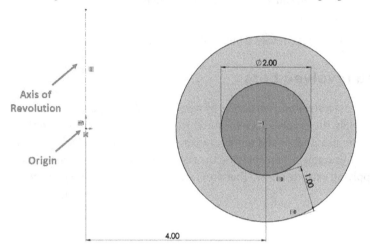

Figure 5.61 – The sketch used to apply the revolved boss feature

3. **Applying the feature**: Here, we are going to apply our revolved boss feature.

To apply the revolved boss feature, follow these steps:

1. Select the **Revolved Boss/Base** command from the **Features** command bar. We don't need to exit the sketch to select the command. However, if we do exit the sketch for whatever reason, we can select the **Revolved Boss/Base** command as shown in *Figure 5.62*, and then select the sketch we want to apply it to:

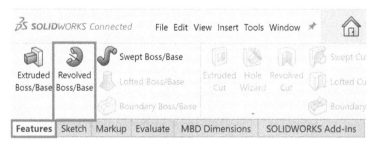

Figure 5.62 – The location of the revolved boss feature

2. Adjust the options in the PropertyManager, as shown in *Figure 5.63*:

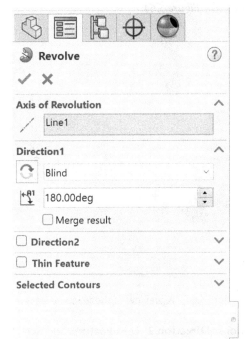

Figure 5.63 – The revolved boss PropertyManager

Here is a brief description of the unique options that can be used with the revolved feature:

- **Axis of Revolution**: This can be any straight line located on the canvas. This axis will be used to apply the feature by rotating the sketch around that axis. In our exercise, we will select the centerline indicated in the preceding figure as the axis of revolution.

- **End Condition**: This provides a few options regarding how the revolution will stop. In this exercise, we will use the **Blind** condition. By doing this, we will decide on the end of the revolution by determining the angle of rotation.

- **Reverse Direction**: This is represented by the two curved arrows next to the end condition field. Clicking on this icon will reverse the rotation direction of the revolved boss.

- **Angle (A1)**: This determines when the revolution stops. In this exercise, the revolution will stop giving the **Blind** end condition, we are stopping it at 180 degrees.

3. Since we're setting the feature's options, we will be able to see a preview of the feature on the canvas. This will look like *Figure 5.64*:

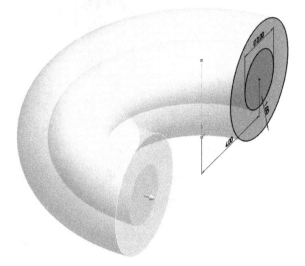

Figure 5.64 – A preview of the revolved boss feature

The rest of the options (**Direction 2**, **Thin Feature**, and **Selected Contours**) were explained when we looked at the extruded boss feature. They have the same functionalities.

4. To apply the revolved boss feature, we can click on the green checkmark at the top of the **Options** panel. The resulting model should look like *Figure 5.65*:

Figure 5.65 – The final shape after applying revolved boss

This concludes how to use the revolved boss feature. In this section, we discussed the following topics:

- How to apply the revolved boss/base feature

- The different options we can use to define the revolved boss/base feature

Now, we can start learning about the revolved cut feature.

Applying revolved cut

The revolved cut feature removes materials from existing bodies by revolving a sketch around an axis of revolution. To show you how to apply the revolved cut feature, we will create the model shown in *Figure 5.66*. Note that the final body is a continuation of the body we just created when we applied the revolved boss feature. Thus, we will use and build upon that model:

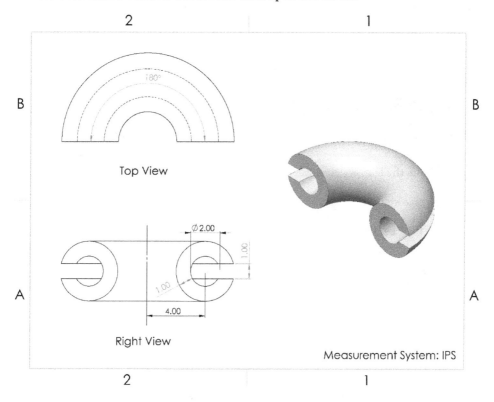

Figure 5.66 – The 3D model we are building using the revolved cut feature

Let's go ahead and complete the model:

1. **Planning**: We'll use the model we created earlier and add the revolved extruded cut feature to create the desired final shape in the drawing.

2. **Sketching**: Here, we will sketch and fully define a rectangle to make the new cut that's shown in *Figure 5.67*. The height of the cut is shown in **Right View** as 1 inch and cuts through the thickness of the wall. You will notice that we picked **1.2** inches for the width of our rectangle since this will be enough to cut through the whole wall. There are more advanced ways to define the cut, but we won't be covering them here. The axis of revolution goes through the center of the model, which also goes through the origin that we provided in the previous exercise:

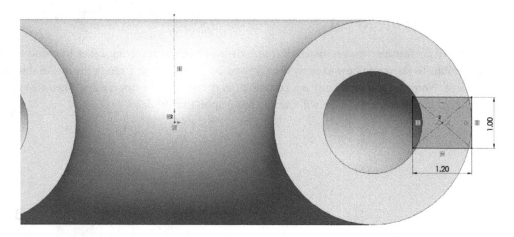

Figure 5.67 – The sketch used to apply the revolved cut feature

3. **Applying the feature**: Here, we will be applying the revolved cut feature.

To apply the revolved cut feature, follow these steps:

1. Select the **Revolved Cut** feature from the command bar, as shown in *Figure 5.68*:

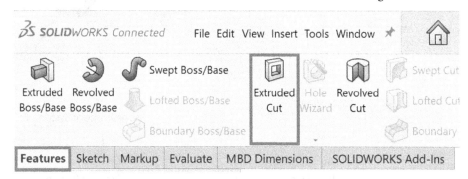

Figure 5.68 – The location of the Revolved Cut feature

2. Set up the **PropertyManager** options that are shown in *Figure 5.69*. We are already familiar with these options. However, here, we're introducing a new end condition, **Up To Surface**. Now, select the other end of the surface, as shown in the following screenshot. After selecting the surface, we will see a preview of our revolved cut:

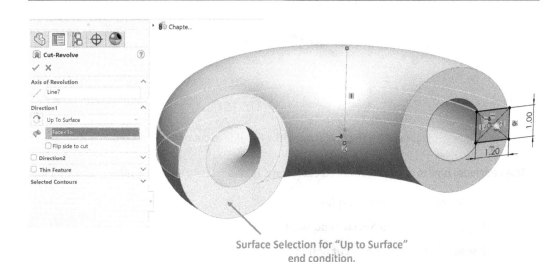

Surface Selection for "Up to Surface" end condition.

Figure 5.69 – The revolved cut PropertyManager and a preview

This end condition means that the cut will start from our sketch and end at a selected surface. Since our revolved cut goes through all of our shapes, it would be more convenient to end it by selecting the end surface rather than writing a numerical value for an angle. Another advantage of selecting this end condition is that it preserves our design intentions. For example, if we wanted to modify the angle of the revolved boss so that it's 270 degrees instead of 180, the revolved cut would update automatically and go through all of our shapes. However, if we use 180 degrees in a **Blind** end condition for our revolved cut and then update the revolved boss to 270 degrees, the cut will stay at 180 degrees.

3. Click on the green check mark to apply the cut. This will result in the model in *Figure 5.70*:

Figure 5.70 – The final 3D model after applying the revolved cut feature

Before we conclude this section on the revolved cut, let us review the **Flip side to cut** option shown in the **PropertyManager** in *Figure 5.69*. Unlike other options, this one is unique to revolved cut. Clicking it will flip the cut, so we keep the highlighted part in the preview and remove the rest. In our example, clicking the **Flip side to cut** option will result in the shape highlighted in *Figure 5.71*:

Figure 5.71 – The final shape if the Flip side to cut option is clicked

This concludes how to apply the revolved cut feature. In this section, we discussed the following topics:

- How to apply the revolved cut feature
- How to use the **Up To Surface** end condition
- How to use the **Flip side to cut** option in the revolved cut feature

Now that we've learned how to apply the revolved boss and revolved cut features, we will learn how to modify them.

Modifying revolved boss and revolved cut

The same procedure that we follow to modify the extruded boss and cut features applies when we modify the revolved boss and cut features. To recap, to edit or delete a revolved boss or cut, we can *right-click* on the feature from the design tree and select **Edit** or **Delete…**, as required. *Figure 5.72* highlights the **Edit** and **Delete…** commands that appear after right-clicking on a feature from the design tree:

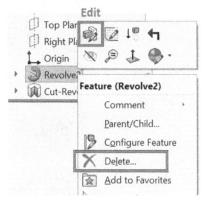

Figure 5.72 – The locations of the Edit and Delete… commands

This concludes this section on revolved boss and cut. In this section, we learned how to apply these features, set them up, and modify them. Those two features will be important when we deal with rounded objects such as shafts and cylinders.

Summary

In this chapter, we learned about our first set of SOLIDWORKS features, all of which allow us to go from creating 2D sketches to 3D models. Here, we learned about the extruded boss/cut and revolved boss/cut features. Each of these feature sets has an additive feature and a subtractive feature, as we learned in this chapter. We also learned about the fillet and chamfer features, which mainly aim to remove sharp edges for our 3D models. For each feature, we learned about what it was, how to apply it, and how to modify it.

In the next chapter, we will explore features that are considered more advanced than the features we explored in this chapter. This includes swept boss/cut and lofted boss/cut, both of which require more than a single sketch to be applied. We'll also address adding more reference geometries, such as new planes and coordinate systems.

Questions

1. What are the features of SOLIDWORKS?

2. What are the extruded boss and cut features?

3. What are the fillet and chamfer features?

4. What are the revolved boss and cut features?

5. Create the model shown in the following figure:

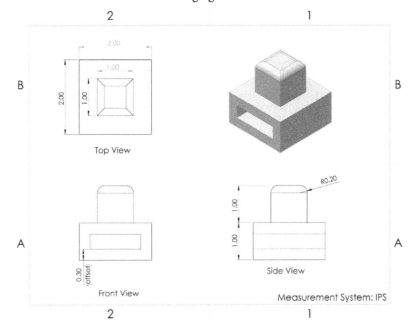

Figure 5.73 – The drawing for question 5

6. Create the model shown in the following figure:

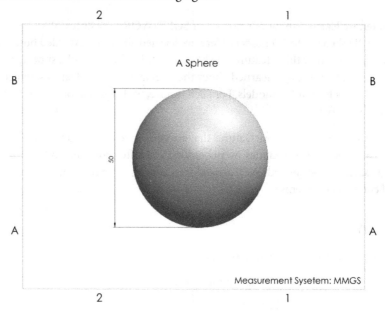

Figure 5.74 – The drawing for question 6

7. Create the model shown in the following figure:

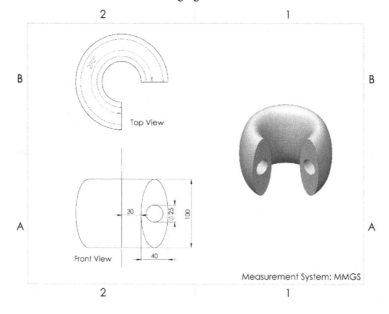

Figure 5.75 – The drawing for question 7

8. Create the model shown in the following figure:

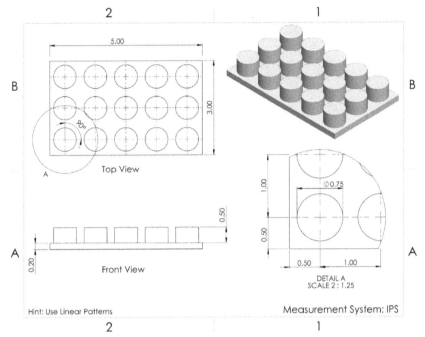

Figure 5.76 – The drawing for question 8

> **Important note**
> The answers to the preceding questions can be found at the end of this book.

Get This Book's PDF Version and Exclusive Extras

6

Basic Secondary Multi-Sketch Features

Many of the simple SOLIDWORKS features require only one sketch to be created, but others need more than that to model more complicated shapes in one go. In this chapter, we will cover essential multi-sketch features, such as Swept Boss and Swept Cut, and Lofted Boss and Lofted Cut. In addition to that, we will define some new planes that are completely different from the default ones.

In this chapter, we will cover the following topics:

- Reference geometries – additional planes
- Understanding and applying Swept Boss and Swept Cut
- Understanding and applying Lofted Boss and Lofted Cut

By the end of this chapter, we will be able to create complex-looking 3D models compared to what we have built already. We will also be able to create irregular shapes compared to those we created in the previous chapter.

Technical requirements

In this chapter, you will need to have access to SOLIDWORKS software.

The CiA video for this chapter can be found at `https://packt.link/pxh4M`

Reference geometries – additional planes

By default, SOLIDWORKS provides us with three planes that we can start sketching on. In addition, we can use any other planner surface as a plane. However, sometimes, we need planes that are different. In this case, we need to introduce our own planes. In this section, we will discuss how we can create additional planes in our 3D space to be used as reference geometries.

Understanding planes, reference geometries, and why we need them

Reference geometries are fundamental elements that are used to anchor our sketches and features in the 3D space. They are used as a base for sketches, features, and coordinate locations. In SOLIDWORKS, reference geometries include planes, coordinate systems, axes, and points. In this section, we will focus on planes.

Whenever we create a sketch, we start by selecting a sketch plane to base our sketch on. Previously, we used the default planes and the planner surfaces resulting from the created solid bodies built by features. However, in some cases, we may need additional planes that do not exist. To illustrate this, take a look at the example model shown in *Figure 6.1*.

The following model consists of a cube and a cylinder that intersects it. The cube is a normal cube, just like the ones we've made in the previous chapters. However, the cylindrical part has been created with an angle. In the following diagram, we used a new plane (shown as *Plane 1*) to sketch and create that cylinder. Note that the plane is different from the default planes and different from all of the other planner surfaces.

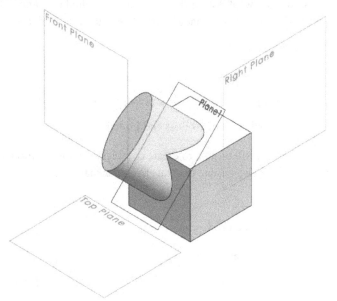

Figure 6.1 – An illustration of the default planes and an additional one

In this section, we will learn how to introduce new planes, such as the one shown in the preceding diagram. But first, we will learn about some of the geometrical principles that define a plane.

Defining planes in geometry

When defining planes in a 3D space, we need to take geometrical principles into account. Hence, before we get practical with SOLIDWORKS, we need to review some of the basic geometrical principles that define a plane. As you may recall, a **plane** can be understood as an infinity-extending surface. To define a plane, we only need to define a piece of it. There are eight common ways that a plane can be defined in space. The following are the ways in which they can be defined, along with an example of each:

1. **Three points**: Any three points in a space can be connected to define a plane, as shown in *Figure 6.2*:

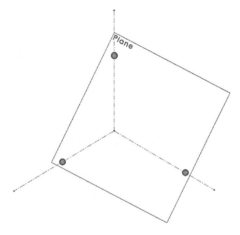

Figure 6.2 – Defining a plane using three points

2. **One point and a line**: Linking a line to a point in space will define a plane, as shown in *Figure 6.3*:

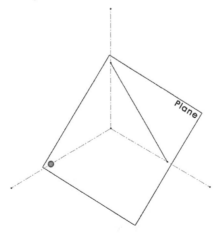

Figure 6.3 – Defining a plane using a point and a line

3. **Two parallel lines**: Any two parallel lines in a space can define a plane, as shown in *Figure 6.4*:

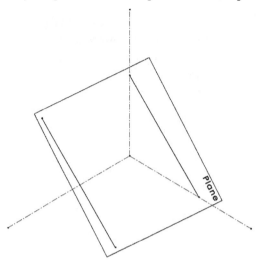

Figure 6.4 – Defining a plane using two parallel lines

4. **Two intersecting lines**: Any two intersecting lines can define a plane, as shown in *Figure 6.5*:

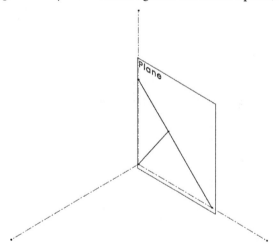

Figure 6.5 – Defining a plane using two intersecting lines

5. **Other planes**: Any existing plane or planner surface can be used to define new planes if we offset it by a certain distance. We can also make the new plane related to the other two. For example, the new plane can be midway between two parallel planes or planner surfaces. *Figure 6.6* highlights a base plane and another plane defined by an offset:

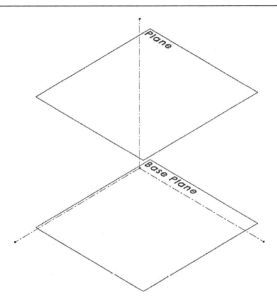

Figure 6.6 – Defining a plane using another plane

6. **A plane and a line**: Mixing an existing plane and a line can result in a new plane. This is done by defining a relationship between the existing plane and a line. These relations can be parallel, perpendicular, or coincident. We can also place an angle or distance in-between them, as shown in *Figure 6.7*:

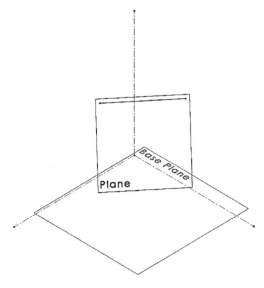

Figure 6.7 – Defining a plane using another plane and a line

7. **A plane and a point**: A plane can define another plane if we offset it. A point can be used to define the offset, as shown in *Figure 6.8*:

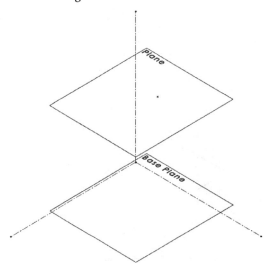

Figure 6.8 – Defining a plane using another plane and a point

8. **A plane and a curve**: Curve tangents can be used to define planes when we want to link the new plane to an existing one, as shown in *Figure 6.9*:

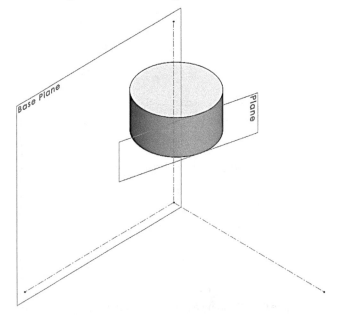

Figure 6.9 – Defining a plane using another plane and a curved surface

Note that these are just the basic ways of defining planes in a space. In most cases, we can manipulate angles, distance, and geometric relations further to generate more diverse planes.

Once we are familiar with how to define a plane from a geometrical perspective, we can easily define a plane in SOLIDWORKS by selecting the different components to define it with. Now, we will learn how to define new planes in SOLIDWORKS.

Defining a new plane in SOLIDWORKS

Now that we know how to define planes in different ways in terms of geometry, we can start learning how to define new planes in SOLIDWORKS. To illustrate this, we will create the model highlighted in *Figure 6.10*:

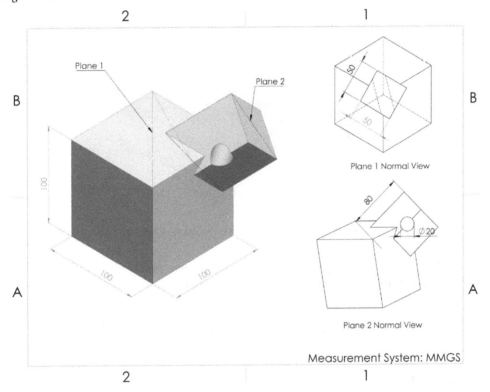

Figure 6.10 – The 3D model we will make in this exercise

Note that, in the preceding model, we have used two additional planes: *Plane 1* and *Plane 2* as indicated in *Figure 6.10*. They are defined as follows:

- *Plane 1*: Three points (vertexes)
- *Plane 2*: Two parallel lines (edges) that come from the side extrusion

To create this model, we need to plan, sketch, and apply features. However, we will also create additional reference planes.

Our plan is to create the base cube first and then create a new reference plane. After that, we'll create the side extrusion using the new plane. Then, we'll create the second reference plane. Finally, we'll create a circular hole. Let's start applying this plan by following these steps:

1. Create the base 100 mm cube to get the model highlighted in *Figure 6.11*:

Figure 6.11 – The first step will be to create a 100 mm cube

2. Now, we can start creating the first reference plane for our model. With the **Features** tab selected, click on the **Reference Geometry** tab and select **Plane**, as in *Figure 6.12*:

Figure 6.12 – The location of the Reference Geometry command

3. The PropertyManager will appear on the left-hand side so that we can select our references. Select three points (vertexes), as shown in the following screenshot. Also, you may need to check the last box under **Options**, that is, **Flip normal**, since we want the normal dimension to be outward, as indicated by the blue arrow. We will get *Figure 6.13*:

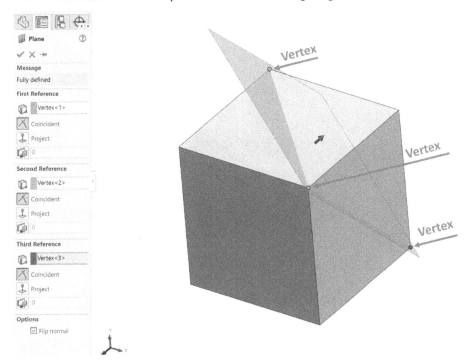

Figure 6.13 – PropertyManager and the plane preview

The options for reference planes are simple. First, we need to select a reference. The reference could be a point, a plane, a planner, or a curved surface. Then, we can select the relation of our new plane to that reference.

In the case of the plane we are creating, the first, second, and third references are all points. Also, the relation of these points to the new plane is that they are all coincident.

> **Note**
> We do not always need three references; the number of references depends on how we define the plane, as we explained in the *Defining planes in geometry* section. To assist with this, the software will let us know if the plane is fully defined or not, as shown on the top of the PropertyManager in *Figure 6.13*.

4. Click on the green checkmark to approve the plane.

5.   We can now start sketching on top of the new plane we just created. In terms of sketching, we can treat the new plane just like the other default planes we have. Select the new plane from the design tree and sketch a 50 mm square that's centered on the edge of the cube, as shown in *Figure 6.14*:

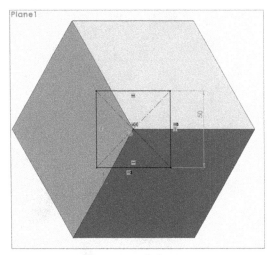

Figure 6.14 – Using the new plane to create new sketches

6.   Apply a feature based on the new sketch. We can apply features to new sketches in the same way we apply features to other sketches: based on default or surface planes.

7.   Apply the extruded boss feature to extrude our sketch by 80 mm. We will get the shape highlighted in *Figure 6.15*:

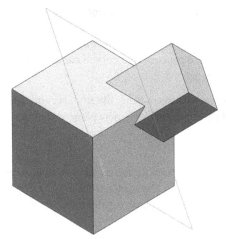

Figure 6.15 – Features can be applied based on sketches from the new planes

8. Now, we will create the second reference plane, which is based on two parallel lines. Select the geometries and select the two parallel edges, as shown in *Figure 6.16*. Then, click on the green checkmark to generate the new plane.

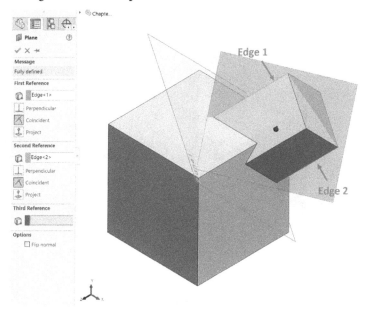

Figure 6.16 – The PropertyManager and preview for our second plane

9. Sketch the new circle with a diameter of 20 mm, as shown in *Figure 6.17*. Note that the center of the circle is located mid-point on the edge.

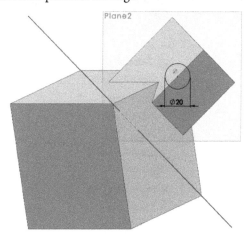

Figure 6.17 – A new circle sketch applied on a new plane

10. Apply the extruded cut feature by setting the end condition to be **Through all - Both**. We are doing this since the hole goes through the whole model and in both directions. We will get the shape highlighted in *Figure 6.18*:

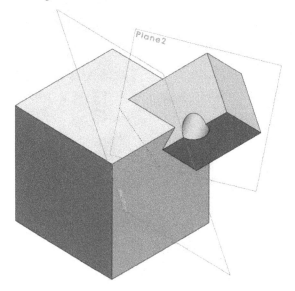

Figure 6.18 – The final shape of our 3D model

This concludes the process of creating the model.

Note that, in our resulting model, we can see that the two planes we introduced are visible. To make the model look cleaner, we can hide the new planes after they have fulfilled our needs. We can do this by right- or left-clicking on the plane listing in the design tree and selecting the eye-shaped option, as shown in *Figure 6.19*:

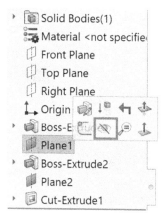

Figure 6.19 – The hide option to hide planes from view

Once we've hidden the visible planes, our cleaner model will look like *Figure 6.20*:

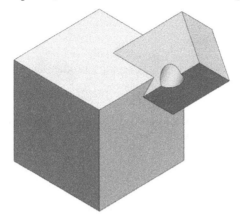

Figure 6.20 – A clean-looking 3D model after hiding the new planes

In this exercise, we created a reference plane with three points and two parallel lines. If we were to create a new plane based on any of the other six methods we looked at earlier, for example, another plane, or two intersecting lines, we could simply follow the same steps. The only difference would be the reference selection and the available selected reference relations.

This concludes how to create planes as additional reference geometry in SOLIDWORKS. In this section, we covered the following topics:

- How to access the command for creating new reference planes
- How to create a reference plane based on three points
- How to create a reference plane based on two parallel lines
- How to use a new reference plane to create sketches and features

Now that we know how to generate new plane reference geometries, we can start learning about our next set of features: Swept Boss and Swept Cut.

Understanding and applying Swept Boss and Swept Cut

Swept Boss and **Swept Cut** allow us to create shapes by sweeping a profile along a path. In this section, we will discuss Swept Boss and Swept Cut in detail. These features are opposites and require more than one sketch if we wish to apply them. We will learn about their definitions, how to apply them, and how to modify them.

What are Swept Boss and Swept Cut?

Swept Boss and Swept Cut are opposing features; one adds materials, while the other removes materials. Let's talk about them in more detail:

- **Swept Boss**: This adds materials by sweeping a profile shape on a designated path
- **Swept Cut**: This removes materials by sweeping a profile shape on a designated path

Figure 6.21 highlights the effect of the Swept Boss and Swept Cut features in a better way:

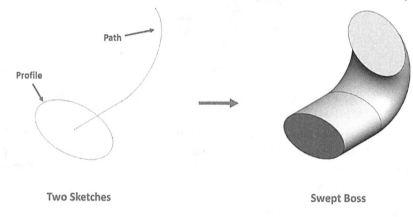

Figure 6.21 – An illustration of what the Swept Boss feature does

The preceding diagram shows Swept Boss, while *Figure 6.22* shows Swept Cut. Note that both features require two sketches before they can be applied:

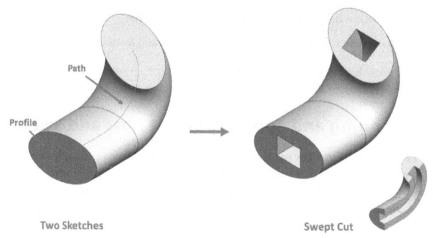

Figure 6.22 – An illustration of what the Swept Cut feature does

Compared to the extruded boss and extruded cut, Swept Boss and Swept Cut provide us with more flexibility when it comes to extruding a shape. While extruded boss and extruded cut add and remove materials directly perpendicular to the sketch plane, Swept Boss and Swept Cut allow us to guide the extrusion as we see fit. Let's start applying them, beginning with Swept Boss.

Applying Swept Boss

In this section, we will demonstrate how to apply the Swept Boss feature. To do this, we will create the model shown in *Figure 6.23*:

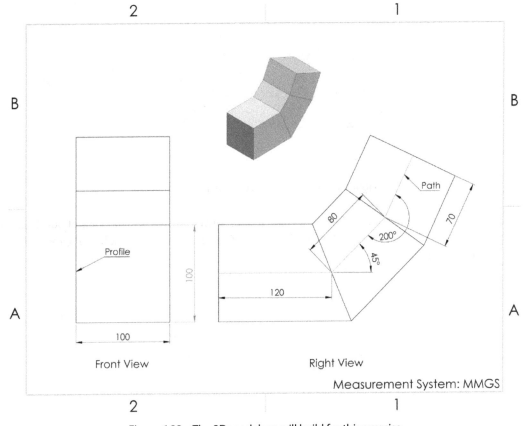

Figure 6.23 – The 3D model we will build for this exercise

To model this shape, we will go through the stages of planning, sketching, and applying features. Our plan will be to create a rectangular profile and then the path. After that, we will apply the Swept Boss feature. Follow these steps to do so:

1. Select the front plane and draw a square centered at the origin. This will represent the profile. Set the side length to 100 mm, as shown in *Figure 6.24*. Then, exit sketch mode.

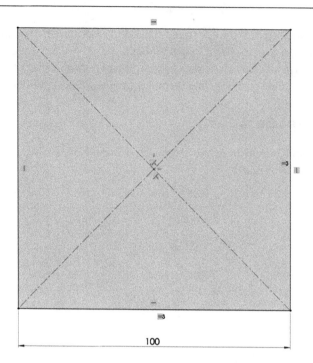

Figure 6.24 – The profile sketch of a 100 mm square

2. Select the right plane and draw the path that's shown in the right-hand view of our diagram. We can make the start of the path the origin. Our sketch should look like *Figure 6.25*. Exit sketch mode after that.

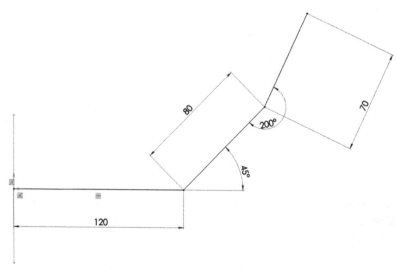

Figure 6.25 – The path sketch

3. Now, we will apply the Swept Boss feature. Go to the **Features** tab in the command bar and select the **Swept Boss/Base** command, as highlighted in *Figure 6.26*:

Figure 6.26 – The location of the sweep features

4. The PropertyManager will appear on the left-hand side so that we can select **Profile and Path**. We can then make the following selections:

I. **Profile**: Select the first rectangle we drew in *Step 1*. We can select it by clicking on it on the canvas.

II. **Path**: Select the curve we sketched in *Step 2*.

Once we've selected **Profile and Path**, we will see a preview of our shape, as shown in *Figure 6.27*:

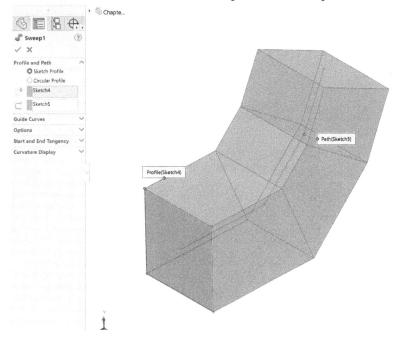

Figure 6.27 – The PropertyManager and a preview of the Swept Boss feature

5. Click on the green checkmark to approve this feature. This will give us the final shape, as shown in *Figure 6.28*:

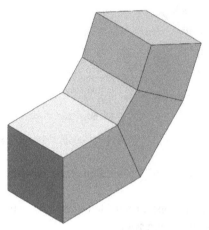

Figure 6.28 – The resulting 3D model after applying Swept Boss

This concludes using the Swept Boss feature. Before we move forward with the Swept Cut feature, let's talk about one of the aspects of the path and explain some of the other options that are available for the Swept Boss feature.

Note that for basic sweeps, the path must either intersect the profile itself or its extension so that it's captured by the feature. In *Figure 6.29*, we have highlighted different types of acceptable and unacceptable paths. The unacceptable paths are indicated with an *X* sign.

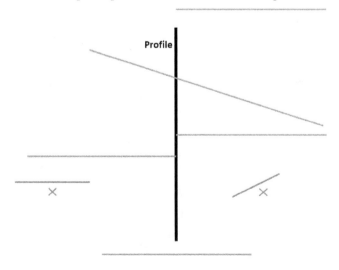

Figure 6.29 – An illustration of the acceptable path location for the sweep features

Now that we know how to create a basic Swept Boss, we can discuss the different options we can use to create more complex Swept Boss applications.

Swept Boss feature options

Now, let's examine some of the other options that are available with the Swept Boss feature that we didn't utilize in the previous exercise. We will cover the options we missed from top to bottom, as shown in *Figure 6.30*:

Figure 6.30 – The PropertyManager sweep feature showing many options

The following is a brief explanation of these options:

- **Circular Profile**: Under **Profile and Path**, we can select the **Circular Profile** option. This makes it easier for us to create the sweep feature when the profile is circular and goes through the middle of the path. When selecting the **Circular Profile** option, a profile sketch will not be needed. This can make it easier to create piping arrangements with fewer sketches.

- **Guide Curves**: With this option, we can add more sketches to help guide our Swept Boss feature. This is helpful when the Swept Boss feature does not have a regular shape that matches the profile. However, it is an advanced feature that we will not use at this level.

- **Profile orientation**: This helps us to determine how the profile moves with the path. There are two options available here:

  - **Follow Path**: This will make the profile tilt along with the path

  - **Keep Normal Constant**: This will keep the profile facing the same direction as it moves on the path

- **Profile Twist**: This can be used to create shapes such as spiral springs or any type of twisted shape since it will allow the profile to twist around the path. *Figures 6.31* and *Figure 6.32* are examples of using **Profile Twist** in two different ways. One was done by twisting a square by 90 degrees, as follows:

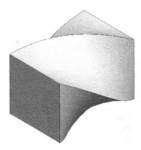

Figure 6.31 – A resulting sweep with the profile twist option

The other was done by twisting a circle by three revolutions around the path, as shown in *Figure 6.32*:

Figure 6.32 – The Profile Twist option can also be used to create spirals

- **Start and End Tangency**: This ensures that the sweep is normal for the path toward the start or endpoint.

- **Thin Feature**: This will allow us to only sweep boss a thin layer around the profile instead of the whole enclosure. This is similar to the Thin Feature option in the extruded boss feature.

- **Curvature Display**: This option allows us to analyze the curvature of our sweep better. This is done by adding visual elements such as mesh preview, zebra stripes, and curvature combs.

This concludes this exercise, which was all about applying the Swept Boss feature. We covered the following topics in this section:

- How to create sketches that can use Swept Boss

- How to apply the Swept Boss feature

- The acceptable paths that can be used with the Swept Boss feature

- The different options that can be used with the Swept Boss feature

Now, we can start learning about the opposite feature of Swept Boss, that is, Swept Cut.

Applying Swept Cut

In this section, we will demonstrate how to use the **Swept Cut** feature. To do this, we will create the model shown in *Figure 6.33* by adding a Swept Cut to the model we created earlier:

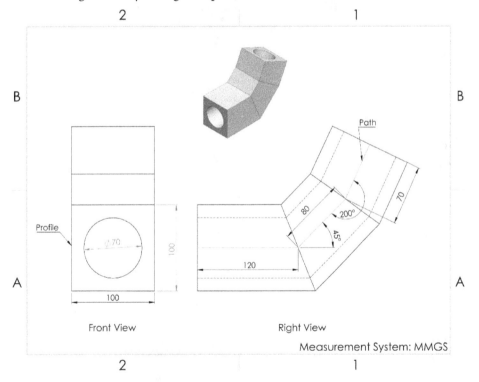

Figure 6.33 – The 3D model we will build in this exercise

To model this shape, we will go through the procedure of planning, sketching, and applying the feature. Our plan will be to create the profile on our existing Swept Boss. After that, we will apply the

Swept Cut feature by following the same path we had for the Swept Boss. To act on this plan, we will follow these steps:

1. Select the face shown in the front view as a sketch surface. Then, sketch a circle with a diameter of 70 mm, as shown in *Figure 6.34*. Exit sketch mode afterward.

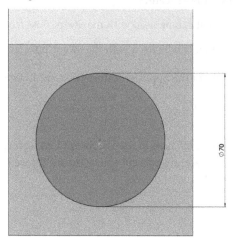

Figure 6.34 – The profile sketch

> **Note**
>
> The path for this cut is the same path that we used to apply the Swept Boss feature. Hence, we don't need to create another path. Instead, we can reuse the one we already have. A reused sketch will be indicated in the design tree with a small icon of an open hand.

2. Now, let's apply the Swept Cut feature. From the **Features** tab, select the **Swept Cut** command, as shown in *Figure 6.35*:

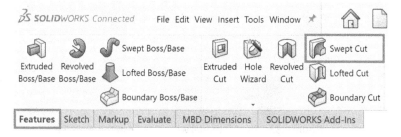

Figure 6.35 – The location of the Swept Cut feature

3. Select the appropriate profile by selecting it from the canvas, just like we did with the Swept Boss feature.

4. We can select the path from the design tree that appears on the canvas. Expand that design tree and look for the sketch we used for the path under the **Sweep** feature. Then, select the sketch that corresponds to the path, as shown in *Figure 6.36*. We will see a preview of the Swept Cut feature.

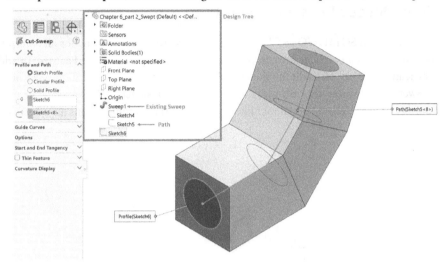

Figure 6.36 – The design tree can be used to select sketches for the path or profile

5. Click on the green checkmark to apply the Swept Cut feature. This will result in the model shown in *Figure 6.37*:

Figure 6.37 – The final 3D model after applying the Swept Cut

This concludes this exercise, which was all about applying the Swept Cut feature. The additional options that are available for the Swept Cut feature are the same as those for the Swept Boss feature. We covered the following topics in this section:

*   How to apply the Swept Cut feature
*   How to reuse a sketch in more than one feature

Now that we know how to apply the Swept Boss and Swept Cut features, we need to know how we can modify them.

Modifying Swept Boss and Swept Cut

Modifying features in SOLIDWORKS is done in the same way that it's done for all features; that is, by right- or left-clicking on the feature in the design tree and selecting **Edit Feature**. In addition to editing the feature, we can also edit the sketches that are guiding the feature. In the case of the Swept Boss and Swept Cut features, this includes the profile and path sketches. To demonstrate this, we will modify our previous model so that it looks like *Figure 6.38*. These modifications have been annotated.

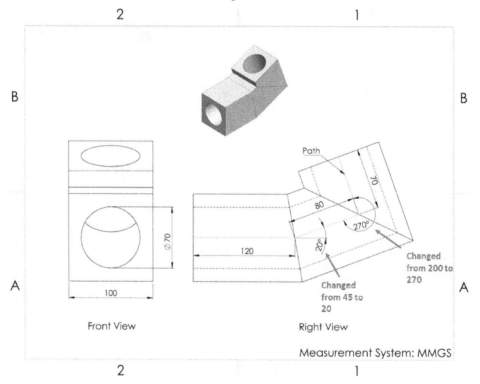

Figure 6.38 – The modifications we will apply for this exercise

Note that this model is very similar to the one we created in the previous exercise. The only difference is the path. Hence, we will only modify the sketch we used for the path. To do that, follow these steps:

1. Find the sketch path in the design tree by expanding the design tree listing of the sweep boss feature. Right-click on the path sketch and select **Edit Sketch**, as highlighted in *Figure 6.39*. Note that if a sketch is being used in more than one sketch, a small hand icon will appear next to it.

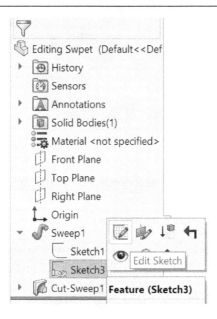

Figure 6.39 – We can edit sketches from the design tree

2. Adjust the sketch by double-clicking on the angle values and changing them so that they match the ones in *Figure 6.40*:

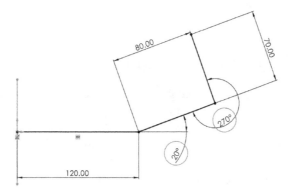

Figure 6.40 – We can adjust the sketches in the canvas once selected from the design tree

3. Exit sketch mode. You will see that the model now looks like *Figure 6.41*. Note that, by editing the sketch, we ended up editing the Swept Boss and Swept Cut features since both are utilizing the same sketch.

Figure 6.41 – The final 3D model after modifying the path sketch

> **Tip**
> A shortcut to edit the dimension is to double-click on the sketch from the design tree and then adjust the displayed dimensions directly without getting into edit mode.

This concludes our exercise on editing Swept Boss and Swept Cut. In this section, we covered the following topics:

- How to edit sketches that are being used for the swept paths
- How to identify sketches in the design tree that are being used in more than one feature

In this section, we covered how to apply and modify the Swept Boss and Swept Cut features. Next, we will cover another set of features: Lofted Boss and Lofted Cut.

Understanding and applying Lofted Boss and Lofted Cut

In this section, we will discuss the Lofted Boss and Lofted Cut features. **Lofted Boss** and **Lofted Cut** allow us to create a shape by sketching and connecting sections of it.Lofted BossLofted Cut These features are opposites, and we require more than one sketch if we wish to apply them. We will learn how to define them, how to apply them, and how to modify them.

What are Lofted Boss and Lofted Cut?

With Lofted Boss and Lofted Cut, we can add or remove materials based on multiple cross-sections. Let's talk about them in more detail:

- **Lofted Boss**: This adds materials by linking different cross-sections together. This includes the start and end of the shape. If we choose to, we can link these cross-sections with guide curves. *Figure 6.42* illustrates the Lofted Boss feature.
- **Lofted Cut**: This removes materials by linking different cross-sections together. This includes the start and end of the cut shape. If we choose to, we can link these cross-sections with guide curves. *Figure 6.43* illustrates the Lofted Cut feature.

These features are polar opposites: one adds materials, while the other removes materials in the same way. The following diagrams illustrate these features:

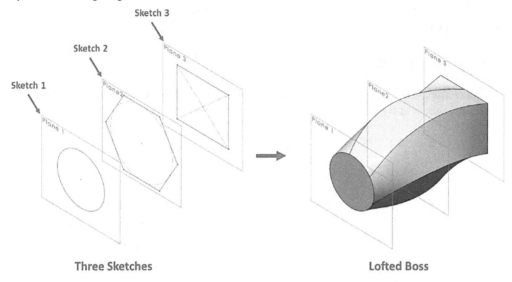

Figure 6.42 – An illustration highlighting the Lofted Boss feature

The preceding diagram shows Lofted Boss, while the following diagram shows Lofted Cut. Note that both features require at least two sketches if we wish to apply them. We can add as many sketches as we wish to define the sections:

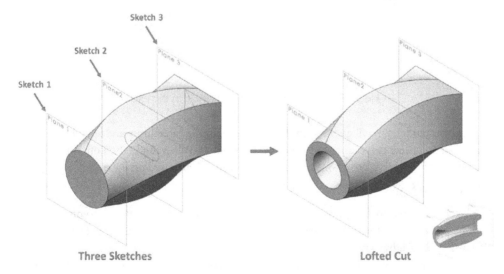

Figure 6.43 – An illustration highlighting the Lofted Cut feature

Lofted Boss and cut can provide us with a unique way of controlling how we add or remove materials compared to other features, such as Swept Boss and Swept Cut and extruded boss and extruded cut. Now that we know what Lofted Boss and Lofted Cut are used for, we can start applying them. Let's start by applying the Lofted Boss feature.

Applying Lofted Boss

In this section, we will cover how to apply the Lofted Boss feature. To illustrate this, we will create the model shown in *Figure 6.44*:

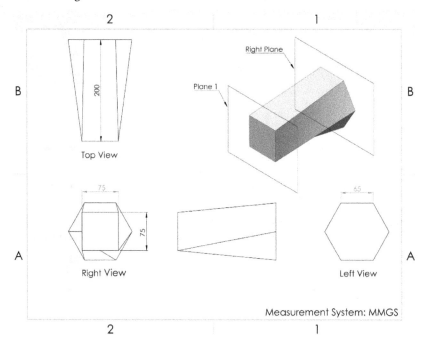

Figure 6.44 – The 3D model we will build in this exercise

To create this model, we will go through the steps of planning, sketching, and applying features. Our plan will be to utilize two planes to create two sketches and then apply the Lofted Boss feature. In this exercise, we'll have to create one additional reference plane based on the default right plane. Let's start by following these steps:

1.   The first step is to define our new reference planes. Select the **Plane** sub-command, which can be found under **Reference Geometries**. Then, create **Plane1** by offsetting a distance of 200 mm from **Right Plane**, as shown in the top view in *Figure 6.45*. The option for plane creation is shown in the following screenshot, and can be found alongside the final shape of the two planes:

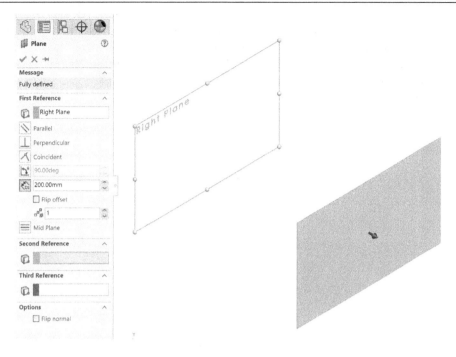

Figure 6.45 – The PropertyManager for our new plane

After approving the new plane, it will appear parallel to **Right Plane**, as shown in *Figure 6.46*:

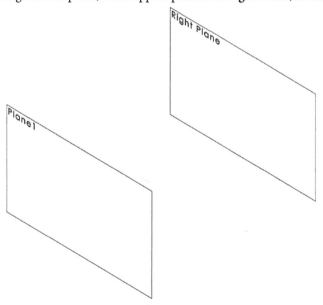

Figure 6.46 – The new plane after being defined

2. Since we are using the Lofted Boss feature, we will create two different sketches in the two planes. Select **Plane1**, which we created previously, and sketch a square whose sides are equal to 75 mm, as shown in the right-hand view of *Figure 6.44*. The resulting sketch will look as follows. Exit sketch mode after that.

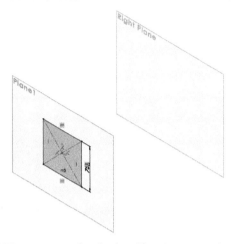

Figure 6.47 – A 75 mm square sketched on Plane1 representing our first profile

Tip

You can show the right plane in the canvas by right- or left-clicking on it in the design tree and selecting **Show**.

3. Select the right plane and sketch a hexagon whose sides are equal to 65 mm, as shown in the left view of *Figure 6.44*. The resulting sketch will look as in *Figure 6.48*. Exit sketch mode after that.

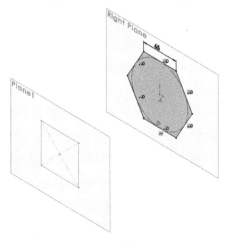

Figure 6.48 – A hexagon sketched on the right plane representing our second profile

4. Now, we will apply the Lofted Boss feature to connect the two sketches. Select the **Lofted Boss/Base** command from the **Features** tab, as shown in *Figure 6.49*:

Figure 6.49 – The location of the Lofted Boss feature

5. After selecting the command, we will get the command's PropertyManager options on the left-hand side, as shown in *Figure 6.50*:

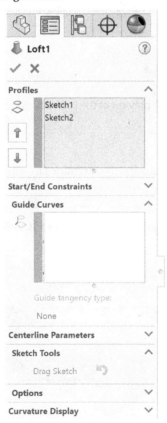

Figure 6.50 – The PropertyManager showing the profile selection

6. For the profiles, select the square first, then the hexagon. Note that the selection is order-sensitive. Once we do that, we will get the following review on the canvas. Note that there is one guideline in the preview to help us to define our loft. This line is controlled by the two endpoints. To adjust it, we can drag the point to another location. This will change the shape of the loft. Take your time and adjust the guideline so that it looks like *Figure 6.51*:

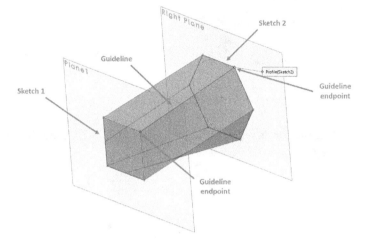

Figure 6.51 – A preview of the Lofted Boss showing a guideline

Note

The initial positioning of the guideline is determined by where you click on the sketch to select the profile.

7. Click on the green checkmark to apply the Lofted Boss feature. The result will be *Figure 6.52*:

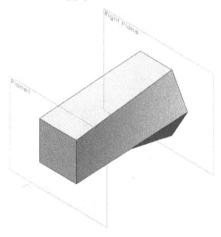

Figure 6.52 – The final resulting shape after applying the Lofted Boss

This concludes our exercise on the Lofted Boss feature. In this exercise, we only covered the most basic lofted application. However, this feature has many advanced options that we did not get around to using. We will explore these options next.

Lofted Boss feature options

In the feature's PropertyManager, we will be able to find all of the feature's options. *Figure 6.53* highlights the PropertyManager for the Lofted Boss feature:

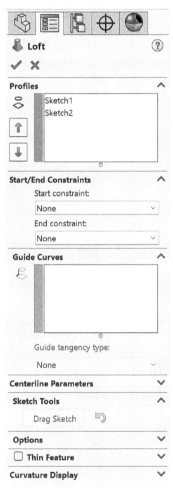

Figure 6.53 – The Lofted Boss's PropertyManager showing the available feature options

The following is a brief explanation of these options:

- **Start/End Constraints**: This gives us more control over the areas that are close to the profile sketches. Under each, we have the following options:

    - **Direction vector**: This pushes the loft toward a specific direction, as per an existing vector. To apply this, we may need to create additional lines to push the loft.

    - **Normal to profile**: This pushes the loft in a direction that's normal/perpendicular to the existing profile we used to build the loft.

- **Guide Curves**: This gives more flexibility in terms of how the loft is constructed. However, we still need to create more curves from different sides to guide the loft. *Figure 6.54* shows our previous loft with multiple guiding curves:

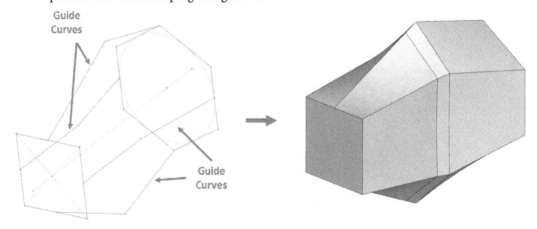

Figure 6.54 – An illustration of the guide curves function

Thin Feature: This changes the loft so that it's only lofting the shell rather than the whole shape.

This concludes the various options for the Lofted Boss feature. In this section, we covered the following topics:

- How to create different profiles with the Lofted Boss feature
- How to apply the Lofted Boss feature
- The different options for the Lofted Boss feature

Now that we know how to apply the Lofted Boss feature, we can start learning how to apply the Lofted Cut feature.

Applying Lofted Cut

The Lofted Cut feature works the same as the Lofted Boss feature, except that it has the opposite effect. To show you how this feature works, we will build on the previous model and create the model shown in *Figure 6.55*:

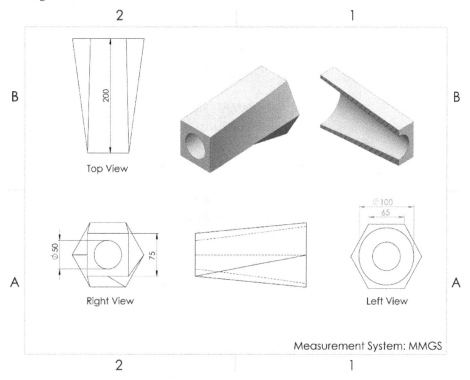

Figure 6.55 – The 3D model we will build in this exercise

Note that, for this model, we will only create an internal cut from the previous model. This cut is governed by two circles on each end. Like we did previously, we will create the model by planning, sketching, and then applying the feature. We will use the existing end faces of our shape to sketch two circles. Then, we will apply the Lofted Cut feature. Follow these steps to implement this plan:

> **Tip**
> Since both of our sketches are located on existing faces, we can hide the two visible planes and sketch on the faces that have been formed from the features instead. To hide a plane, we can right-click on it from the design tree and select the **Hide** option, which is the small eye icon.

1. Select the square face as a sketch plane and sketch a 50 mm circle, as shown in *Figure 6.56*. Use the origin as the center of the circle. Exit sketch mode after that.

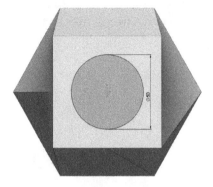

Figure 6.56 – The sketch of our first profile

2. Select the hexagonal face as a sketch plane and sketch a 100 mm circle, as shown in *Figure 6.57*. Use the origin as the center of the circle. Exit sketch mode after that.

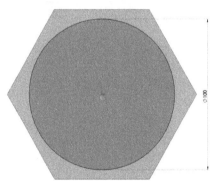

Figure 6.57 – The sketch of our second profile

3. Now, let's apply the feature. Select **Lofted Cut** from the **Features** tab, as shown in *Figure 6.58*:

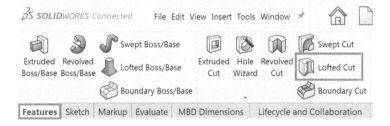

Figure 6.58 – The location of the Lofted Cut command

4. The **PropertyManager** options for the Lofted Cut are the same as they are for the Lofted Boss feature. Under **Profiles**, select the two circles we sketched earlier. We will get the preview shown in *Figure 6.59*. Note that since our two profile sketches are circles, shifting the guideline will make the cut look more like a sandglass figure. Thus, to maintain a uniform cut from the smaller to the wider circle, we can either maintain a straight guideline or use additional guide curves. For this exercise, we use the earlier option, as we are covering guide curves next.

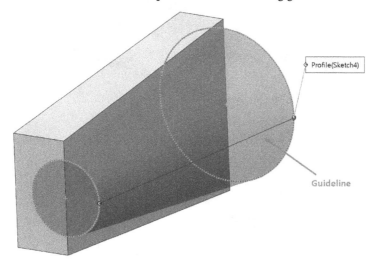

Figure 6.59 – The preview of the Lofted Cut

5. Click on the green checkmark to apply the Lofted Cut feature. The final shape will look like *Figure 6.60*:

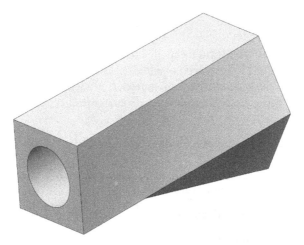

Figure 6.60 – The final 3D model after applying the Lofted Cut

We can use the cross-section viewing feature at the top of the canvas to view the shape from the inside as well. The result of doing this is shown in *Figure 6.61*:

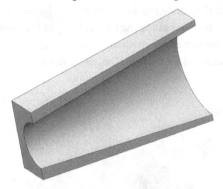

Figure 6.61 – A cross-section of the model showing the cut

6.   To create a section view, click on the icon shown in *Figure 6.62*. Then, adjust the **Section View** cross in the PropertyManager.

Figure 6.62 – The Section View command on top of the canvas

> **Note**
> Even though we created the list cut using the Lofted Cut feature, we can reach the same result with other features, such as revolved cut and draft. Keep in mind that there are often many different paths that can reach the same geometrical model. However, we might choose one path over another depending on our design intent or ease of application.

This concludes our exercise of applying the Lofted Cut feature. In this section, we covered the following topics:

- How to apply the Lofted Cut feature
- The effects that the provided guidelines have on circular profiles
- How to make a section view

Now that we know how to apply these two features, we need to know how we can modify them.

Modifying Lofted Boss and cut

Modifying the Lofted Boss and Lofted Cut features follows the same procedure that we follow to modify any other feature. We can right-click on the feature to modify its options. We can also modify the sketches of the profile to adjust the shape of the loft.

Before we conclude this section, let's cover a key aspect of the Lofted Boss and Lofted Cut features: guide curves.

Guide curves

Guide curves provide us with more flexibility when it comes to lofted features that we lack when applying the feature without them. Because of that, we will create the shape shown in *Figure 6.63* to learn about one way of using guide curves:

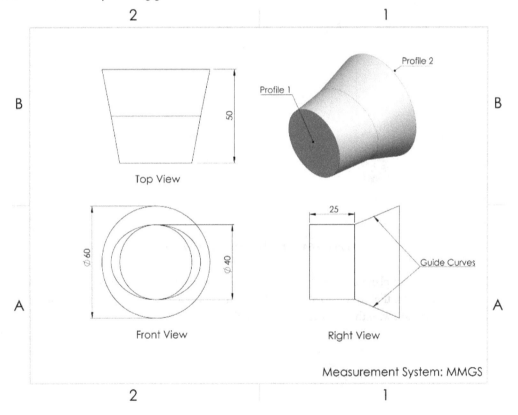

Figure 6.63 – The 3D model we are building in this exercise

Let's go through our usual procedure of creating models: planning, sketching, and applying features. Our plan will be to create the two profiles and then create the two guidelines. After that, we will apply the Lofted Boss feature. Follow these steps to implement this plan:

1. Create an extra plane by offsetting the front plane by 50 mm. Then, create the two circle profiles that are shown in *Figure 6.64*. The smaller diameter is 40 mm, while the larger one is 60 mm.

> **Tip**
>
> We can hide **Front Plane** and **Plane1** to make our canvas clearer and less crowded.

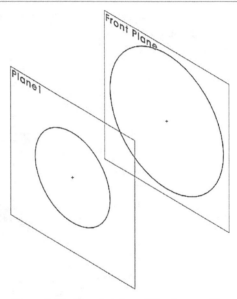

Figure 6.64 – The sketches of the two profiles

2. Using the right plane as a sketch plane, sketch the upper guide curve, as shown in the following diagram. Note that to fully define the sketch, the endpoints of the guide curve should have a pierce relation with the profile, as highlighted in *Figure 6.65*:

> **Note**
>
> The guide curves must intersect with the profiles.

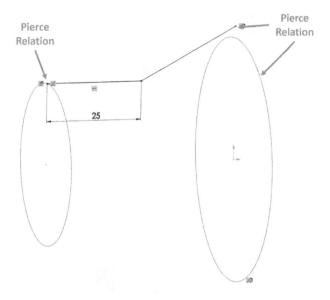

Figure 6.65 – The sketch of the guide curve

3. Mirror the first guide curve for the other side, as shown in *Figure 6.66*. We are doing this because the lower part of the curve is the same as the top one. Then, exit sketch mode.

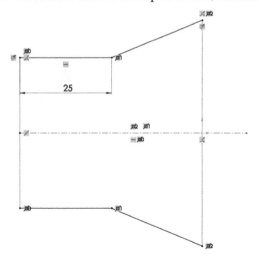

Figure 6.66 – The final two guide curves required for the 3D model

4. Now, let's apply the Lofted Boss feature. We will use all of the profiles and the guide sketches to apply it. Select the **Lofted Boss** command. Then, select the two circular profiles.

5. Under the **Guide Curves** section, select the two guide curves we created. Note that we created the two guide curves under one sketch. Hence, when selecting one curve from the canvas, we will see the window that's shown in *Figure 6.67*. As we can see, we can select the **Open Loop** option and then click on the green checkmark to apply it.

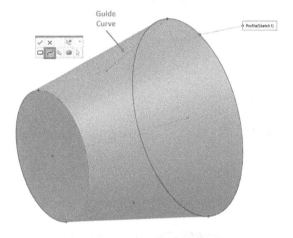

Figure 6.67 – The preview and selection procedure of one guide curve

Note

Our loft will shift toward the guide curve.

6. Following the same procedure that we completed in *Step 5*, select the other guide curve. After that, our preview will look like *Figure 6.68*:

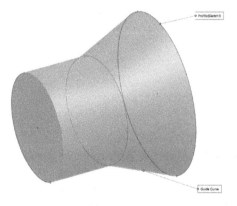

Figure 6.68 – The final preview after both guide curves are applied

7. Click on the green checkmark to apply the loft. The resulting model will look like *Figure 6.69*:

Figure 6.69 – The final resulting shape

Note that we have only applied two guide curves to govern the loft from the upper and lower sides. Hence, from the unguided side, we'll notice that the resulting shape is more elliptical. If we need more guidance when it comes to shape, we can increase the number of guide curves as we see fit. There is no limit to the number of profiles and guide curves we can have.

This concludes our section on lofted guide curves, where we learned how to apply guide curves to control our lofts and how to select parts of a sketch to included in the guide curves.

In this section, we covered the Lofted Boss and Lofted Cut features, how to apply them, and how to modify them. We also learned about guide curves.

Summary

In this chapter, we learned about a set of features that allow us to create more complex 3D models than what we were able to create in the previous chapters. We learned about plane reference geometries, which allow us to add new reference planes in addition to the default ones. We also covered the Swept Boss, Swept Cut, Lofted Boss, and Lofted Cut features. Each feature set requires more than one sketch to apply. For each, we learned what they are, how to apply them, and how to modify them. The features that we covered in this chapter allow us to generate 3D models, such as flexible tubing and irregularly shaped casings.

In the next chapter, we will learn about mass properties, which allow us to assign materials and calculate different properties, such as the mass of our 3D models.

Questions

1. What are the eight methods of defining new planes?

2. Why do we need to define new planes?

3. What are Swept Boss and Swept Cut?

4. What are Lofted Boss and Lofted Cut?

5.   Create the following model:

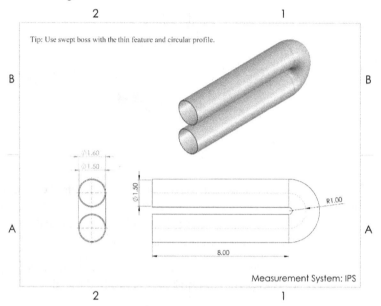

Figure 6.70 – The 3D model for Question 5

6.   Create the following model:

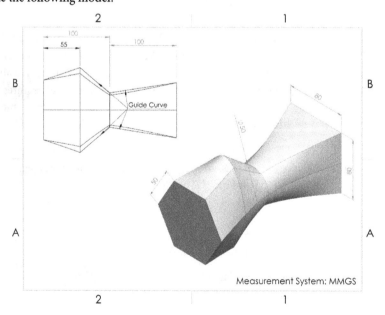

Figure 6.71 – The 3D model for Question 6

7.   Create the following model:

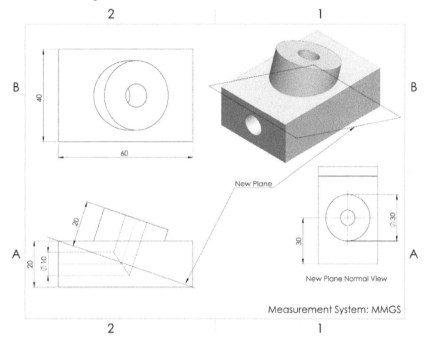

New Plane

New Plane Normal View

Measurement System: MMGS

Figure 6.72 – The 3D model for Question 7

> **Important note**
> The answers to the preceding questions can be found at the end of this book.

Get This Book's PDF Version and Exclusive Extras

Part 4:
Basic Evaluation and
Assemblies – Associate Level

After building 3D models of parts, we might be required to assign them materials, evaluate their mass properties, and put them together to form an assembly. This part will introduce all that and cover all that is expected for the Associate level. This includes assigning materials, evaluating mass properties, and applying standard mates to link different parts in an assembly.

This part has the following chapters:

- *Chapter 7, Materials and Mass Properties*
- *Chapter 8, Standard Assembly Mates*

7

Materials and Mass Properties

Whenever we design or model an object, we have to consider the structural materials to work with. In other words, should the object be made of steel, iron, plastic, wood, or any other materials? SOLIDWORKS provides a library with a variety of materials that we can choose from. It also provides tools that we can use to find the mass properties of the object at hand, such as the volume, mass, or center of mass.

The following topics will be covered in this chapter:

- Reference geometries – defining a new coordinate system
- Assigning materials and evaluating and overriding mass properties

By the end of this chapter, you will be able to assign materials to your parts and evaluate different associated mass properties. This will include finding the mass and the volume of a 3D model. We will also cover how to introduce a new coordinate system to your model. These skills will help you decide on what materials to choose for the products you are designing. It will also help you estimate different costs associated with materials and their related aspects, such as transportation and storage.

Technical requirements

This chapter will require that you have access to the SOLIDWORKS software.

The CiA video for this chapter can be found at `https://packt.link/WxQIi`

Reference geometries – defining a new coordinate system

By default, SOLIDWORKS provides us with one coordinate system. This system is centered on the origin. The origin is also the starting point of the three axes: **X**, **Y**, and **Z**. In certain cases, we require another base of a coordinate system. In this section, we will explore why we need new coordinate systems and how to define them. This will help us calculate different properties, such as the center of mass, from a different perspective, which is a key skill that's required for collaborative work and a SOLIDWORKS professional.

What is a reference coordinate system and why are new ones needed?

In SOLIDWORKS, we can understand **reference geometries** as ones we can use as a base or as a reference for something else. For example, a plane is considered a reference geometry that we use as a base for sketches. A coordinate system is a reference geometry since we can use it as a base to determine specific locations within the canvas. In this section, we will learn what a coordinate system is and how to define a new one in SOLIDWORKS.

In our sketch creation process, we always link our sketch to the origin. The origin is the base point for our coordinate system, which extends through the three axes: **X, Y, and Z**. In the lower-left corner of the SOLIDWORKS canvas, we can notice the direction of the three axes, as shown in *Figure 7.1*:

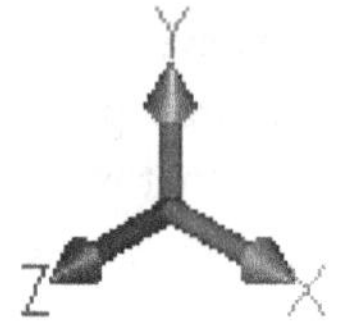

Figure 7.1 – The direction of the axes found in the canvas

Using the coordinate system, we can locate any point in the canvas according to its **X**, **Y**, and **Z** coordinates. When in sketch mode, the lower-right corner of the canvas will show us the location of the cursor according to the **X**, **Y**, and **Z** locations. *Figure 7.2* highlights where we can find these coordinates:

Z	Y	X				
-5.07in	-1.2in	0in	Under Defined	Editing Sketch1	IPS	

Figure 7.2 – The location of the cursor is indicated according to the axes below the canvas

> **Note**
> The order of the **X**, **Y**, and **Z** coordinates, as shown in *Figure 7.2*, will change according to the sketch plane.

Since SOLIDWORKS already provides us with a default coordinate system, why do we need additional ones? Here are two reasons why we may need a new coordinate system:

- **To calculate some mass properties in relation to a different point of reference**: For example, if we measure the center of mass of an object, the measurement will be relative to the **X**, **Y**, and **Z** locations for the default coordinate system. With additional coordinate systems, we can determine the center of mass according to a different coordinate system of our choosing. We will explore this later in this chapter.

- **To switch our views of directions**: In certain applications, we may need to redefine our axes to change our sense of direction as we are creating a model. Additional coordinate systems can help us accomplish that. Also, when working with different people on a specific design, pointing toward different coordinate systems can ease communication.

We have just covered reference coordinate systems and why we may need additional ones for certain applications. Now, we will start learning how to define a new coordinate system within SOLIDWORKS.

How to create a new coordinate system

In this section, we will learn how to introduce a new coordinate system. To highlight this, we will create a simple triangular prism and introduce a new coordinate system, as shown in *Figure 7.3*:

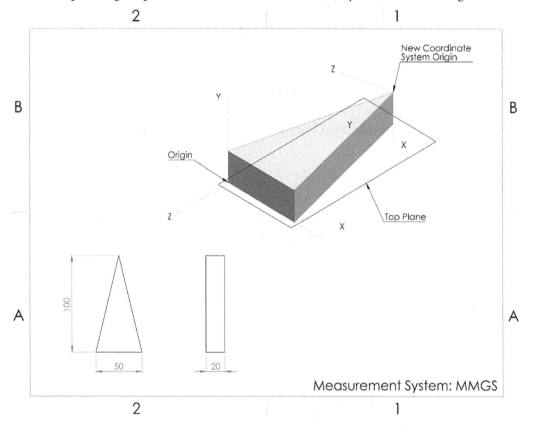

Figure 7.3 – The drawing we will build in this exercise

We will start creating this model by planning, sketching, and applying features. However, note that since we are familiar with the steps and with creating various shapes already, we will only explain this process in brief. Our plan for this model will be to start by creating the overall simple shape first and then introduce the new coordinate system.

Follow these steps:

1. Sketch the isosceles triangular base using the top plane and apply an extruded boss with the dimensions indicated in *Figure 7.3*. We should have the shape shown in *Figure 7.4*:

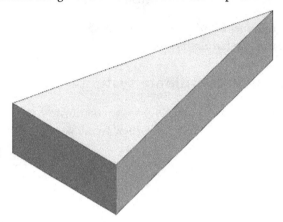

Figure 7.4 – The first step is to build the 3D model

2. Under the **Features** tab, select **Coordinate System** under **Reference Geometry**, as shown in *Figure 7.5*:

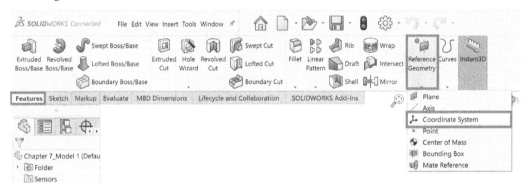

Figure 7.5 – The location of the new Coordinate System command

3. The **PropertyManager** will appear on the left, showing the options that we need to define our new coordinate system. To define a new coordinate system, we have to specify the origin and two axes. Set the origin by selecting the vertex indicated in *Figure 7.6*.

4. We can do the same thing with **X axis** and **Y axis** by selecting edges. Note the arrows next to each axis selection of **PropertyManager**; they allow us to switch the direction of the axis. We can flip the direction of the axis until we get the right orientation, as shown in *Figure 7.6*:

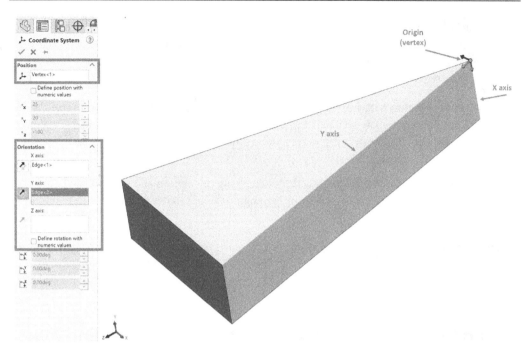

Figure 7.6 – The preview of the new coordinate system and its PropertyManager

5. Click on the green checkmark to approve the new coordinate system. Once we apply the new system, it will be shown on the model as **Coordinate System1**, as shown in *Figure 7.7*:

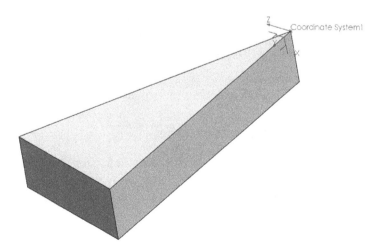

Figure 7.7 – The new coordinate system will be shown on the canvas after being applied

This concludes this exercise of creating the new coordinate system we needed. We will use this model again in the next part of this chapter to study mass properties. However, before moving on, let's explain the **Define position with numeric values** and **Define rotation with numeric values** options shown in the PropertyManager in *Figure 7.6*:

- **Define position with numeric values**: Checking this option allows us to position the origin of our new coordinate system using numerical distance values in the X, Y, and Z axes as they relate to the absolute origin and axes. In contrast, the example we followed earlier positioned the new origin using an existing vertex.

- **Define rotation with numeric values**: Checking this option allows us to rotate the axes of the new coordinate system by inputting angular values that are measured in relation to the absolute axes.

Both of these options can be helpful when we're looking to define a new coordinate system that is independent of the existing geometries of the part. With that, we can conclude our discussion on defining new coordinate systems. Next, we will start assigning a structural material and evaluate the mass property of our model.

Assigning materials and evaluating and overriding mass properties

Whenever we design physical objects, we have to think about what materials we will choose to build those objects. These objects can be made out of plastic, steel, iron, and so on. SOLIDWORKS provides us with an array of different materials to choose from. It also provides us with the properties of each of those materials, such as their density, strength, thermal conductivity, and other properties related to the specific material. Also, once we assign a material to our model, we can evaluate the different **mass properties** of our objects, such as the mass and the center of mass.

In this section, we will learn how to assign materials and how to evaluate the mass properties of our models. We will also learn how to override the evaluated mass properties.

Assigning materials to parts

In this section, we will discuss how to assign specific materials to our parts. To do that, we will assign the **Steel: AISI 304** material to the model shown in *Figure 7.8*. Note that we created this model earlier in this chapter, as shown in *Figure 7.4*:

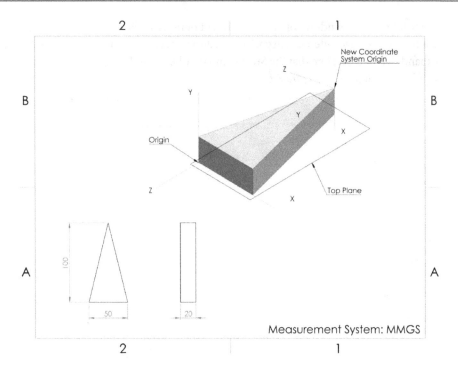

Figure 7.8 – The 3D model we will use for material assignment

To assign materials to our model, follow these steps:

1. On the design tree, right-click on the **Material <not specified>** entry. Then, click on the **Edit Material** option, as in *Figure 7.9*:

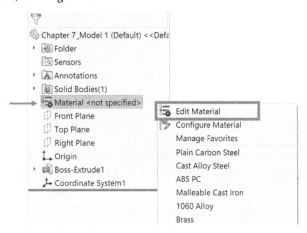

Figure 7.9 – The location of the Edit Material command

2. We will see a new window containing different options, as shown in the following screenshot. On the left, we have different categories of materials to choose from. Since our material is **Steel**, expand that option. Note that the **Steel** menu may be expanded by default when you get to the window highlighted in *Figure 7.10*:

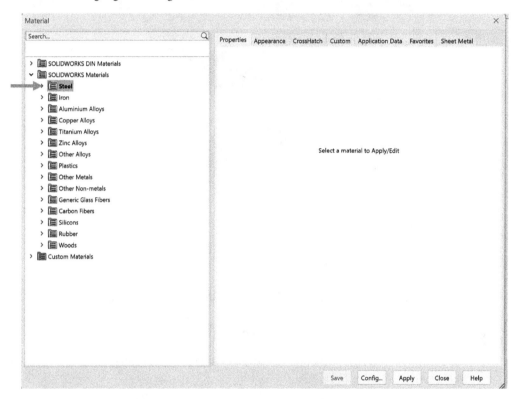

Figure 7.10 – The material library with a selection of different materials to choose from

> **Tip**
> At the top of the materials list, there is a search box that can help you find the required material faster.

3. Expand the **Steel** menu and select **AISI 304**. Our window will look as shown in *Figure 7.11*. Note the highlighted options, which show **Unit of Measurement** and **Material Properties**:

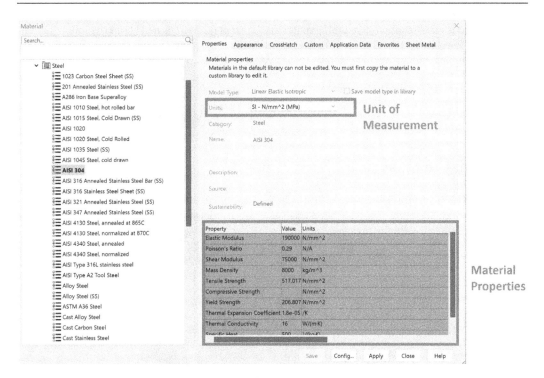

Figure 7.11 – We can choose the unit of measurement to view the material properties

Note

We can change the display unit to other units, such as **Imperial** and **Metric**. This will change the unit that's displayed for the rest of the properties. The listed material properties include **Mass Density**, **Tensile Strength**, and **Poisson's Ratio**. We can click on other materials to find out what their properties are.

4.   Click on **Apply** to apply the material to our part.

After applying the material, we will notice that the color of the part in the canvas will change to match the assigned material. For example, when assigning **AISI 304** stainless steel, the visual part in the canvas will change color to light gray. We can choose not to apply the material appearance from the **Appearance** tab before clicking **Apply**.

Tip

We can add our frequently used materials to the **Favorites** tab by right-clicking on the material and adding it to our favorites. Also, if we are looking for a material that does not exist in the SOLIDWORKS library, we can add a custom material to our local material library.

This concludes this section on assigning materials to our parts. We learned about the following topics:

- How to assign different materials to an existing part
- How to find different material properties for each material

At this point, we have a material assigned to our part. Next, we will learn how to view the mass properties of our part.

Viewing the mass properties of parts

In this section, we will learn how to view the different mass properties of our existing model. To demonstrate this, we will complete tasks relating to the model shown in *Figure 7.12*. Note that this is the same model as from the previous section:

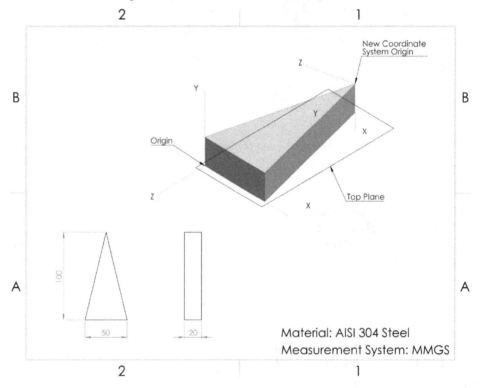

Figure 7.12 – The 3D model we will use to view the mass properties

In this exercise, we will complete the following four tasks:

1. Find the mass of the model in grams

2. Find the center of mass concerning the origin in millimeters

3. Find the center of mass concerning the new coordinate system indicated in the diagram in millimeters

4. Find the mass of the model in pounds

Note that all these tasks involve finding different mass properties, such as the mass and the center of mass. Hence, before we can accomplish these tasks, we need to discuss how to view the mass properties.

Viewing mass properties

To view mass properties, we can click on **Mass Properties** from the **Evaluate** tab, as highlighted in *Figure 7.13*:

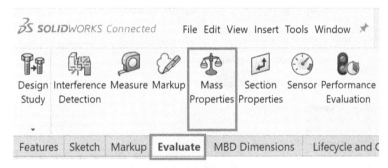

Figure 7.13 – The location of the Mass Properties command

This will show us the window shown in *Figure 7.14*, which contains different mass properties related to our object. These include **Density**, **Mass**, **Center of mass**, **Volume**, **Moments of inertia**, and other properties. If you cannot find the **Mass Properties** option that's shown in the preceding screenshot, you can click on the **Tools** menu, and then go to **Evaluate**, where you will also find the **Mass Properties** option.

> **Note**
>
> The units of measurement for the mass properties are the same as the units of measurement for the document. We will learn how to change that when we tackle the fourth task.

We will be able to find most of the information we need to resolve our tasks in the **Mass Properties** window. We will cover these tasks one by one:

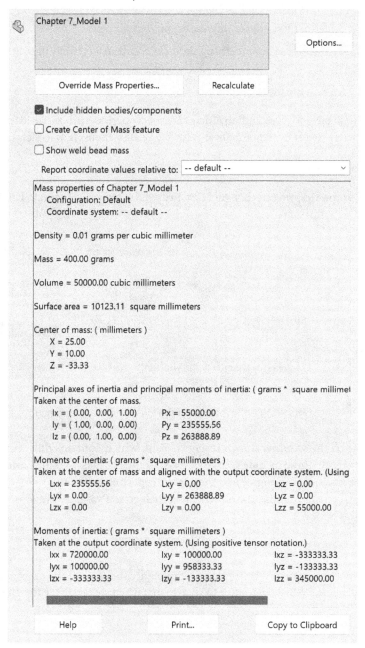

Figure 7.14 – The Mass Properties window will display different properties for the 3D model

Now, we can start accomplishing our four tasks by finding the mass of the model in grams.

Finding the mass of the model in grams

We can find the mass of the model in grams directly from the list of mass properties. Since the unit of measurement for the document is MMGS, the displayed mass is in grams. As shown in *Figure 7.15* of the **Mass Properties** window, **Mass** is listed as **400.00 grams**:

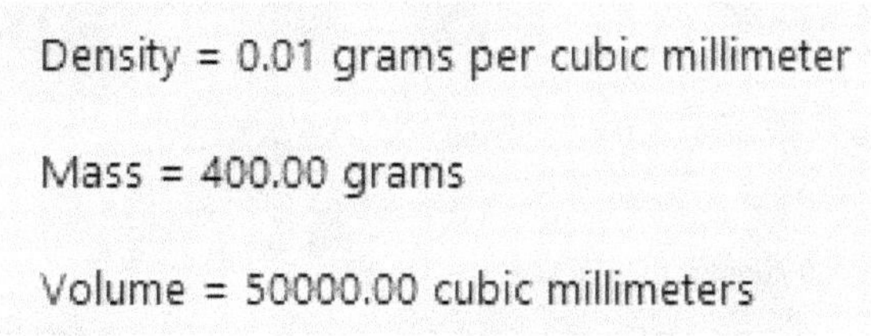

Density = 0.01 grams per cubic millimeter

Mass = 400.00 grams

Volume = 50000.00 cubic millimeters

Figure 7.15 – Density, Mass, and Volume are among the displayed mass properties

Now, let's find the center of mass concerning the origin in millimeters.

Finding the center of mass concerning the origin in millimeters

From the **Mass Properties** window, we can find **Center of mass: (millimeters)**, as shown in *Figure 7.16*:

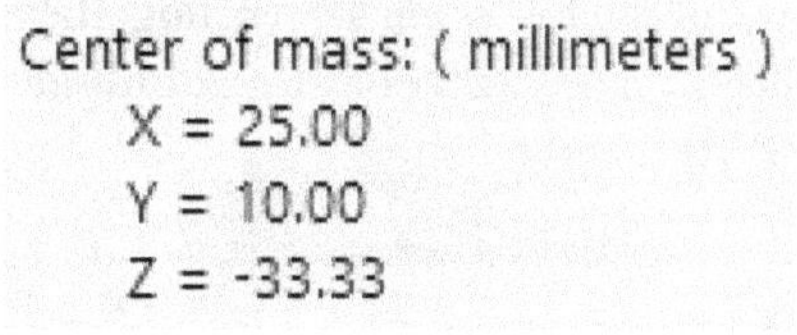

Center of mass: (millimeters)
X = 25.00
Y = 10.00
Z = -33.33

Figure 7.16 – Center of mass, as shown in the Mass Properties window

Note that **Center of mass** is a relational value. In other words, the X, Y, and Z coordinates are concerned with the origin and the default coordinate system. By default, SOLIDWORKS calculates all of the relational values in relation to the default coordinate system. We can adjust that to another coordinate system, which we'll do in the next task.

Next, we will find the center of mass concerning the new coordinate system.

Finding the center of mass concerning the new coordinate system in millimeters

In this task, we need to find the same information that we found in the second task, but using a different coordinate system. We created the new coordinate system for the model earlier in this chapter. To change the calculations so that they relate to the new coordinate system, you can change the field next to **Report coordinate values relative to** to the other coordinate system, as shown in *Figure 7.17*:

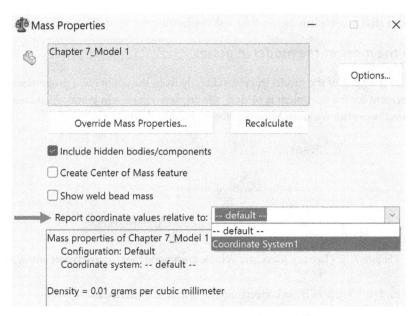

Figure 7.17 – We can evaluate the mass properties according to different coordinate systems

After selecting the new coordinate system, we will notice that the relational values, including **Center of mass**, will change, as shown in the following screenshot. The absolute values such as **Mass** and **Volume** are the same. Note that the new **Center of mass: (millimeters)** is **X = 10.00, Y = 64.68,** and **Z = 16.17**, as shown in *Figure 7.18*:

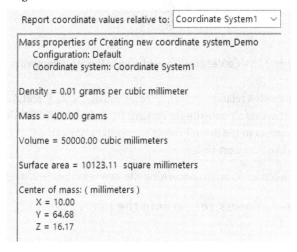

Figure 7.18 – Location-based values can change according to the coordinate system being used

Lastly, we will find the mass of the model in pounds.

Finding the mass of the model in pounds

For this task, we need to change the units of measurement for mass from grams to pounds. To do this, we can follow these steps:

1. Click on **Options...** in the **Mass Properties** window, as shown in *Figure 7.19*:

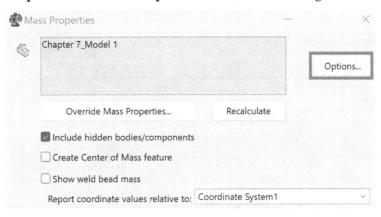

Figure 7.19 – The location of the Options... command

2. We will be taken to the **Mass/Section Property Options** menu, which will allow us to customize the measurement units for our mass properties. To change the mass unit, click on **Use custom settings**. Then, under **Mass**, select **pounds**, as shown in *Figure 7.20*:

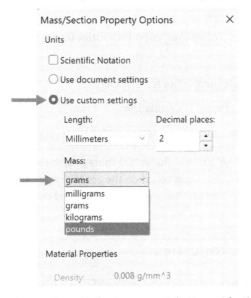

Figure 7.20 – We can change the unit of measurement that's used for the mass properties

3. Click **OK** to confirm our unit selection. This will apply the new mass unit to the **Mass Properties** window, as shown in *Figure 7.21*:

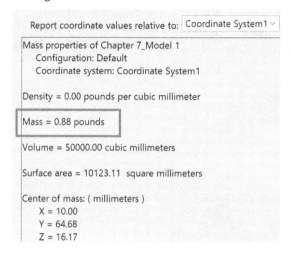

Report coordinate values relative to: Coordinate System1 ⌄

Mass properties of Chapter 7_Model 1
    Configuration: Default
    Coordinate system: Coordinate System1

Density = 0.00 pounds per cubic millimeter

Mass = 0.88 pounds

Volume = 50000.00 cubic millimeters

Surface area = 10123.11 square millimeters

Center of mass: (millimeters)
    X = 10.00
    Y = 64.68
    Z = 16.17

Figure 7.21 – Displayed units for mass and density switched to pounds

Note that the mass is now calculated in pounds with a value of **0.88 pounds**.

> **Note**
>
> You can change the displayed decimal places using the same **Options...** window.

This concludes this section on viewing mass properties for existing models. We learned about the following topics:

- How to view the mass properties for a model
- How to adjust mass property calculations for a new coordinate system
- How to customize the unit of measurement for mass property evaluation

So far, we have learned how to view the actual mass properties of our parts. These are based on calculations that SOLIDWORKS does based on the geometry and the material of our part. However, we also have the option to override those values with manual entries. We will learn about this in the next section.

Overriding mass properties

In this section, we will learn how to override mass property values. By default, SOLIDWORKS calculates the mass properties based on the material that's assigned, as well as the design itself. However, in some instances, we might want to override those calculated values. For example, we might have a part in an assembly where we know its final mass but not the exact design that will result in the mass. In this case, we can simply override the mass to our required value.

To demonstrate this, we will override the mass of our model from 400 grams to 500 grams. To accomplish this, follow these steps:

1. Open the **Mass Properties** window and click on **Override Mass Properties...**, as shown in *Figure 7.22*:

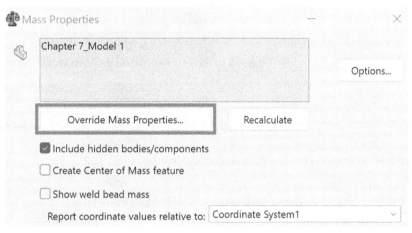

Figure 7.22 – The location of the Override Mass Properties… command

2. Check the **Override mass** box and input a value of 500, as shown in *Figure 7.23*:

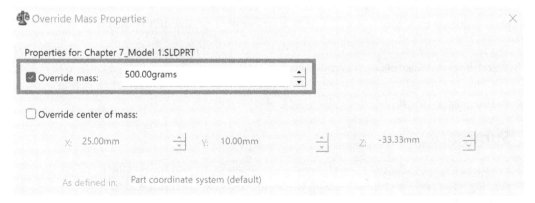

Figure 7.23 – We can change the mass properties values by overriding them

3. Click **OK** to implement the adjustment.

This will redefine the mass as **500 grams**. When viewing the mass properties, SOLIDWORKS will note the mass with **user-overridden**, as shown in *Figure 7.24*:

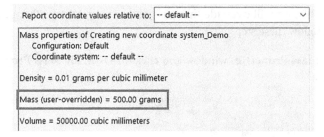

Figure 7.24 – The overridden mass will be indicated as such in the Mass Properties window

> **Tip**
>
> When overriding the mass, we can state the unit of measurement in the text input and SOLIDWORKS will automatically convert it to the default unit. For example, in *Figure 7.23*, we can type 1 pound, and it will automatically change to 453.59 grams.

To change the mass back, we can go back to the **Override Mass Properties…** command and uncheck the **Override mass** box. If we do this, our mass will go back to 400 grams as a calculated value. In addition to overriding the mass, we can also override the values for the center of mass and moments of inertia.

In practice, overriding the mass properties is more common when we're working with assemblies rather than parts. We do this if we want to keep a certain effect of a part in its interaction with other parts in an assembly. This is especially the case if our parts are not fully refined yet.

This concludes this section on assigning materials and evaluating mass properties. Being able to assign materials and evaluate mass properties for our model is essential in deciding what material we should use. Also, it plays an important role in calculating the cost of production, materials, transportation, and other applications that can relate to the evaluated mass properties.

Summary

In this chapter, we discussed reference coordinate systems and mass properties. We started by defining what a reference coordinate system is, why we need new ones, and how to define new ones in SOLIDWORKS. Then, we learned about mass properties.

We learned how to access the materials library we have in SOLIDWORKS and how to assign a particular material to our parts. Then, we learned how to find properties such as mass and volume for our models. Being able to determine mass properties will help us decide on what structural materials we can pick for the products we design. It will also help us to determine the costs associated with production, transportation, and so on.

In the next chapter, we will start working with assemblies. Assemblies are more than one part that's joined together in one file. Most of the products we use in our everyday lives, such as laptops, cars, and pens, consist of multiple parts that have been put together to form the final product or assembly. This makes these tools key to designing products that are used in our everyday lives.

Questions

Answer the following questions to test your knowledge of this chapter:

1. What are coordinate systems?

2. What parameters do we need to define a new coordinate system in SOLIDWORKS?

3. What information do we get by evaluating the mass properties of a part?

4. Create the following model and define the indicated **Coordinate System 1**:

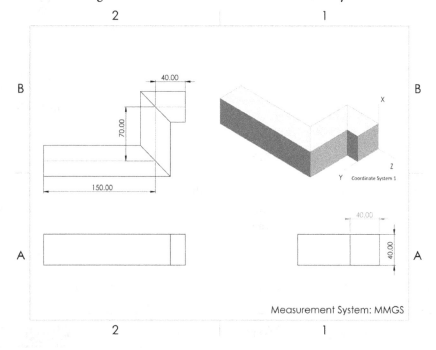

Figure 7.25 – The drawing for question 4

5. Assign the **Aluminum Alloy: 1060 Alloy** material to the model we created in *Question 4*. What is the mass in grams? What is the center of mass (in millimeters) concerning **Coordinate System 1**, as indicated in the preceding diagram?

6. Create the following model and define the indicated coordinate system:

Figure 7.26 – The drawing for question 6

7. Assign the **Plain Carbon Steel** material to the model we created in *Question 4*. What is the mass in pounds? What is the center of mass (in inches) concerning **Coordinate System 1**, as indicated in the preceding diagram?

> **Important note**
> The answers to the preceding questions can be found at the end of this book.

8

Standard Assembly Mates

Most of the products we interact with in our daily lives, such as laptops, phones, and cameras, are made up of many different components that have been put together; that is, they have been assembled. One of the major elements of mastering SOLIDWORKS is being able to use **SOLIDWORKS assemblies**, which allow us to put multiple parts together to create a single artifact. In this chapter, we will cover basic SOLIDWORKS assemblies, in particular, **standard mates**.

In this chapter, we will cover the following topics:

- Opening an assembly file and adding parts
- Understanding and applying non-value-oriented standard mates
- Understanding and applying value-driven standard mates
- Utilizing materials and mass properties for assemblies

By the end of this chapter, we will be able to put different parts together to form an assembly. The objective of this chapter is to get us to generate complex artifacts by linking different parts together and creating an assembly using standard mates.

Technical requirements

In this chapter, you will need to have access to the SOLIDWORKS software. The project files for this chapter are available at the following GitHub repository: `https://github.com/PacktPublishing/Learn-SOLIDWORKS-2025-Third-Edition`

The CiA video for this chapter can be found at `https://packt.link/OM7ba`

Opening assemblies and adding parts

In this section, we will take our first steps toward working with SOLIDWORKS assemblies. We will cover what SOLIDWORKS assemblies are, how to start an assembly file, and how we can add a variety of components to our assembly file. Opening an assembly file and adding different parts to it is the first step we need to take when we start any assembly.

Defining SOLIDWORKS assemblies

There are three main sections of SOLIDWORKS: **parts**, **assemblies**, and **drawings**. For each type, SOLIDWORKS creates a different file type with different file extensions. For assemblies, the file extension is .SLDASM, while a part has a file extension of .SLDPRT. In *Chapter 10*, we will cover drawings, which have .SLDDRW as the file extension.

With an assembly file, we can link more than one part file together to form one product. *Figure 8.1* and *Figure 8.2* highlight two examples of assembly files. *Figure 8.1* shows a simple assembly that consists of only three parts. We can see each and every part since they are highlighted by solid lines and borders that have been filled in with a variety of colors:

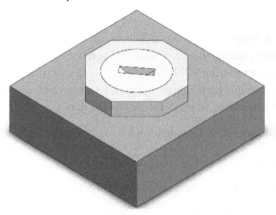

Figure 8.1 – A simple assembly of three parts

Figure 8.2 highlights a more complex assembly than the first one. It is a mechanical assembly that consists of over 50 different parts:

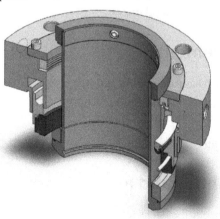

Figure 8.2 – An assembly of 50 parts

The different parts interact with each other via mates. **Mates** are very similar to the relations we used in sketching, for example, coincident, perpendicular, and tangent. However, they work for assemblies. Now that we know what assemblies are, we can move on and create our first assembly file.

Starting a SOLIDWORKS assembly file and adding parts to it

To demonstrate how to start an **assembly file** and add parts to it, we will start an assembly file and add the following parts to it. Make sure that you download the files that accompany this chapter. The parts you will download are as follows:

- A small triangular part, as shown in *Figure 8.3*:

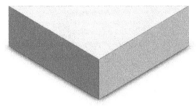

Figure 8.3 – The rectangular prism is included in this chapter's downloads

- A larger base part, which will house the smaller triangular part, as shown in *Figure 8.4*:

Figure 8.4 – The part file for this shape is included in the chapter's downloads

Now that you've downloaded these parts onto your computer, we can start opening our assembly file.

Starting an assembly file

To start an assembly file, follow these steps:

1. Select **New** from the top of the SOLIDWORKS interface, as shown in *Figure 8.5*:

Figure 8.5 – Where to start a new file

2. Select **Assembly** and click **OK**, as highlighted in *Figure 8.6*:

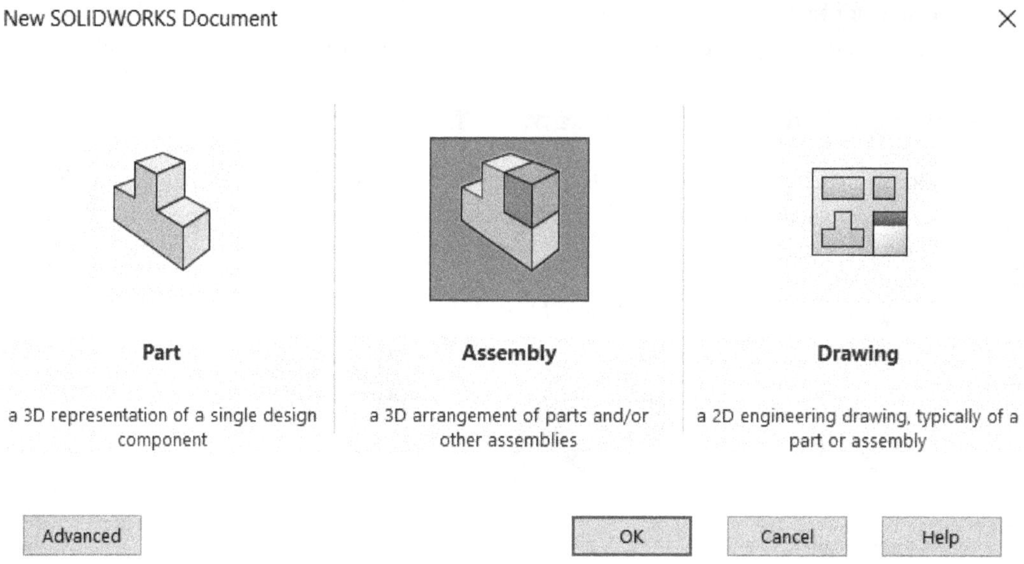

New SOLIDWORKS Document   ✕

Part	**Assembly**	**Drawing**
a 3D representation of a single design component	a 3D arrangement of parts and/or other assemblies	a 2D engineering drawing, typically of a part or assembly

Advanced		OK	Cancel	Help

Figure 8.6 – A window showing the different new file types you can start with

Now that we have opened our assembly file, we can add the two parts we downloaded to it. We will do that next.

Adding parts to the assembly file

To add our two parts to the assembly file, follow these steps:

1. In the **Assembly** tab, select **Insert Components**, as shown in *Figure 8.7*.

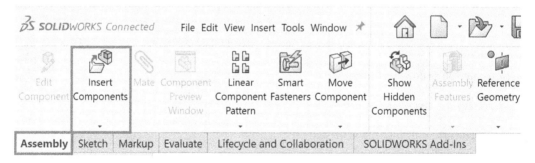

Figure 8.7 – The Insert Components option to add parts to your assembly

2. Navigate to the `Base.SLDPRT` file, which can be found in the SOLIDWORKS parts attached to this chapter. Click **Open** after selecting the file, as shown in *Figure 8.8*. Alternatively, we can double-click on the file to open it:

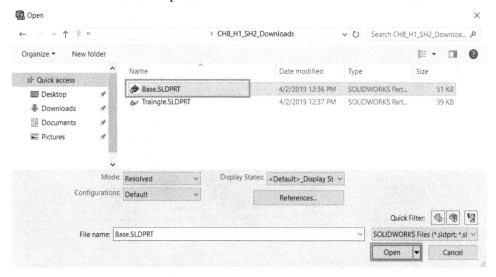

Figure 8.8 – The browser window to open a part in an assembly

3. The part will appear on the assembly's canvas. At the bottom of the page, we will see some options that can help us orient the part if needed. Once we are satisfied with the part's orientation, we can left-click on the canvas using the mouse to place the part. Alternatively, we can click on the green checkmark, which can be found in the top-right or the top-left corner of the page, as shown in *Figure 8.9*:

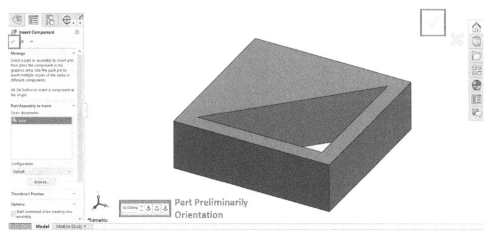

Figure 8.9 – The part appearing in the canvas after inserting it in the assembly file

4. Following *Steps 1-3* again, add the other part, `Triangle.SLDPRT`.

5. The triangle part will appear on the canvas as well. We can left-click to place the part in the assembly. We will end up with the two parts on the canvas, as shown in *Figure 8.10*. You may have a different placement for these two parts than what's shown here, though. This is not an issue at this point.

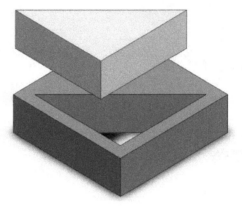

Figure 8.10 – The two parts will be in the assembly canvas

This concludes this exercise of adding parts to our assembly file. Before we move on, however, let's mention three key points when it comes to adding parts to our assembly:

- **Fixed parts**: The first part you insert into the assembly file is a fixed part by default. Fixed parts don't move in the assembly environment and are fully defined. In the design tree, a fixed part is annotated with (**f**) next to its name.

- **Floated parts**: The second part that's inserted is a floating part by default. Floating parts can be moved around the assembly environment since they are not defined. We will use mates to define floating parts later in this chapter.

- **Dragging parts**: You can click and hold the second part (triangle) and move it around the canvas. The first part is fixed by default so it cannot be moved.

We can change any part's status from fixed to floating and vice versa by right-clicking on the part and selecting the **Float** or **Fix** command. The **Float** command is highlighted in *Figure 8.11*:

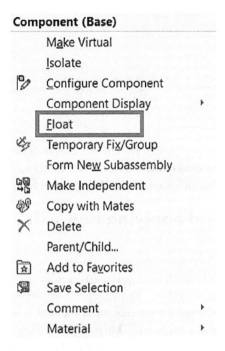

Figure 8.11 – The Float command location

> **Important note**
> If the part is fixed, we will see the **Float** command, as shown in *Figure 8.11*. However, if the part is floating, then we will see the **Fix** command instead to make the part fixed.

At this point, we have inserted our two parts into an assembly file. However, the parts hold no linkage to each other. Next, we will look at mates, which will enable us to interlink the different parts we have.

Understanding mates

Mates are similar to sketch relations, but they act on assemblies. They govern how different parts interact with each other or move in relation to each other. As an example, examine the keys on a computer keyboard. Each key is stationed in a specific location and restrained by specific movements, such as up and down. We can think of this positioning and movement as being governed by an assembly's mates.

There are three categories of mates in SOLIDWORKS: **standard mates**, **advanced mates**, and **mechanical mates**. We will only cover standard mates in this chapter. Standard mates provide the following options:

- Coincident
- Parallel

- Perpendicular
- Tangent
- Concentric
- Lock
- Distance angle

Some standard mates require us to input a numerical value, such as the mate's distance and angle. We can refer to these as **value-oriented mates**. The rest of the mates do not require a numerical value. We can refer to these as **non-value-oriented mates**. We will learn more about these next.

Understanding and applying non-value-oriented standard mates

In this section, we will start linking different parts together in an assembly using the non-value-oriented standard mates. We will learn about the coincident, parallel, perpendicular, tangent, concentric, and lock mates. These mates don't need a numerical value input to be applied to them; instead, they are constructed based on their relationship with different geometrical elements. In this section, we will learn what those mates are and how to apply them. We will also cover the different levels of defining an assembly. We will start by defining each of those non-value-oriented standard mates.

Defining the non-value-oriented standard mates

The non-**value-oriented standard mates** are coincident, parallel, perpendicular, tangent, concentric, and lock. These are special in that they don't require any numerical input to be applied to them or defined for them. They are very similar to the sketching relations we applied while sketching. Here is a brief explanation of each standard mate:

- **Coincident**: This allows a coincident relation between two surfaces, a line and a point, and two lines.
- **Parallel**: This allows us to set two surfaces, two edges, or a surface and an edge so that they're parallel to each other.
- **Perpendicular**: This allows us to set two surfaces, two edges, or a surface and an edge so that they're perpendicular to each other.
- **Tangent**: This allows two curved surfaces to be tangent to each other. This can also happen between a curved surface and an edge, as well as a straight surface.
- **Concentric**: This allows two curves to have the same center.
- **Lock**: This locks two parts together. When two parts are locked, they will copy each other's movements.

Now that we know what these standard mates do, we will start applying them to link different parts of our assembly. We will start with the coincident and perpendicular mates.

Applying the coincident and perpendicular mates

To explore how to make the mates coincident and perpendicular, we will make the assembly shown in *Figure 8.12*. The assembly is made out of two parts – a **Base** part and a **Triangle** part. We opened these parts in an assembly file earlier in this chapter. We will continue from there:

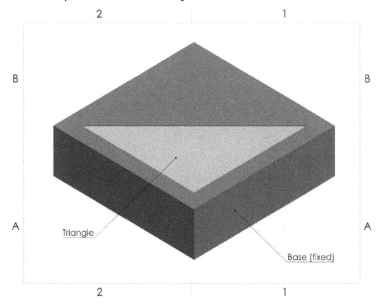

Figure 8.12 – The final assembly we will reach in this exercise

If you are starting over, you can download the two parts from the files for this chapter and open them in an assembly file. Our starting point will be having these two parts arranged on an assembly canvas, as shown in *Figure 8.13*:

Figure 8.13 – The two parts placed in an assembly without mates

Now that we have two parts in our assembly, we will apply the mates. We will start with the coincident mate, followed by the perpendicular mate.

Applying the coincident mate

To apply the coincident mate to our assembly, follow these steps:

1. To apply mates, select the **Mate** command, which can be found under the **Assembly** tab, as shown in *Figure 8.14*:

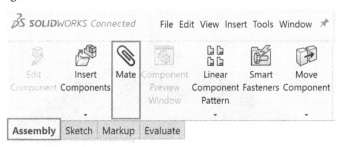

Figure 8.14 – The location of the Mate command

2. After selecting **Mate**, we will see a PropertyManager on the left-hand side of the screen. At this point, we will be asked to select which elements we want to mate. This can include surfaces, edges, and points. For this selection, we'll select the top surfaces of the base and the triangle, as shown in *Figure 8.15*. These will fill in the **Mate Selections** space, highlighted on the left side in *Figure 8.15*:

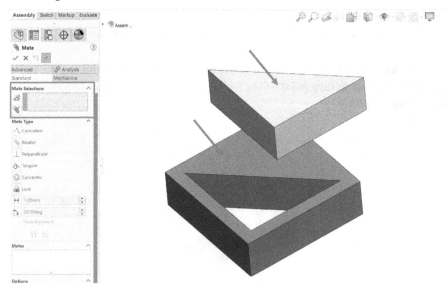

Figure 8.15 – The surfaces we can mate together

After the selection, we will see that the two parts move in relation to each other. Also, one mate will be selected automatically. In this case, the mate will be coincident. We can change this mate if we want to use another one. However, in this case, the coincident mate will work for us. Our canvas will look as follows. To apply the mate, click on the green checkmark on top of the mate's `PropertyManager`, as highlighted in *Figure 8.16*:

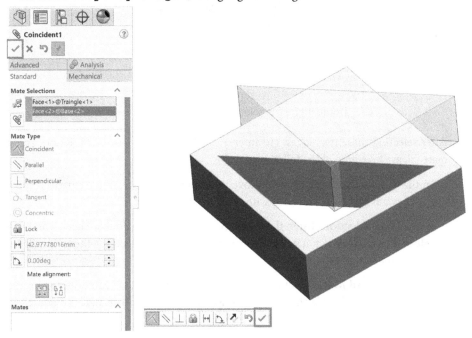

Figure 8.16 – The mate will be previewed in the canvas after application

> **Tip**
>
> To check what effect the applied mate has, click and hold the triangle and move it around. We will see that the movement of the part has been restricted due to the applied mate.

When we try moving the triangle around the canvas, we will notice that it can only move sideways to the base. However, it won't move up and down relative to the base. Now, let's apply another coincident mate to restrain the triangle more. Note that the `Mate` command is still active, which means we can apply more mates.

3. Select the edges shown in *Figure 8.17*. These are the outer edges of the triangle and the inner edge of the base. Again, SOLIDWORKS will interpret that we want the coincident mate and automatically apply it:

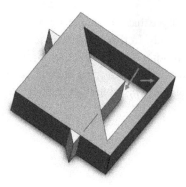

Figure 8.17 – The selection of the coincident mate

After selecting these two edges, our interface will look like *Figure 8.18*. Before applying the mate, please take note of the following:

- In the canvas, the two parts will move to preview the mate. Make sure that the preview matches our needs, as indicated with *A* in *Figure 8.18*.

- In the **Mate Selections** list in the PropertyManager, we can double-check whether we have selected two edges. If not, we can delete the undesired selections there. This is indicated with *B* in *Figure 8.18*.

- In the **Standard Mate** selection in the PropertyManager, we can double-check that the selected mate is what we want to apply. If we want to use another mate, we can select it from there. This check is indicated with *C* in *Figure 8.18*:

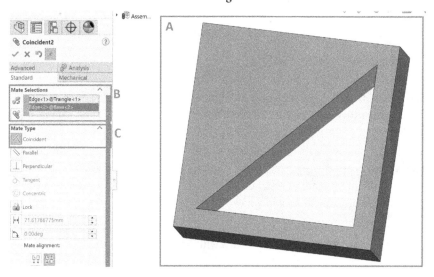

Figure 8.18 – Different checks to ensure we applied the correct mate

4. After conducting all the checks, we can click on the green checkmark to apply the mates.

> **Tip**
> The pin icon next to the red cross allows you to keep the mate PropertyManager visible after applying the mate, making it faster to apply multiple mates one after the other.

This concludes our application of the coincident mate. Next, we will apply the perpendicular mate to our assembly.

Applying the perpendicular mate

To find out what restraints are missing from our assembly, we can click and hold the triangle and drag it. We will see that the triangle is hinged at the corner we just mated. To restrain this movement, we can apply the perpendicular mate between the faces, as shown in *Figure 8.19*:

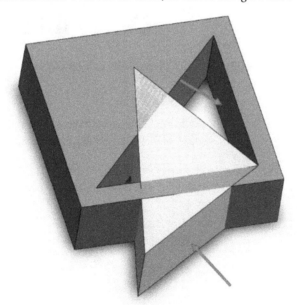

Figure 8.19 – The surface selection to apply the perpendicular mate

To apply the mate, follow these steps:

1. Select the **Mate** command and then select the **Perpendicular** mate.
2. Select the two faces under **Mate Selections**. Our view will look like *Figure 8.20*. Again, before applying the mate, note the position of the parts, the mate selection, and the selected standard mates:

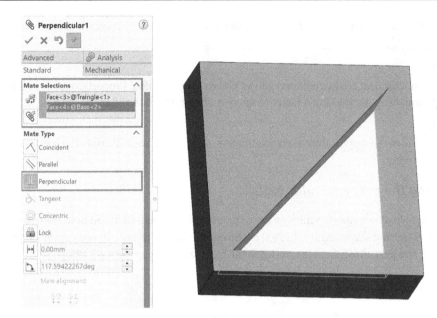

Figure 8.20 – The mate's PropertyManager and a preview of the perpendicular mate

3. Click on the green checkmark to apply the mate.

4. Since we don't need to apply any more mates, we can close the mate's PropertyManager by pressing the *Esc* key on the keyboard or clicking the red cross at the top of the PropertyManager.

At this point, our assembly will look as follows. Note that if we try to drag the triangle in any direction, it won't move. This indicates that our assembly is now fully defined, as in *Figure 8.21*:

Figure 8.21 – The final look of our assembly

Important note

We can end up with the same assembly using different mates. For example, in the prior example, we can have the same result by applying more coincident relations or perpendicular relations. Sometimes, there are no right or wrong mates to apply while at other times, it can depend on the design intent behind our product.

This concludes the application of the coincident and perpendicular mates to our assembly. We will follow the same procedure to apply all the other mates. While applying the coincident and perpendicular mates, we learned about the following:

- How to access the mate command

- How to select different elements and apply the mates to restrain them

- How to check for unrestrained movements by holding and dragging parts

We have just finished fully defining our assembly by using the coincident and perpendicular mates. Next, we will work on another assembly, which will involve the parallel, tangent, concentric, and lock mates.

Applying the parallel, tangent, concentric, and lock mates

In this section, we will explore how to apply the parallel, tangent, concentric, and lock mates. To do that, we will apply the mates that are shown in *Figure 8.22*. We will refer to this drawing as we apply the different mates:

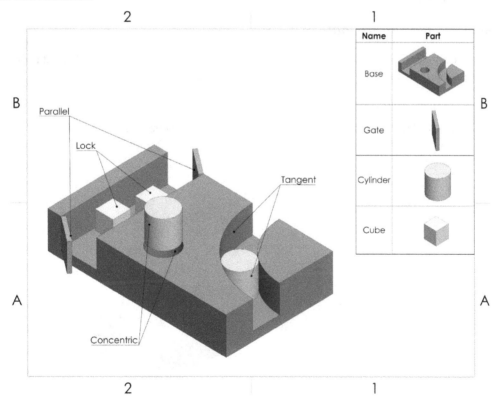

Figure 8.22 – The assembly we will work on in this exercise

To start, download the parts and the assembly file attached to this chapter. Our starting point, which is where we will apply all the mates, will be from the provided assembly file, which looks like *Figure 8.23*:

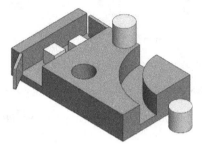

Figure 8.23 – The starting point of this exercise from the downloadable files

The attached assembly file already has a few coincident mates applied to it. The procedure of applying all the standard mates is similar, so we won't go into too much detail regarding the next four mates we will apply: parallel, tangent, concentric, and lock.

Applying the parallel mate

In this section, we will apply the parallel mate to the two gates that were shown in the initial drawing. These are also highlighted in the following figure. To do this, follow these steps:

1.   Go to the **Mate** command and select the **Parallel** mate.

2.   Under **Mate Selections**, select the two faces shown in *Figure 8.24*:

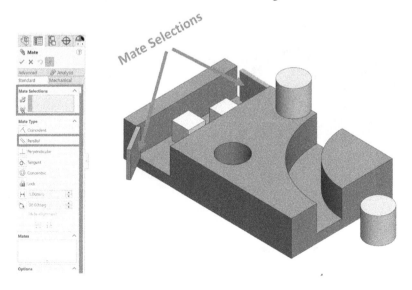

Figure 8.24 – The mate selection for the parallel mate

> **Important note**
>
> The positions of the gates will shift as the mate takes effect.

3. Apply the mate by clicking on the green checkmark.

After applying the mate, it's good practice to drag the gates to see what effect the new mate has. You will notice that, as we move one gate, the other gate will also move to keep the two faces parallel to each other. In this example, we applied the parallel mate to two faces, but we can also apply the mate in the same way to two straight edges or an edge with a face.

This concludes this exercise on applying the parallel mate. Next, we will start applying the tangent mate.

Applying the tangent mate

In this section, we will apply the tangent mate to the cylinder and base that were highlighted in the initial drawing. To do this, follow these steps:

1. Go to the **Mate** command and select the **Tangent** mate.
2. Under **Mate Selections**, select the two faces, as shown in *Figure 8.25*. Note that the position of the cylinder will shift as the mate takes effect:

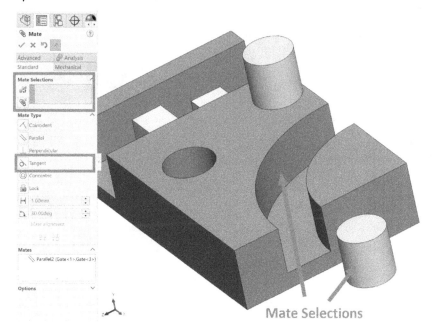

Figure 8.25 – The mate selection for the tangent mate

3.   Apply the mate by clicking on the green checkmark.

After applying the mate, it's good practice to drag the cylinder to see what effect the new mate has. You will notice that the cylinder will move while keeping a tangent relation with the side we selected in the base part. In this example, we applied the tangent mate to two faces, but we can also apply the mate in the same way to two edges or an edge with a face.

This concludes this exercise on applying the tangent mate. Next, we will start applying the concentric mate.

Applying the concentric mate

In this section, we will apply the concentric mate to the cylinder and base that are highlighted in the initial drawing. To do this, follow these steps:

1.   Go to the **Mate** command and select the **Concentric** mate.

2.   Under **Mate Selections**, select the two faces shown in *Figure 8.26*. Note the position of the cylinder will shift as the mate takes effect:

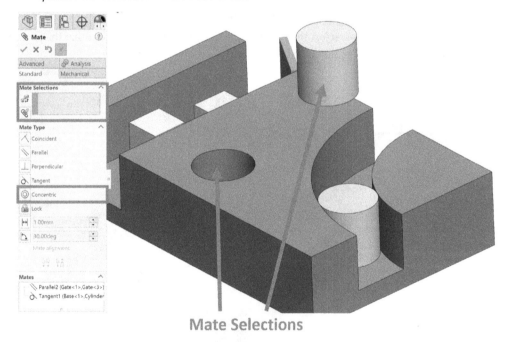

Mate Selections

Figure 8.26 – The mate selection for the concentric mate

3.   Apply the mate by clicking on the green checkmark.

After applying the mate, it's good practice to drag the cylinder to see what effect the new mate has. You will notice that the cylinder will only move vertically, that is, up and down, so that the two rounded faces share the same center. In this example, we applied the concentric mate to two faces, but we can also apply the mate in the same way to two arc edges or an arc edge with an arc face.

This concludes this exercise on applying the concentric mate. Next, we will start examining the lock mate.

Applying the lock mate

In this section, we will apply the lock mate to the cubes that were highlighted in the initial drawing. To do this, follow these steps:

1. Go to the **Mate** command and select the **Lock** mate.

2. Under **Mate Selections**, select the two cubes shown in *Figure 8.27*. Note that the position of the cubes will not change after we apply the lock mate:

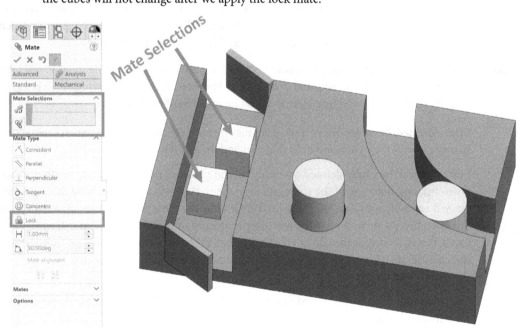

Figure 8.27 – The mate selection for the lock mate

3. Apply the mate by clicking on the green checkmark.

After applying the mate, it's good practice to drag the cubes to see what effect it has. You will notice that as we move one cube, the other cube will move in the same way. This includes linear movements, as well as rotational movements. The lock mate can only be applied to whole parts.

This concludes the application of the lock mate. At this point, we have covered how to apply all the non-value-oriented standard mates. Next, we will learn what *fully defined* means in the context of assemblies. We will also learn about other types of assembly definitions.

Under-defining, fully defining, and over-defining an assembly

When we finished the first assembly exercise, which is where we used the coincident and perpendicular mates, we noticed that all the parts were restrained from moving in any direction. Thus, the triangle will not move in any direction when it's dragged. This indicates that the assembly is now *fully defined* since both parts were fully restrained. This status is shown in the lower right-hand corner of the canvas, as shown in *Figure 8.28*:

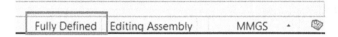

Figure 8.28 – The state of assembly is defined at the bottom of the canvas

However, in the second assembly exercise, where we looked at the parallel, tangent, concentric, and lock mates, we noticed that we could still drag the parts around, even after keeping certain movement constraints. When this happens, the status of the assembly will be **Under Defined**.

Similar to sketching, there are three different statuses and terms when it comes to defining an assembly. Those are under-defined, fully defined, and over-defined. However, the way we interpret them is slightly different in the context of assemblies. Here is a brief description of each status:

- **Under-defined**: There are one or more parts that are not fully constrained in terms of movement. In other words, if we click and hold that part and drag it, it will move.

- **Fully defined**: In a fully defined assembly, all the parts are fully restrained. In other words, if we click and hold any of the parts and drag them, they will not move. Note that, in a fully defined assembly, we can apply more mates that serve the same purpose without over-defining the assembly. As such, even if the assembly is already fully defined, we can still apply more mates, that is, as long as they do not contradict each other. This aspect is different from defining sketches. In sketching, any relation that's added after fully defining a sketch will make the sketch over-defined.

- **Over-defined**: In an over-defined assembly, we have mates that contradict each other. Thus, we will need to delete or redefine some of the existing mates.

The indication at the bottom of the interface refers to the definition status of the whole assembly. Let's learn how we can find out the status of each individual part.

Finding the definition statuses of the parts

To find the definition status of each part in the assembly, we can look at the assembly design tree. At the beginning of each part's listing, SOLIDWORKS indicates the statuses of the different parts with symbols such as (**f**), (-), and (+). The meaning of each symbol is as follows:

- (**f**): Fixed

- (-): Under-defined

- **No symbol**: Fully defined
- (+): Over-defined

Figure 8.29 shows each of the symbols in the design tree:

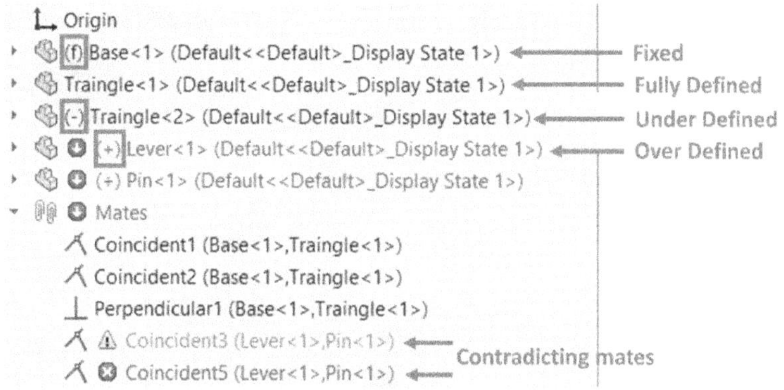

Figure 8.29 – The parts' statuses as shown in the assembly design tree

Finding out the status of each part will help us when we need to define our assemblies. With this, we can decide which part needs more mates or which mate we should reconsider. However, an important question is that in the context of assemblies, which definition status is better? We will discuss that question next.

Which assembly definition status is better?

When defining assemblies, we should avoid having an over-defined assembly. However, there are certain advantages of having our assembly under-defined or fully defined. Here are some scenarios for both cases:

- If the assembly has a moving part, a common practice will be to have the assembly under-defined so that the desired movement is visible if we were to drag and move the part around. For example, if we assemble a windmill, we can choose to have the blades under-defined to show how they move and how all the parts interact with each other during that movement.
- If all the parts in the assembly are fixed, a common practice will be to have the assembly fully defined. A common example of fully fixed assemblies is tables, which don't have any moving parts.

As we can see, keeping our assemblies fully defined or under-defined has certain advantages. As designers or draftsmen, we will have to weigh up the advantages of each and adapt our own approach. Next, we will learn how to view and adjust active mates.

Viewing and adjusting active mates

In the assembly design tree, we will see a list of all the parts in the assembly, as well as the mates that have been applied to those parts. The lowest part of the assembly design tree shows the mates. We can expand this list to view all the mates that were involved in making the assembly. *Figure 8.30* shows the mates we used to make the assembly we constructed earlier, that is, two **Coincident** mates and one **Perpendicular** mate:

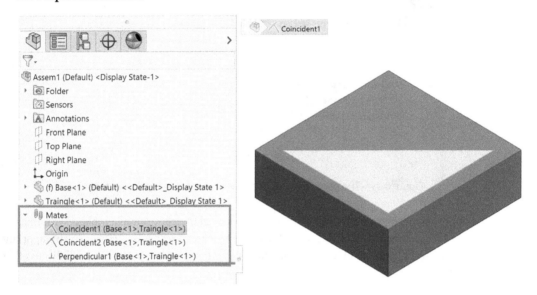

Figure 8.30 – Existing mates are listed in the assembly design tree

> **Tip**
> To see which elements of the parts are involved in the mates, we can click on the mate in the design tree. Then, the involved elements will be highlighted in the canvas, as shown in *Figure 8.30*.

Now, we know how to view the active mates that we have in our assembly. Next, we will learn how to modify them.

Modifying existing mates

To modify a particular mate, we can right- or left-click on the mate from the design tree. We will see the menu in *Figure 8.31*. Here, we can choose to edit, delete, or suppress the selected mate. Modifying mates follows the same procedure as modifying features:

Figure 8.31 – A menu appears after right-clicking a mate, giving us different options

So far, we have learned how to use all the non-value-driven standard mates. We have also learned about the different statuses of assemblies, in addition to how to view and modify existing mates. Now, we can start learning about value-driven standard mates.

Understanding and applying value-driven standard mates

This section covers the standard mates that are defined by numerical values, that is, the distance and angle mates. We will learn about what they do and how to apply them to an assembly. By the end of this section, we will be familiar with applying all the standard mates, which is our first step when it comes to working with SOLIDWORKS assembly tools.

Defining value-driven standard mates

Value-driven standard mates are those that depend on numerical values so that they can be set. They include two standard mates – distance and angle. Here is a brief definition of these two mates:

- **Distance**: This sets a certain fixed distance between two entities, such as edges and planner surfaces.

- **Angle**: This sets a fixed angle between two planner surfaces or edges.

Whenever we define one of these mates, we need to input a number that indicates the desired distance or angle. Now that we know what the distance and angle mates are, we can start applying them.

Applying the distance and angle mates

Here, we will apply the distance and angle mates to create the assembly shown in *Figure 8.32*. You can download all the indicated parts from the download files that are linked to this chapter:

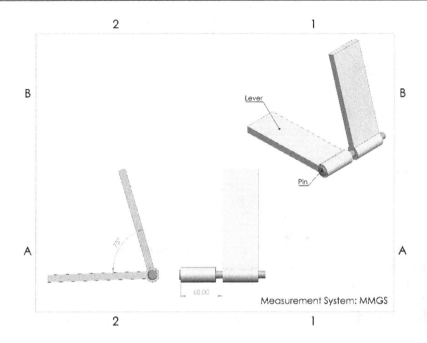

Figure 8.32 – The assembly we are making in this exercise

Note that, in the preceding drawing, the two levers are separated by a set distance of 60 mm. Also, the two levers have an angle of 75 degrees between them. This indicates that we can utilize the distance and angle relations to complete the assembly. To complete this assembly, we need to download the files that are attached to this chapter and open the Lever-Pin Assembly.SLDASM assembly file. The assembly looks like *Figure 8.33*. Note that the assembly already has mates that are restraining the three parts. However, we still need to add the distance and angle mates in order to achieve the assembly shown in the preceding figure. We can move the parts around in the assembly to find out how are they restrained:

Figure 8.33 – The initial status of the downloadable assembly for this exercise

Now that we have downloaded our parts, we can start applying the mates. We will start with the distance mate.

Applying the distance mate

To apply the distance mate, follow these steps:

1. Select the **Mate** command. In **Mate Selections**, select the two planner surfaces, as shown in *Figure 8.34*.

2. Select the **Distance** mate and select 60.00mm in the distance space, as highlighted in *Figure 8.34*.

3. Apply the mate by clicking on the green checkmark.

4. Note the **Flip dimension** checkbox below the distance value. Checking this box will switch the distance from being toward the left to being toward the right and vice versa:

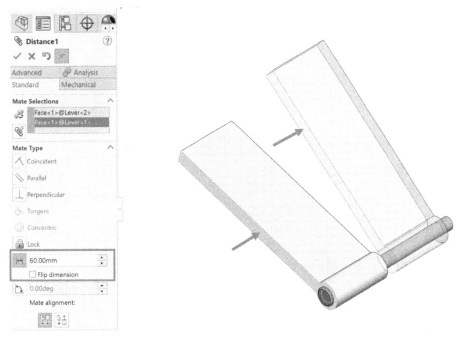

Figure 8.34 – The selection and the PropertyManager for the distance mate

After applying the mate, it's good practice to test its effect. We can do that by dragging the different parts to find out how the new mate is taking effect. In this case, we will notice that the levers can still rotate; however, they cannot move away from each other, that is, along the pin. Now, we will apply our next mate so that we can set the angle between the levers.

Applying the angle mate

To apply the angle mate, follow these steps:

1. Select the **Mate** command. In **Mate Selections**, select the two planner surfaces, as shown in *Figure 8.35*.

2. Select the standard mate **Angle** and input 75 degrees for the angle, as stated in the initial drawing and as illustrated in *Figure 8.35*.

3. Apply the mate by clicking on the green checkmark.

Similar to the distance mate, there is a **Flip dimension** checkbox to flip the dimension that the angle is measured by. Try checking the box to see what effect this has on the assembly. Once checked, the output will be reflected in the preview on the canvas.

Note that, when applying mates, we may get a shortcut menu showing the various mates. We can use this menu in the same way we use the PropertyManager:

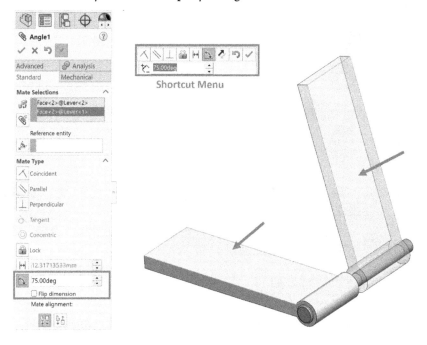

Figure 8.35 – The selection and the PropertyManager for the angle mate

Once we have applied the angle mate, the final shape will look like *Figure 8.36*. Note that, if we drag any of the levers, the other lever will rotate with it while keeping the angle between them equal to 75 degrees. Also, note that the assembly is still under-defined; however, we will keep it that way to show a simple simulation of how the different parts in the assembly move together:

Figure 8.36 – The final status of the assembly

This concludes this exercise on the distance and angle mates. We have learned about the following topics:

- What the distance and angle mates are and what they do
- How to apply the distance and angle mates

At this point, we have covered all the standard mates that can be used in a SOLIDWORKS assembly. These allow us to model products that consist of different parts that interact with each other. Next, we will start looking at materials and mass properties within the context of an assembly.

Utilizing materials and mass properties for assemblies

When creating assemblies, we may need to determine the mass, volume, center of mass, and other related mass properties. This information is necessary, as it helps us understand our product from a physical perspective. This information can help us develop or modify our product further in case we ever want to achieve a certain mass, volume, or other properties to meet a specific requirement. Similar to when working with parts, we can evaluate mass properties within the context of assemblies. In this section, we will learn about setting up new coordinate systems, editing materials for the parts within the assembly, and how to evaluate the different mass properties for our assembly.

Setting a new coordinate system for an assembly

In many cases, we may need to reorient models ourselves directionally and find the center of mass from different locations. These are more common practices when we're working with assemblies compared to when we're working with parts. This is due to it being less intuitive to build upon the default coordinate system within the assemblies' environment. The previous chapter examined this topic in more detail.

To define a new coordinate system, we can follow the same procedure that we followed when we defined coordinate systems for parts. To access the command, we can go to the **Assembly** commands category and select **Coordinate System** under **Reference Geometry**, as highlighted in *Figure 8.37*:

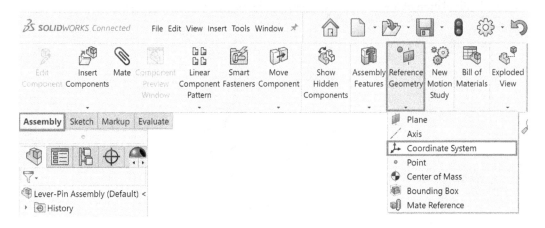

Figure 8.37 – The location of the new Coordinate System command

To define a new coordinate system, we have to define the origin, as well as the direction of the axes, similar to defining a coordinate system in parts. Introducing a new coordinate system in assemblies is a common practice when we're measuring coordinate-orientated mass properties, such as the center of mass. Next, we will address how to edit materials within assemblies. Refer to *Chapter 7* for more information about coordinate systems. The procedure of setting and dealing with new coordinate systems is the same for both parts and assemblies.

Material edits in assemblies

There is no material assignment for the assembly. Instead, each part will carry its own material assignment. If the part was assigned a material when it was created, then this assignment will simply be transferred to the assembly. Within the assembly environment, we can still edit and assign materials to individual parts. We will learn how to do this here.

Assigning materials to parts in the assembly environment

We can assign a material to individual parts from the assembly environment. Follow these steps to do so:

1. Decide which part you would like to assign a material to.

2. Expand the part from the design tree and right-click on **Material**. Then, select **Edit Material**, as shown in *Figure 8.38*:

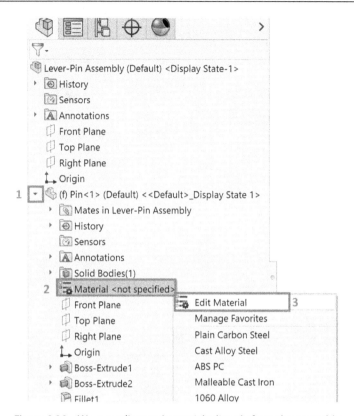

Figure 8.38 – We can edit parts' materials directly from the assembly

3. Assign or adjust the assigned material as needed.

Important note

Once we assign the material to the part from within the assembly, it will be updated in the original part file since they are now linked.

We can assign the same material to more than one part in one go by highlighting the parts, right-clicking, and then selecting the desired material, as shown in *Figure 8.39*:

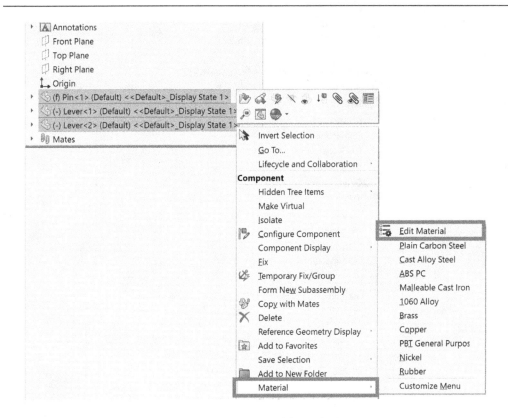

Figure 8.39 – We can edit the material for multiple parts at once

This concludes editing a part's material assignments within an assembly file. When editing materials in an assembly, take note of the following points:

- If we change the material within the assembly and save it, the part's material assignment will be updated as well. This is because the parts and the assembly files are connected.

- If we edit the material assignment for a repeated part (that is, we have more than one copy in the assembly), all the part's materials will be updated to match the new edit.

Now that we have materials assigned to our parts, we can start evaluating our mass properties.

Evaluating mass properties for assemblies

To evaluate the mass properties for an assembly, we can click on **Mass Properties** under the **Evaluate** commands category, as shown in *Figure 8.40*:

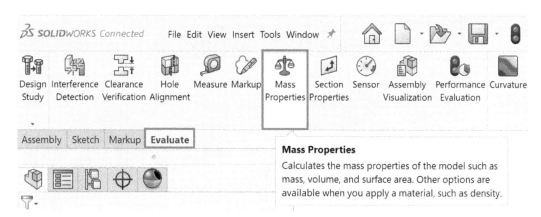

Figure 8.40 – The location of the Mass Properties command

This will show us the same information we received when we evaluated mass properties for parts. Refer to *Chapter 7* for more information. The only difference is that the mass properties here will be a reflection on the whole assembly rather than on individual parts.

Note that the center of mass is calculated based on the position of the different parts that make up the assembly. If the assembly is under-defined and we move the parts, the center of mass will change. This is in addition to all the other properties that are calculated based on the coordinate system's position, such as the moment of inertia. When moving the assembly, we may need to click on **Recalculate** in order to recalculate the mass properties, as shown in *Figure 8.41*:

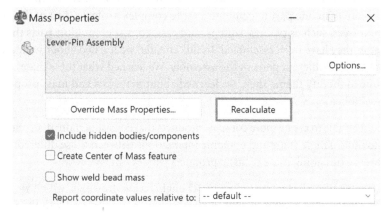

Figure 8.41 – The Recalculate command updates the values in the PropertyManager

Within assemblies, we can also calculate the mass properties of a specific part in relation to the assembly's coordinate system. To do this, we can select that part in the **Mass Properties** selection window, as highlighted in *Figure 8.42*. This allows us to show the mass properties of a specific part within the assemblies' environment:

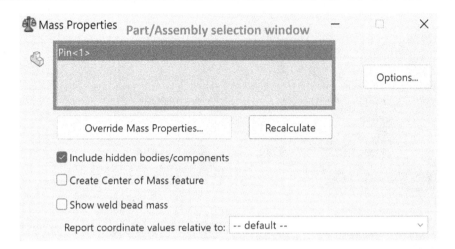

Figure 8.42 – We can find the mass properties of specific parts within the assembly environment

In this section, we have learned about mass properties in the context of assemblies. We have learned about setting new coordinate systems, adjusting the material assignments for our parts, and evaluating the mass properties of our assemblies.

Summary

In this chapter, we started working with assemblies. In SOLIDWORKS assemblies, we are able to put together more than one part to generate a more complex artifact. Most of the products we use in our everyday lives, such as phones, laptops, and cars, consist of multiple parts that have been put together; that is, they have been assembled. In this chapter, we learned about standard mates, which help us create links to different parts of the assembly. We learned what these mates do, how to apply them, and how to modify them. Then, we learned about materials and mass properties within the context of assemblies.

Now, we should be able to create more complex products that consist of more than one part. We should also be able to build simple static and dynamic interactions between those different parts. All of this brings us closer to designing more realistic products with SOLIDWORKS.

In the next chapter, we will start introducing 2D engineering drawings, which we will use to share our 3D models with individuals and organizations outside our circle or with those who don't have access to SOLIDWORKS. We will cover what engineering drawings are, why we need them, and how to interpret them.

Questions

1. What are the SOLIDWORKS assemblies?

2. What are mates? What are the three different types of mates?

3. What are standard mates?

4. Download the parts linked to this question and assemble them to form the drawing shown in *Figure 8.43*. The assembly should be fully defined:

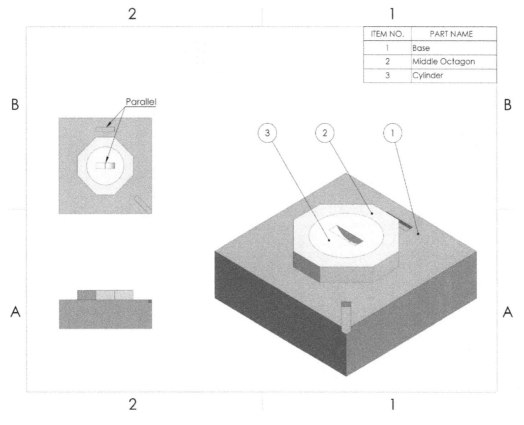

Figure 8.43 – The drawing for Question 4

5. Using the assembly from the previous question, adjust the material for each part and define the coordinate system shown in the following drawing in *Figure 8.44*. Determine the mass in grams and the center of mass in millimeters according to the newly defined coordinate system:

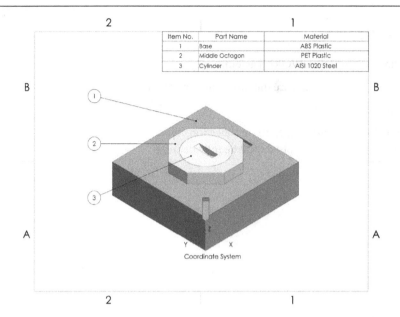

Item No.	Part Name	Material
1	Base	ABS Plastic
2	Middle Octagon	PET Plastic
3	Cylinder	AISI 1020 Steel

Figure 8.44 – The drawing for Question 5

6. Download the parts linked to this question and assemble them to form the following drawing in *Figure 8.45*. The assembly should be fully defined:

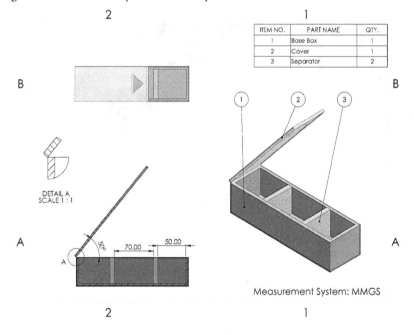

ITEM NO.	PART NAME	QTY.
1	Base Box	1
2	Cover	1
3	Separator	2

Figure 8.45 – The drawing for Question 6

7. Using the assembly from the previous question, adjust the material for each part and define the coordinate system for the following drawing in *Figure 8.46*. Determine the mass in pounds and the center of mass in inches, according to the newly defined coordinate system:

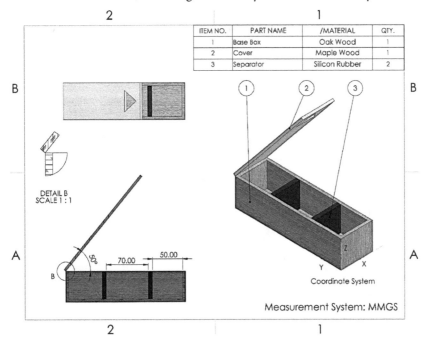

ITEM NO.	PART NAME	/MATERIAL	QTY.
1	Base Box	Oak Wood	1
2	Cover	Maple Wood	1
3	Separator	Silicon Rubber	2

DETAIL B
SCALE 1 : 1

Coordinate System

Measurement System: MMGS

Figure 8.46 – The drawing for Question 7

Important note

The answers to the preceding questions can be found at the end of this book.

Get This Book's PDF Version and Exclusive Extras

Part 5: 2D Engineering Drawings Foundation

2D engineering drawings are foundational to the design workflow. They enable you to document and communicate your work. This part will cover the foundations of engineering drawing interpretations, how to construct simple engineering drawings, and how to generate and adjust bills of materials in the SOLIDWORKS drawing environment.

This part has the following chapters:

- *Chapter 9, Introduction to Engineering Drawing*
- *Project 1 – 3D Modeling a Pair of Glasses*
- *Chapter 10, Basic SOLIDWORKS Drawing Layout and Annotations*
- *Chapter 11, Bills of Materials*

9

Introduction to Engineering Drawings

Whenever we want to communicate a specific design to others, for example, makers, manufacturers, or evaluators, a 2D engineering drawing is often required. Such drawings usually communicate the shape, materials, and dimensions of any product. This chapter will cover the basic knowledge of engineering drawings and how to interpret different layouts found in them. Interpreting drawings is an essential skill for us to be able to generate our own drawings and to collaborate with other people by interpreting their drawings.

The following topics will be covered in this chapter:

- Understanding engineering drawings
- Interpreting engineering drawings

By the end of this chapter, you will have gained knowledge about different engineering drawing concepts. In addition, you will be able to interpret the different types of lines and views often found in engineering drawings.

For this chapter, there is no Code in Action (CiA) video available.

Understanding engineering drawings

Engineering drawings are what we use to communicate designs to other entities. Whenever we produce a design for a specific product, we are often required to present an engineering drawing with it to communicate the design. Within an engineering drawing, we can communicate the shape of the design, the materials, the suppliers, and any other information we want to communicate.

Also, when engineers and technicians maintain a certain plant or facility, they interact with engineering drawings in their day-to-day jobs. This is to identify what the machine is comprised of, how to maintain it, and the materials required for that. A couple of examples of engineering drawings are shown in *Figure 9.1* and *Figure 9.2*. The following figure is of a simple part, communicating only the shape and overall dimensions of the part:

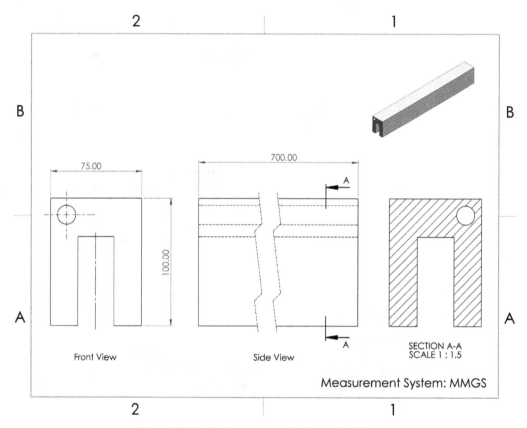

Figure 9.1 – A drawing communicating a simple part

Figure 9.2 is a more complex assembly. Note that this drawing does not communicate dimensions; rather, it communicates the different parts included in the assembly. In other words, it communicates the bill of materials:

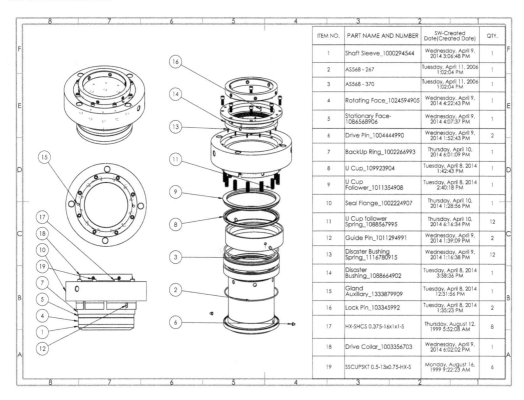

ITEM NO.	PART NAME AND NUMBER	SW-Created Date(Created Date)	QTY.
1	Shaft Sleeve_1000294544	Wednesday, April 9, 2014 3:06:48 PM	1
2	AS568 - 267	Tuesday, April 11, 2006 1:02:04 PM	1
3	AS568 - 370	Tuesday, April 11, 2006 1:02:04 PM	1
4	Rotating Face_1024594905	Wednesday, April 9, 2014 4:22:43 PM	1
5	Stationary Face-1086568906	Wednesday, April 9, 2014 4:07:37 PM	1
6	Drive Pin_1004444990	Wednesday, April 9, 2014 1:52:43 PM	2
7	BackUp Ring_1002266993	Thursday, April 10, 2014 6:01:09 PM	1
8	U Cup_109923904	Tuesday, April 8, 2014 1:42:43 PM	1
9	U Cup Follower_1011354908	Tuesday, April 8, 2014 2:40:18 PM	1
10	Seal Flange_1002224907	Thursday, April 10, 2014 1:28:56 PM	1
11	U Cup follower Spring_1088567995	Thursday, April 10, 2014 6:16:34 PM	12
12	Guide Pin_1011294991	Wednesday, April 9, 2014 1:39:09 PM	2
13	Disaster Bushing Spring_1116780915	Wednesday, April 9, 2014 1:16:38 PM	12
14	Disaster Bushing_1088664902	Tuesday, April 8, 2014 3:58:36 PM	1
15	Gland Auxiliary_1333879909	Tuesday, April 8, 2014 12:31:56 PM	1
16	Lock Pin_103345992	Tuesday, April 8, 2014 1:35:23 PM	2
17	HX-SHCS 0.375-16x1x1-S	Thursday, August 12, 1999 5:52:08 AM	8
18	Drive Collar_1003356703	Wednesday, April 9, 2014 6:02:02 PM	1
19	SSCUPSKT 0.5-13x0.75-HX-S	Monday, August 16, 1999 9:22:23 AM	6

Figure 9.2 – A drawing showing an assembly and its bill of materials

Engineering drawings vary in complexity according to what they communicate. Also, the information displayed on a drawing sheet can vary from one organization to another. However, all drawings follow the same standards in terms of communicating different aspects of the drawing. Engineering drawings became an essential tool for communication due to their flexible distribution. They can be printed on paper or sent as images or PDFs for viewing with common software, such as an image viewer or a PDF reader.

One major element of SOLIDWORKS is drawings. This enables us to create engineering drawings for our parts and assemblies. To be able to create drawings in SOLIDWORKS, it is important for us to have some understanding of basic drawing standards and communication practices.

In this section, we have learned what engineering drawings are and what their purpose is. Now, we can start learning about some key standards used when creating drawings to help us interpret them.

Interpreting engineering drawings

Being able to interpret engineering drawings is an essential part of creating them. In this section, we will cover essential drawing competencies, including how to interpret different types of lines and different types of drawing views. We will start by understanding lines, then views, and then projections. Interpreting drawings is a skill that grows with time as we are exposed to more drawings. To start, we will learn key standards that are followed when generating drawings. Those common standards will help us interpret drawings regardless of their source.

So first, we will start by interpreting the most essential drawing element – lines.

Interpreting lines

In simple terms, we can look at **drawings** as different lines connected together. However, the shape of a line gives it a different meaning. *Figure 9.3* shows the most common types of lines found in engineering drawings:

Type and shape of the line	Line indication
Visible object lines 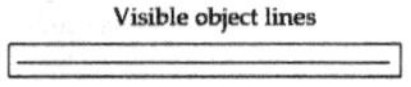	Visible object lines show the visible outline of the object.
Hidden line 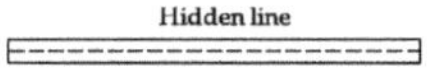	This shows the hidden outline of the object from the drawing's viewpoint. This includes any details that are at the back of the object.
Centerline 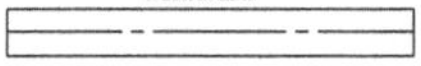	This indicates the center of any two entities. For example, the center of two edges and the center of a circle.
Dimension line 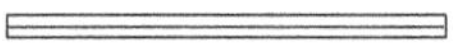	This line is not part of the object; rather, it indicates the dimension of a drawing entity. Note that dimension lines are much lighter in comparison to visible object lines.
Break line 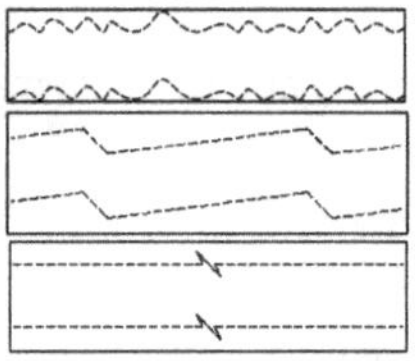	This indicates a break in the object in the drawing. This is often used to fit relatively long objects within a drawing, such as long construction beams. Note that there are many different types of break lines. The three types shown here are the most common ones. They are jagged cut, zig-zag cut, and small zig-zag cut, as seen from top to bottom.
Section / hash lines 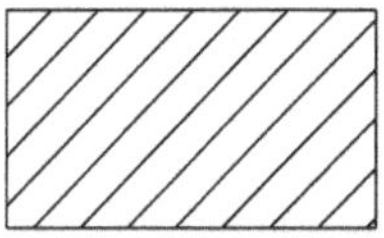	Section lines are inclined lines indicating a cut in a section. We will see them in section views.
Section cutting line 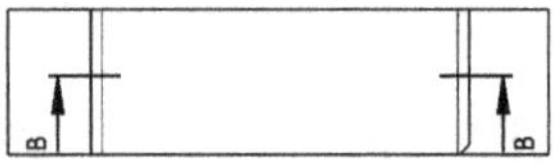	Section cutting lines highlight the viewing location and angle of a section cut.

Figure 9.3 – Different common lines found in engineering drawings

The next figure highlights a model and its 2D engineering drawing. All the lines in *Figure 9.3* are highlighted in the following drawing for easy reference:

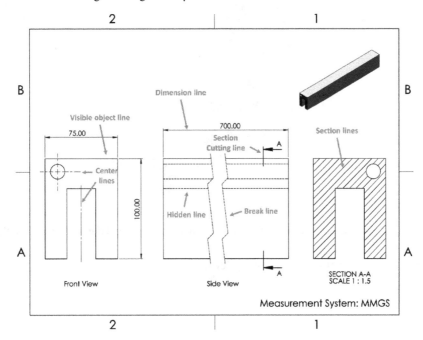

Figure 9.4 – An engineering drawing utilizing different types of lines

Note that when we create drawings with SOLIDWORKS, the software will generate all those lines according to the international standard. However, it is important for us to be able to identify the different types of lines when we see them. Now that we know how to interpret lines, we can start learning how to interpret views.

Interpreting views

In a general sense, a drawing consists of different views of a specific object. Each of the views can give us a deep insight into the shape of the object. As views are also indicated with lines, there is a lot of common knowledge between lines and views. We will look at the most common views at this level and how we can interpret them. We will briefly discuss auxiliary views, section views, detail views, broken-out section views, and crop views.

To investigate all the views, we will examine them using the model shown in *Figure 9.5*:

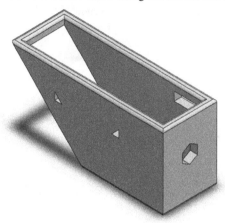

Figure 9.5 – Different drawing views can be used to communicate different elements of this part

In *Figure 9.5*, we will explore the following views:

- Orthogonal views
- Auxiliary views
- Section views
- Detail views
- Broken-out section views
- Crop views

For each view, we will define its purpose and highlight how it looks in relation to the preceding model.

Orthogonal views

Orthogonal views are the most common views we will come across in engineering drawings. They are basically a combination of the front, side, top, bottom, and back views. There are two common standards in constructing orthogonal views. Those are **first angle projections** and **third projections**. Those projection standards are different, based on the interpretation of the top, right, left, and bottom views surrounding the base front view. *Figure 9.6* shows the orthogonal view for our model based on a third angle projection standard:

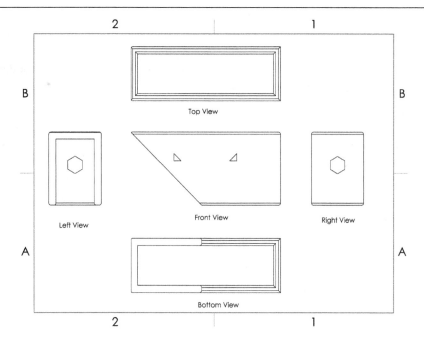

Figure 9.6 – A drawing with a third angle orthogonal view projection

Figure 9.7 shows the same orthogonal views following the first angle projection standard:

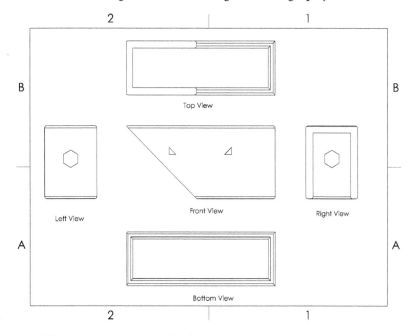

Figure 9.7 – A drawing with a first angle orthogonal view projection

The two standards are more prominent in different countries. For example, first-angle projections are more common in Europe and India, while third-angle projections are more prominent in the United States of America and Japan. You can follow the link in the *Further reading* section at the end of the chapter for more information about those standards.

Auxiliary view

The **auxiliary view** shows the view of the model if we look at it from a selected surface or edge, that is, a perpendicular projection of a surface. *Figure 9.8* highlights the front view of the model as well as the auxiliary view from a tilted angle. Note that the indicated arrow shows the view angle of the auxiliary view. Auxiliary views are often used to show the true size of a specific angle or to communicate specific details that are not clear enough from basic orthographic views:

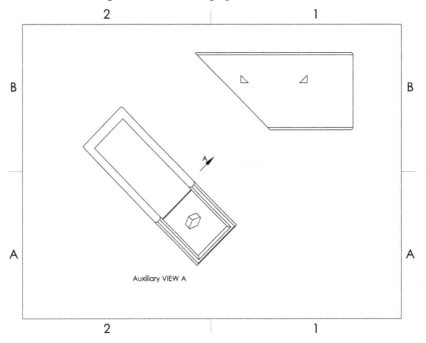

Figure 9.8 – A drawing with an auxiliary view

Section views

Section views allow us to see cross-sections of our models on a selected plane. This type of view allows us to see details that otherwise would be hidden from normal orthogonal views. *Figure 9.9* shows the front view as well as a section view of our model.

The section line in the front view indicates where the cut was made, while the arrow indicates which side we are looking at:

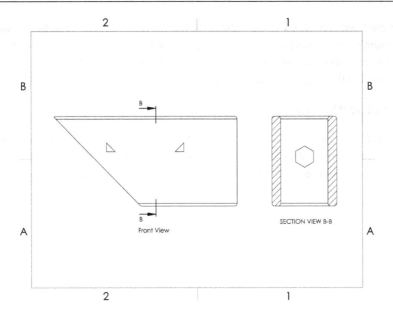

Figure 9.9 – A drawing with a section view

Detail views

Detail views allow us to see and note small details that are otherwise harder to notice. *Figure 9.10* shows the front view as well as a detail view of the small triangle:

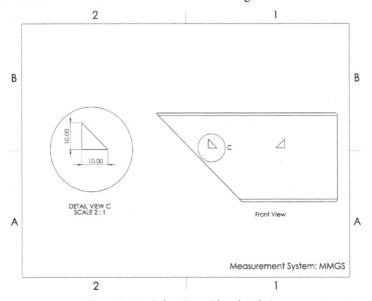

Figure 9.10 – A drawing with a detail view

Note that the front view has a circle indicated by *C* to show the area/zone from where the view is taken. The same letter is used to name the view; thus, our detail view is also named *C*. Also, note that the detail view has its own scale to show how big it is. Generally, the detail view would have a larger scale compared to the original view.

Broken-out section views

Broken-out section views are local section views that do not require a section line. They allow us to see what is behind a specific surface. *Figure 9.11* highlights the front view with and without a broken-out section. Broken-out section views allow us to see details that are hidden from view:

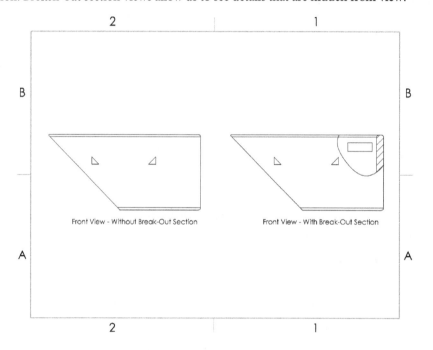

Figure 9.11 – A drawing with a broken-out section view

Crop views

In a **crop view**, we can crop a specific part of a drawing and show it as a standalone view. *Figure 9.12* highlights the full right view of our model as well as a crop view of it:

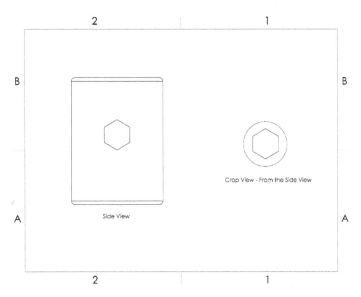

Figure 9.12 – A drawing with a crop view

By now, we have learned the major types of views that we will get exposed to when working with SOLIDWORKS, or when interpreting drawings provided to us. Next, we will learn about axonometric projections.

Axonometric projections

Simply put, we can understand **axonometric projections** as 3D views of the model we are creating the drawing for – in other words, the object tilted to a certain degree in comparison to the plane of projection, which is our drawing sheet. There are three common types of axonometric projections: **isometric**, **dimetric**, and **trimetric**. To understand the difference between the three projections, we will examine them in relation to a simple cube. *Figure 9.13* illustrates the same cube, shown in the three types of projection:

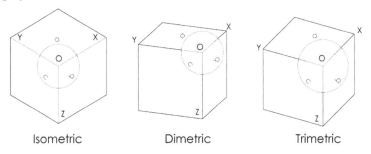

Isometric   Dimetric   Trimetric

Figure 9.13 – A cube shown in isometric, dimetric, and trimetric views

Here are short descriptions illustrating the difference between the three types of axonometric projection:

- **Isometric**: Here, angles *a*, *b*, and *c* are equal. Also, lines O–X, O–Y, and O–Z are equal. This projection is the most common projection we will see and use.

- **Dimetric**: Here, angles *a* and *b* are equal. Also, lines O–Y and O–Z are equal.

- **Trimetric**: All the indicated angles and lines are unequal.

This concludes our brief introduction to axonometric projections. At the level of this book, we will not learn how to generate all of the highlighted views and projections we have covered. However, it is important for us to be able to interpret them.

Summary

Engineering drawings are essential to communicate our designs to manufacturers, maintenance teams, or any other entity. Engineering drawings not only communicate dimensions; they can also communicate materials, part specifications, tolerances, and whatever an organization/individual takes as a practice or a standard to follow. In this chapter, we learned what engineering drawings are and what their purpose is. We also learned how to interpret different standards related to lines and drawing views. Being able to interpret drawings is a fundamental skill for working with SOLIDWORKS. However, as with other skills, it takes time and practice to master.

Next, we will work on a 3D modeling project to create a pair of glasses. The project will provide you with comprehensive practice of all the topics covered from the beginning of the book to this chapter.

Questions

1. What are 2D engineering drawings?

2. What does the following line indicate?

Figure 9.14 – The line referred to in Question 2

3. What is the difference between visible object lines and dimension lines?

4. What are hidden lines, what do they indicate, and what do they look like?

5. What are detail views and when do we use them?

6. What is a crop view?

7. In the following figure, what is the name of the view shown on the right? What is its purpose?

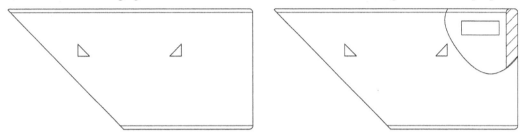

Figure 9.15 – The drawing view referred to in Question 7

> **Important note**
> The answers to the preceding questions can be found at the end of this book.

Further reading

More information about the first and third angle projections in orthogonal views can be found here: https://en.wikipedia.org/wiki/Multiview_orthographic_projection.

Project 1
3D Modelling a Pair of Glasses

SOLIDWORKS is a 3D design tool. Just like all tools, the more you use it, the better you become at it. In this project chapter, you will be provided with work that you can complete to hone your skills. In this project, you will be 3D modeling and assembling a pair of glasses from a set of engineering drawings.

This project chapter will cover the following topics:

- Understanding the project
- 3D modeling the individual parts
- Creating the assembly

By the end of this chapter, you will have more confidence in using the different SOLIDWORKS tools for practical projects.

Technical requirements

You will need to have access to SOLIDWORKS to complete the project.

Understanding the project

Understanding what the project entails is essential before starting the work. This will allow you to draw a plan and manage your work expectations when completing the project. For this exercise project, you will be 3D modeling a pair of glasses, as shown in the following figure:

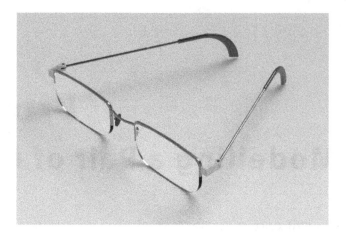

Figure P1.1– The pair of glasses you will 3D model in this project

The pair consists of 15 parts, 8 of which are unique parts. The following figure highlights the bill of materials, showing the names of the parts, their quantity, and their position in the assembly:

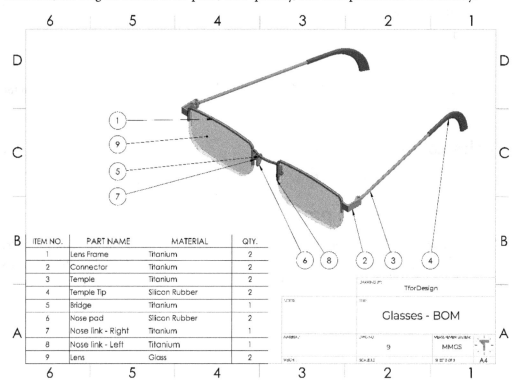

ITEM NO.	PART NAME	MATERIAL	QTY.
1	Lens Frame	Titanium	2
2	Connector	Titanium	2
3	Temple	Titanium	2
4	Temple Tip	Silicon Rubber	2
5	Bridge	Titanium	1
6	Nose pad	Silicon Rubber	2
7	Nose link - Right	Titanium	1
8	Nose link - Left	Titanium	1
9	Lens	Glass	2

Figure P1.2 – The pair and the bill of material

At this point, you already have an idea of the project's outcome and the complexity of the needed parts and assembly. In this next section, we will provide you will the engineering drawings needed to replicate all the parts and assembly. Now that we have an idea about the project's final output, we can discuss how you can tackle it in the context of this writing.

> **Note**
>
> The drawings and 3D models presented in this project are for practice purposes rather than for manufacturing purposes.

There are two ways in which you can tackle this project depending on your 3D modelling level. They are as follows:

- **Moderate level**: Take a look at the drawings and provided hints to complete the project.
- **Advanced level**: Only take a look at the drawings without using the hints.

Other than the two suggested approaches, you can also design your own way to 3D model the glasses utilizing the provided drawings and selected hints.

> **Tip**
>
> You can treat the project as your own and customize the provided glasses to end up with your own unique design.

In this write-up, we will first explore the individual parts, then move into the assembly. So, let us get started with the parts. We will also provide you with hints that can assist you with your work.

3D modeling the individual parts

In this section, we will explore the different part drawings to make up the pair of glasses. The provided drawings have enough information for you to replicate all the parts to end up with an identical result to the one shown in *Figure P1.1*.

Thus, one option you have for going about the project is to create an exact replica of the given drawings. However, you can also choose to customize and adjust different elements of the design to make it your own. Keep in mind that this is your project, so feel free to treat it as such.

Creating the individual parts

The provided pair of glasses consists of 15 parts. However, only 8 parts are unique, which you will need to 3D model as highlighted in *Figure P1.2*. The parts you will need to 3D model are as follows:

- Lens frame
- Lens
- Bridge
- Nose link
- Nose pad
- Connector
- Temple
- Temple tip

> **Note**
> The names of the parts presented in the bill of materials might be different than practiced names in different parts of the world.

Your task is to use the presented drawings to 3D Model the individual parts. As you are 3D modeling the different parts of the glasses, keep in mind that there is no one correct way of 3D modeling any of the parts. However, we will provide you with some hints that can push you forward if you find yourself getting stuck. You can also feel free to customize your own design using the given drawings as a base of inspiration.

> **Note**
> The order in which the drawings are listed is arbitrary.

Let us start exploring the drawings one after the other. The first drawing is for the **lens frame**:

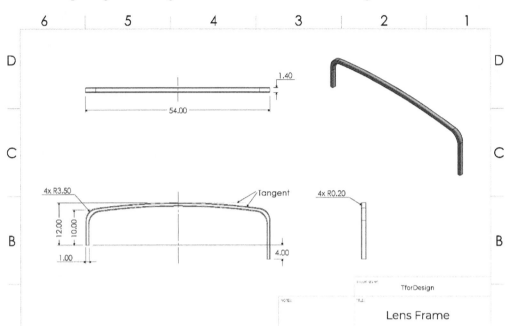

Figure P1.3 – Detailed drawing for the lens frame

Here is a sample procedure for 3D modeling the lens frame:

1. You can start with an extruded boss to create the main shape of the frame, as shown in *Figure P1.4*:

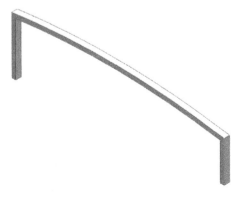

Figure P1.4 – Extruded cut can be used to start the lens frame

2. Apply fillets to get the rounded edges.

Next, we can look at the **lens** itself, which will be held by the lens frame:

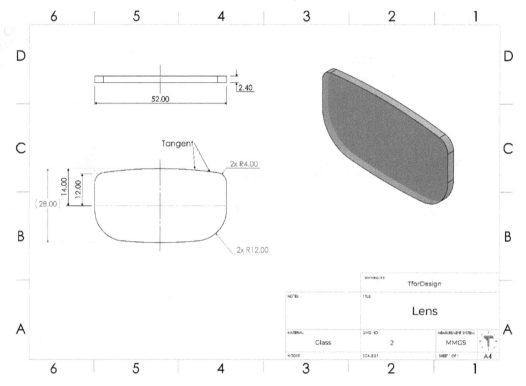

Figure P1.5 – Detailed drawing for the lens

Here is a sample procedure for 3D modeling the lens:

1. Use an extruded boss to build the following shape:

Figure P1.6 – A possible first step in creating the lens using extruded boss

2.   Use fillets to round the corners.

After the lens, we can explore the **bridge**, which will be at the center of the glasses:

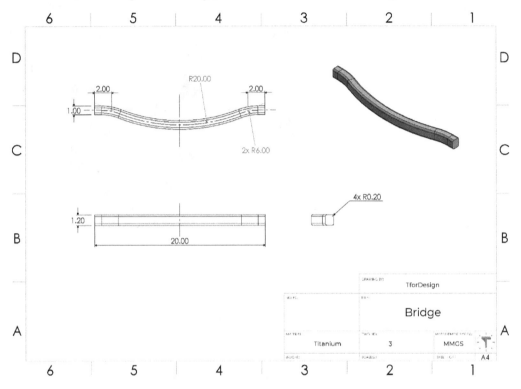

Figure P1.7 – Detailed drawing for the bridge

Here is a sample procedure for 3D modeling the bridge:

1.   Use a swept boss with a square profile to create the basic structure shown in the following figure:

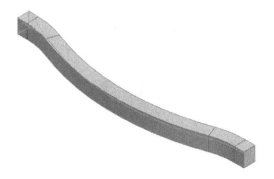

Figure P1.8 – Swept boss can be used to create the bridge

2.   Use the fillet feature to round the rounded corners, as shown in *Figure P1.7*.

After the bridge, we can take a look at the nose pad, which will be connected to the nose link and resting on the nose of the user:

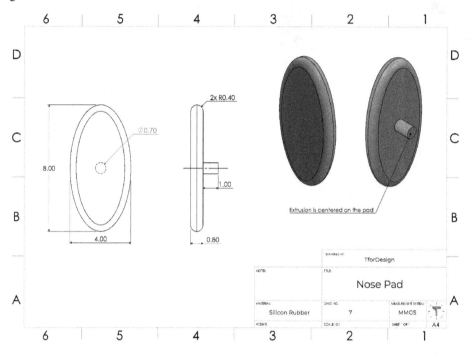

Figure P1.9 – Detailed drawing for the nose pad

Here is a sample procedure for 3D modeling the nose pad:

1.   Use an extruded boss with an elliptical sketch to create the following shape.

Figure P1.10 – Extruded boss can create the bulk of the nose pad

2. Apply another extruded boss to create the cylindrical extrusion, as shown in the following figure.

Figure P1.11 – A second extruded boss can be used to create the small step on the nose pad

3. Use the fillet feature to create the rounded R0.4mm edges.

Next, we can have a closer look into the **connector**, which will be linked to the lens frame on one side and the temple rod on the other side:

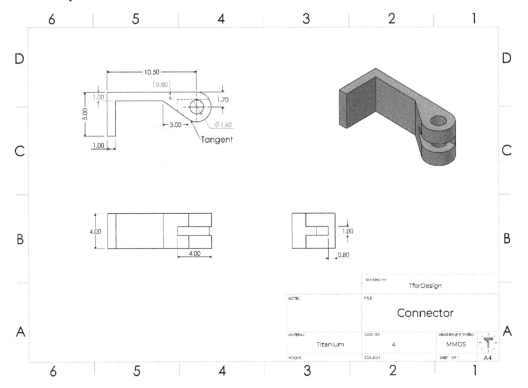

Figure P1.12 – Detailed drawing for the connector

Here is a sample procedure for 3D modeling the connector:

1. Use an extruded cut to create the main shape of the connector, as shown in the following figure.

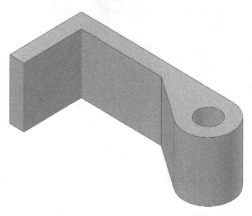

Figure P1.13 – Most of the connector can be done with one extruded boss

2. Use an extruded cut to create the middle cut to end up with the following shape:

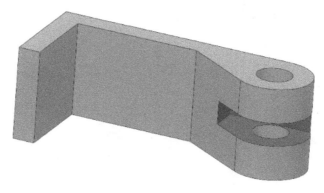

Figure P1.14 – The slot in the connector can be created with an extruded cut

After the connector, we start working on the temple, which is the long part that links to the connector and helps hold the glasses against one's ears:

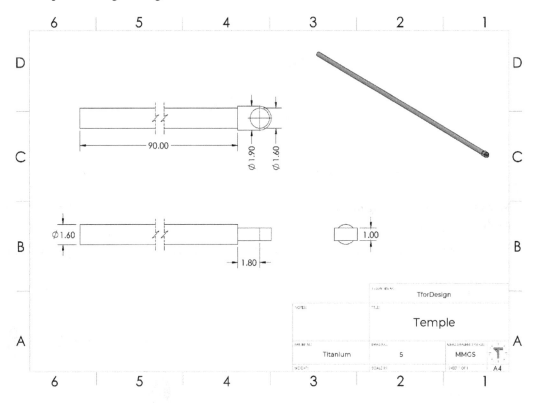

Figure P1.15 – Detailed drawing for the temple

Here is a sample procedure for 3D modeling the temple:

1. Use the extruded boss feature to create the following shape.

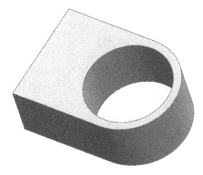

Figure P1.16 – An extruded boss can generate the ring end of the temple

2. Apply another extruded boss to create the long cylindrical rod ending with the final temple as shown:

Figure P1.17 – The 90 mm long cylindrical part can be created with an extruded boss

Note that the drawing in *Figure P1.15* utilizes break lines to shorten the length of the temple rod in the drawing sheet.

Next, we can start looking at the last part, which is the temple tip, or the relatively softer part touching the user's ear:

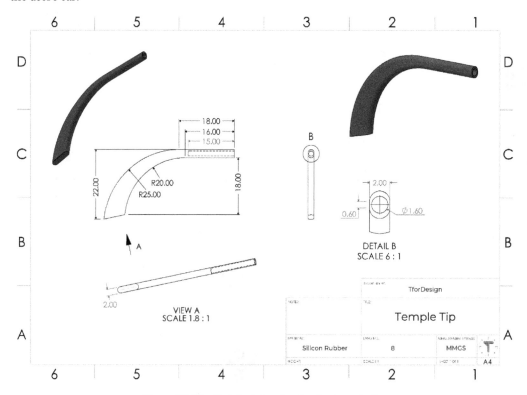

Figure P1.18 – Detailed drawing for the temple tip

Here is a sample procedure for 3D modeling the temple tip:

1. We can use the lofted boss feature with guide curves to create the shape of the temple tip. We can use sketches that look as follows.

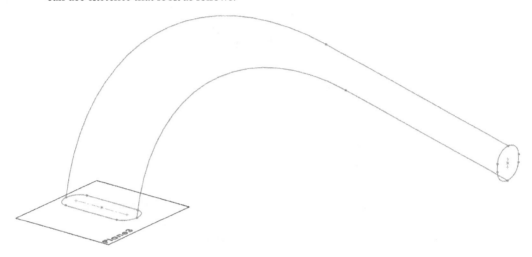

Figure P1.19 – The profiles and guide curves used to create the base lofted boss

2. Using the sketched profiles and guide curves, use the lofted boss feature to end up with the following shape.

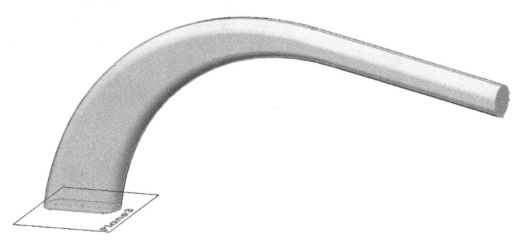

Figure P1.20 – The resulting shape after applying the lofted boss

3. Use extruded cut to create the 1.6 mm diameter hole, as shown in the following figure:

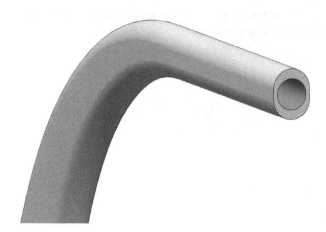

Figure P1.21 – The hole in the temple tip can be made with an extruded cut

Next, we can have a closer look at the nose link, which will link the frame of the glasses to the nose pad:

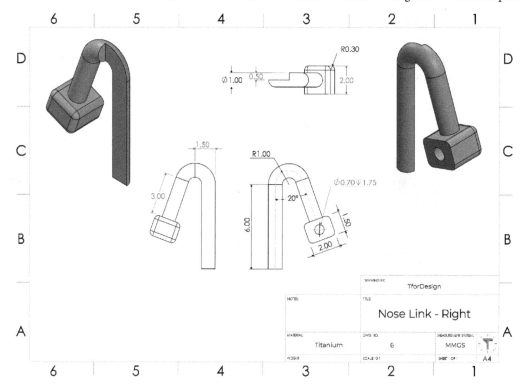

Figure P1.22 – Detailed drawing for the nose link

Here is a sample procedure for 3D modeling the nose link:

1. Use the swept boss feature with a circular profile to create the tube-looking shape of the nose link shown in the following figure.

Figure P1.23 – Swept boss be used to start creating the nose link

2. Use an extruded boss to create the cubical block shown in the following figure.

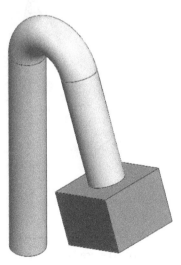

Figure P1.24 – An extuded boss built using a face resulted from the swept boss

3. Use the extruded cut feature to generate the 0.7 mm diameter circular hole on the block, as shown in the following figure:

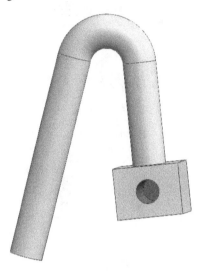

Figure P1.25 – The circular hole in the nose link can be created with an extruded cut

4. Use an extruded cut to cut off a rectangular part of the nose link, as indicated in the following figure.

Figure P1.26 – A rectangular sketch can be used with an extruded boss to create the shown cut

5. Use the feature fillet to create the R 0.3 mm rounded edges.

Now, we are done 3D modeling all the unique parts of our glasses. Feel free to treat the glasses as your own by adjusting the sizes and changing the design. There are no right or wrong answers to how to 3D model. Next, we will create the mirrored part for our nose link.

Creating a mirrored part

A mirrored part has the same features as the original part but is flipped around a specific plane. They are similar to our right and left arms. Both arms have the same features. However, they mirror one another. Other common examples around us are right and left earphones, right and left shoes, right and left casings, and so on.

The nose link part has right and left configurations in the pair of glasses we are creating for this project. We have already created the right nose link. We can create the left part by following these steps:

1. Open up the *nose link* part we created earlier.

2. Select the mirroring plane as shown in the following screenshot:

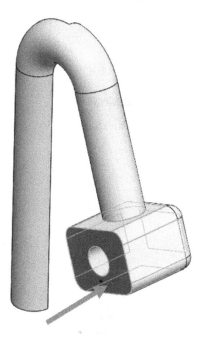

Figure P1.27 – The face to use as a mirror plane to mirror the part

Tip
Any plane, including new reference planes, can be used to create the mirrored part.

3.   With the mirror plane selected, go to **Insert**, then select **Mirror Part…**:

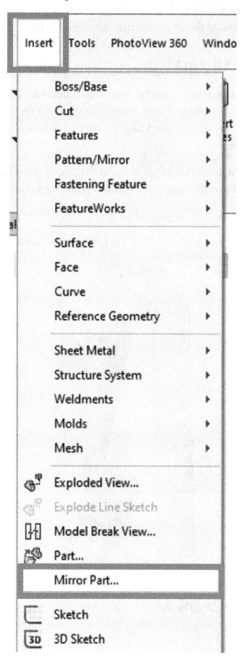

Figure P1.28 – The location of the Mirror Part command

4. This will open up a new part with the menu shown in *Figure P1.29*. This will allow us to pick the elements we want to transfer to the new mirrored part. In our case, we need to transfer **Solid bodies**. Make sure that is checked as in the following screenshot.

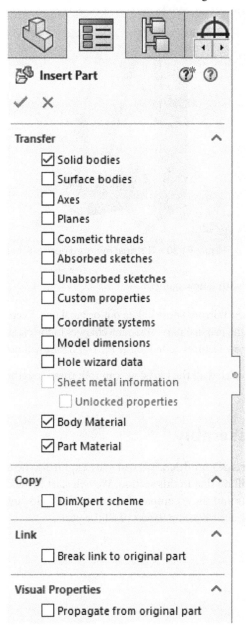

Figure P1.29 – The initial transfer options when creating a mirrored part

5.   Click on the *green checkmark* to create the new mirrored part:

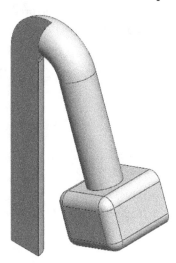

Figure P1.30 – The final mirrored nose link part

6.   Save the new part with a new name.

By doing this, we have a new mirrored nose link to our original one. Keep in mind that the mirrored part takes all its features from the original part. Thus, any changes to the original part will be reflected in the mirrored one. However, new features added to the mirrored part will not be added to the original one.

Now that we have 3D modeled all the parts, we can join them together in an assembly to form the full pair of glasses.

Creating the assembly

Now, we have all the parts 3D modeled. We can start exploring the assembly and start joining all the parts together. We will do that in this section. We will start by talking about the fixed part, then explore the different parts and the accompanying mates. The following drawing highlights the fully assembled pair of glasses, as well as the names of all the parts.

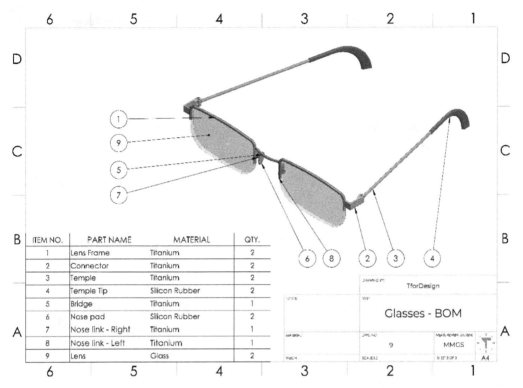

ITEM NO.	PART NAME	MATERIAL	QTY.
1	Lens Frame	Titanium	2
2	Connector	Titanium	2
3	Temple	Titanium	2
4	Temple Tip	Silicon Rubber	2
5	Bridge	Titanium	1
6	Nose pad	Silicon Rubber	2
7	Nose link - Right	Titanium	1
8	Nose link - Left	Titanium	1
9	Lens	Glass	2

Figure P1.31 – The pair assembly and its arrangement of parts

The first part we insert into an assembly becomes a fixed part by default. There is no one answer for which part should be used as a fixed part in the assembly. However, a good practice is to select an inherently non-moving part. A good candidate is the *bridge*.

The following figure highlights major mates for the pair. You may add more mates as required to get the desired result:

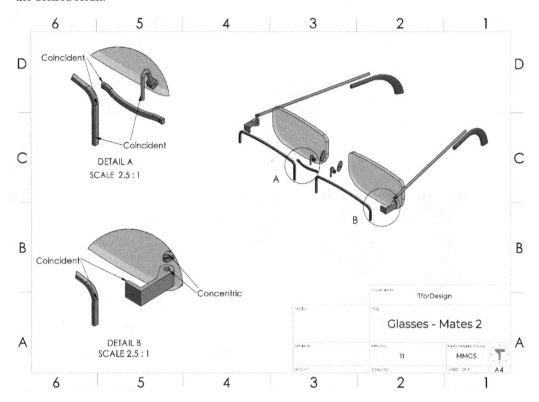

Figure P1.32 – Detailed drawing showing the major assembly metrics

The next drawing shows some extra details on the assembly, such as the angle of the nose pad in relation to the lens frame:

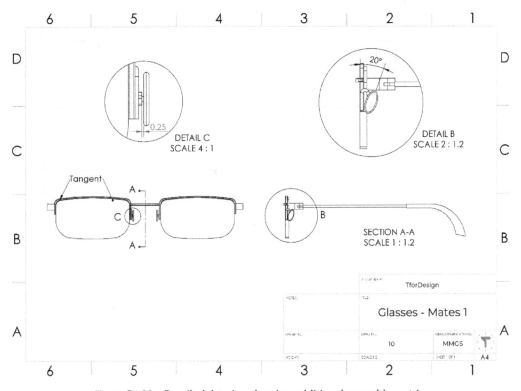

Figure P1.33 – Detailed drawing showing additional assembly metrics

Putting an assembly together is more open-ended than 3D modeling a part. So, you can formulate your own way and sequence for joining the different parts together. However, the following figure presents a possible sequence that you can follow if you are unsure where to start. Note that where the mate coincident is the only one mentioned, it was applied more than once.

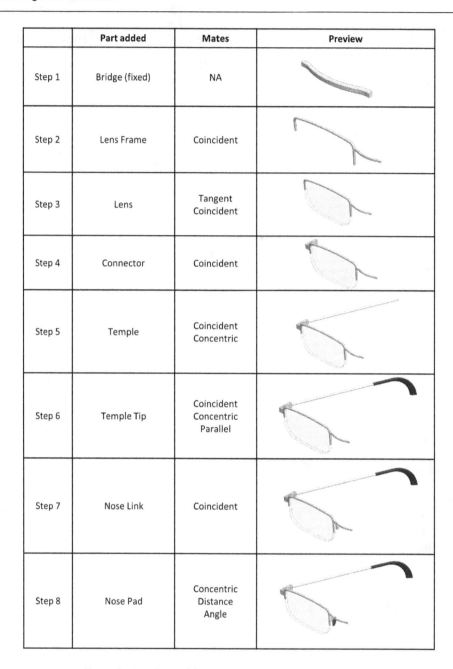

	Part added	Mates	Preview
Step 1	Bridge (fixed)	NA	
Step 2	Lens Frame	Coincident	
Step 3	Lens	Tangent Coincident	
Step 4	Connector	Coincident	
Step 5	Temple	Coincident Concentric	
Step 6	Temple Tip	Coincident Concentric Parallel	
Step 7	Nose Link	Coincident	
Step 8	Nose Pad	Concentric Distance Angle	

Figure P1.34 – A possible sequence in building the assembly

By the end of *Step 8*, half the glasses would be assembled. You can follow *Steps 2-8* again to assemble the other half. Alternatively, you can use the **Mirror Components** command to duplicate applicable parts to the other side.

As you are assembling the parts in the assembly file, note that the given mates will not fully define the assembly. Rather, they will allow some movements that can easily demonstrate the functionality of the glasses. For example, the temple will rotate, keeping one end linked to the connector in a similar mechanism to a standard pair of glasses. You can choose to add additional mates to lock the assembly in a specific position.

By completing the assembly, you have completed the project work to 3D model a pair of glasses.

Summary

In this project chapter, you worked towards 3D modeling a set of glasses. To make that happen, you had to interpret engineering drawings, 3D model different parts, then join them together in an assembly. The skills used to complete the project include the essential 3D modeling skills you will use for any project.

In the coming chapters, we will start addressing more advanced commands and features that will allow us to optimize our 3D modeling approach and 3D model more complex geometries faster.

10

Basic SOLIDWORKS Drawing Layout and Annotations

Designers and engineers need to be able to explain their 3D models to other teams or manufacturers. This could be to review the design at hand or manufacture it. In this chapter, we will learn how to use SOLIDWORKS' drawing tools to do that. We will cover how to generate simple drawings with orthogonal views, how to communicate dimensions and drawing information, and how to export drawings as shareable images or PDF files.

In this chapter, we will cover the following topics:

- Opening a SOLIDWORKS drawing file
- Generating orthographic and isometric views
- Communicating dimensions and design
- Utilizing the drawing sheet's information block
- Exporting the drawing as a PDF file or image

By the end of this chapter, you will be able to generate simple engineering drawings to explain your design to individuals or groups that are not SOLIDWORKS users. You will also be able to produce drawings that can be used for manufacturing, documentation, and archiving.

Technical requirements

In this chapter, you will need to have access to the SOLIDWORKS software.

The project files for this chapter are available at the following GitHub repository: `https://github.com/PacktPublishing/Learn-SOLIDWORKS-2025-Third-Edition`

The CiA video for this chapter can be found at `https://packt.link/y9wqf`

Opening a SOLIDWORKS drawing file

In practice, whenever we want to create a drawing in SOLIDWORKS, the first thing we will do is open a new SOLIDWORKS drawing file. This will have a different format than that of parts and assemblies files. In this section, we will learn how to open a drawing file. This will be our first step when we start working with SOLIDWORKS drawings. To open a new SOLIDWORKS drawing file, follow these steps:

1. Click on the new document icon at the top of the interface, as shown in the following *Figure 10.1*:

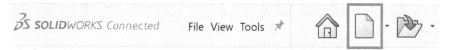

Figure 10.1 – The location of the button that allows you to open a new file

> **Tip**
> You can press *Ctrl + N* on the keyboard for a shortcut to open a new file.

2. Select **Drawing** and click **OK**, or double-click on **Drawing**, as highlighted in the following *Figure 10.2*:

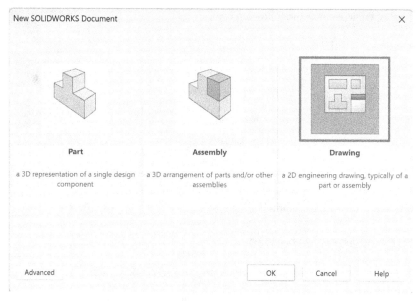

Figure 10.2 – Drawing file selection

3. This will open a drawing file, as shown in *Figure 10.3*. The first window will prompt us to select the size of the drawing sheet we want to use. The list shown here contains all the major standard sheet sizes that are used in the industry. In this exercise, we will select the first one, **A (ANSI) Landscape**, and then click **OK**:

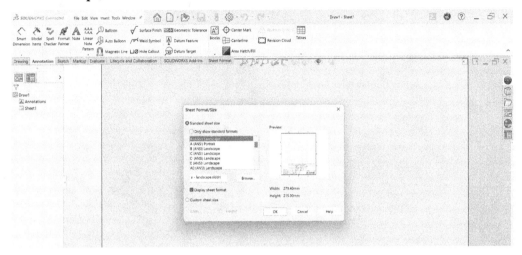

Figure 10.3 – The many available standard sheets

This will open the sheet and interface highlighted in the following *Figure 10.4*. We will be working with this throughout this chapter:

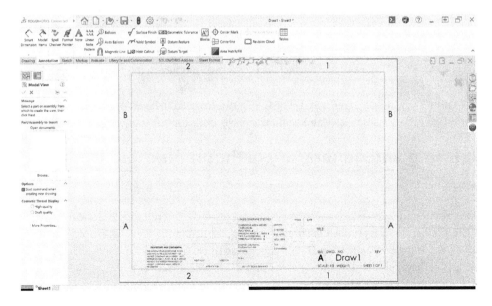

Figure 10.4 – The A (ANSI) Landscape sheet is ready for us to work on

In the rest of this chapter, we will work together to create a simple engineering drawing and export it as a PDF file so that it can be shared. To follow along, make sure that you download the SOLIDWORKS part file that comes with this chapter. *Figure 10.5* shows the drawing we'll have produced by the end of this chapter:

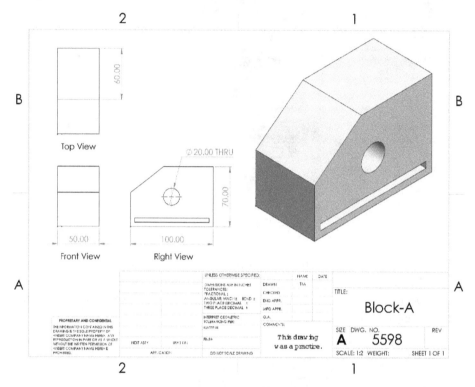

Figure 10.5 – The final drawing we will produce in this chapter

Now that we know how to open a SOLIDWORKS file, we need to insert our model into it to generate the standard orthographic views and isometric projections for our model.

Exploring and generating orthographic and isometric views

Orthographic and isometric drawing views are the most common forms of drawing views. They are also the simplest to interpret. All the drawings we presented in the previous chapters contained these types of drawing views. Now, we will start generating them. In this section, we will start a drawing file and input orthographic and isometric views into it. We will also cover adjusting the scale and display style for our drawing. We will start by selecting our targeted model. Then, we will generate and adjust our orthographic and isometric drawing views.

> **Note**
> Another common term used for orthographic views is orthogonal views.

Selecting a model to plot

SOLIDWORKS' drawing tools are based on parts or assemblies that have already been modeled. A drawing file will be linked to the parts or assemblies it communicates with. Thus, after opening a drawing file, our first step is to select the part or assembly file that we want to include in the drawing. Throughout this practical exercise, we will use the following model to create the drawing. To follow along, download the model that's linked to this chapter, which is shown in the following *Figure 10.6*:

Figure 10.6 – The 3D model we will use to generate a drawing

To begin creating your drawing, follow these steps:

1. Open a new drawing file. We covered this process in the previous section.

2. After opening a new drawing file, you will notice that **Model View** is shown on the left, as highlighted in *Figure 10.7*. If not, we can select it by clicking on **Model View** in the **Drawing** tab.

3. Click on **Browse...** to find the model you want to use as shown in *Figure 10.7*.

Figure 10.7 – The Model View button for inserting a 3D model

4. Then, search for the model linked to this exercise, select it, and click **Open**, as shown in the following *Figure 10.8*:

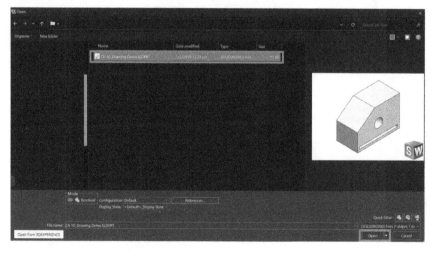

Figure 10.8 – The 2D drawing that is generated from a 3D model

This will create a link between our part file and our drawing file. Next, we will generate our orthographic and isometric views.

Generating orthographic and isometric views

Once we've selected the model from the model view, we can automatically input our orthographic and isometric views. Note that orthographic views are third-angle projections. After generating the third-angle orthographic projections, we will cover how to change them into first-angle projections. To generate the orthographic views, we can follow these steps:

1. Move the cursor onto the body of the sheet. A rectangular outline will appear. By default, this represents the front view of the model. This is shown in the following *Figure 10.9*:

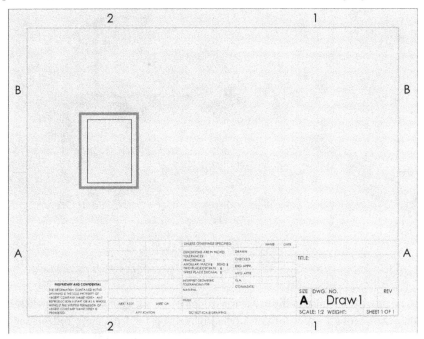

Figure 10.9 – The front view that will appear when dragging a part into the drawing

---

Note

This front view represents the view of the 3D model from its front plane.

---

2. Left-click the mouse to input the front view. After that, move the mouse to the right of the view. You will see the right view appear, as shown in *Figure 10.10*. Left-click the mouse again to input that view.

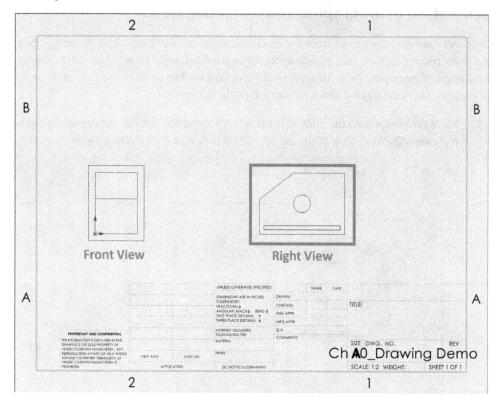

Figure 10.10 – The orthographic views that are generated dynamically

3. Then, move the mouse to the top of the front view to see the top view. Left-click again to input that. Then, move the mouse to the upper-right side of the front view. You will see the isometric view appear. Left-click again to input that view. Our drawing will be something similar to that shown in *Figure 10.11*.

Note

This is a dynamic way of inputting orthographic and isometric views. If we move the mouse in any direction, we will notice that the view changes. These views are referred to as projected views.

4. After inputting the front, right, top, and isometric views, we can simply press *Esc* on the keyboard or click on the green checkmark at the top right to confirm that the views are correct, as shown in the following *Figure 10.11*:

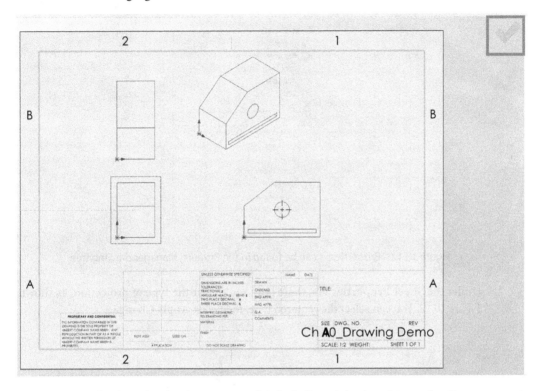

Figure 10.11 – After inserting the views, we can click the indicated checkmark to confirm they are correct

At this point, we should have a few orthographic views and an isometric view in our drawing canvas. Before we make further adjustments to our drawing views, we need to address how to change our third-angle orthographic projections to first-angle orthographic projections and vice versa. In addition, we will address some principles that are related to our initial views, that is, the parent and child views; another way we can insert views; and how to delete views. We will start with how to change our third-angle projections to first-angle projections.

Changing from third-angle projections to first-angle projections

The orthographic projections created earlier in this exercise were third-angle projections. However, you might be requested to produce drawings in first-angle projections. If that is the case, we can simply change the projection style to match the requirements. To adjust the projection style, follow these steps:

1. Right-click on the drawing sheet from the PropertyManager drawing tree found on the left of the drawing interface. Select **Properties…**, as shown in the following *Figure 10.12*:

Figure 10.12 – Properties… can be found in the PropertyManager drawing tree

2. Under the **Sheet Properties** tab, there is an option for the type of projection, as shown in *Figure 10.13*. You can select the required type, then click **Apply Changes**:

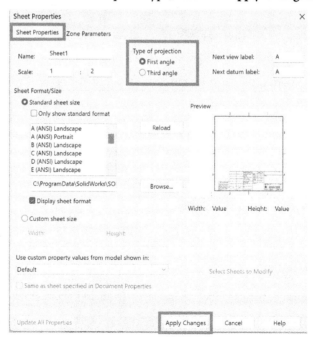

Figure 10.13 – The Sheet Properties tab allows you to adjust an active sheet

Once the changes are applied, the existing views will switch from being third-angle projections to first-angle projections. The same procedure can be followed to change the projection type from a first angle to a third angle. Now, we can move on to exploring other drawing principles, starting with the parent and child views. For the remainder of this demonstration, we will use the third-angle projection.

The parent and child views

In the preceding drawing, the right, top, and isometric views were created based on the front view. As such, we can understand the front view as the parent view of all the other views that are generated using it. *Figure 10.14* highlights the parent view and child views of the drawings we just created. This allows us to implement changes in more than one view at a time. When we move the parent view, we will see that all the child orthographic views move with it. Also, note that when changing parameters, such as the drawing scale in the parent view, those adjustments will be copied to all the child views:

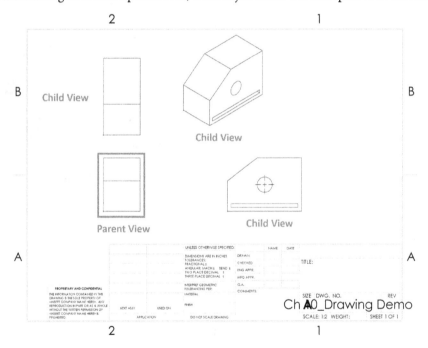

Figure 10.14 – Parent view and child views

By default, child views will copy the features of the parent view. However, prevent this by changing the PropertyManager settings in the drawing tree. Next, we will explore another way of adding views: using the View Palette.

Adding views via the View Palette

Other than adding views via the **Model View** feature, we can add separate views more flexibly via the View Palette. We can access the View Palette via the Task Pane on the right-hand side of the screen. To insert views via the View Palette, we can follow these steps:

1. Click on the View Palette from the Task Pane shortcuts menu on the right-hand side of the screen, as shown in the following *Figure 10.15*:

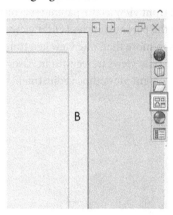

Figure 10.15 – The location of the View Palette

2. If the model is already opened, we will be able to see it in the drop-down menu at the top of the options, as highlighted in the following *Figure 10.16*. If not, we can browse to select a part or assembly:

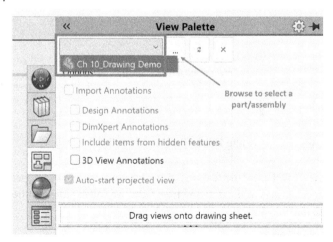

Figure 10.16 – A drop-down menu showing the active 3D model in the drawing

3. Once we've selected the model, we will have several viewing options to choose from, as highlighted in the following *Figure 10.17*. To add a view, we can simply drag it onto the drawing sheet:

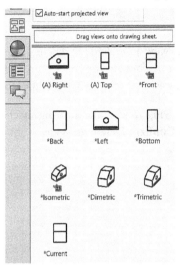

Figure 10.17 – Different views that can be directly dragged onto the drawing sheet

4. Drag the **Trimetric** view from the View Palette onto the drawing canvas. This will insert that view into the drawing. This will make our drawing appear similar to that shown in the following *Figure 10.18*:

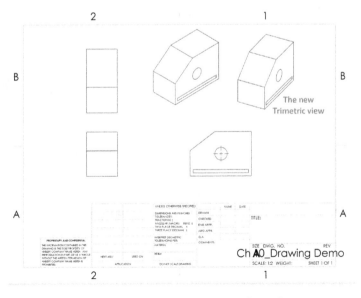

Figure 10.18 – The resulting drawing with an additional trimetric view

Using the View Palette to add views provides us with a quicker way to directly drag and drop specific drawing views from the side of the interface. Next, we will learn how to delete views.

Deleting views

Often, we may input a drawing view and decide to delete it later. For example, if after inserting the trimetric view, as we did earlier, we come to the conclusion that it adds no value, we would want to delete it. There are two methods we can follow when it comes to deleting a drawing view:

- **Method 1**: Select the drawing view on the drawing sheet and then press the *Delete* key on the keyboard. We will get a **Confirm Delete** message, as shown in the following *Figure 10.19*. To delete the view, click **Yes**:

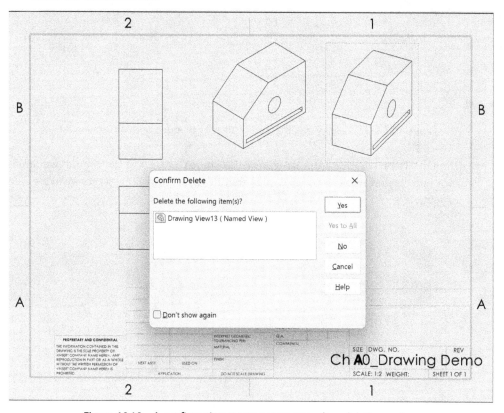

Figure 10.19 – A confirmation message appears when deleting a view

- **Method 2**: Right-click on the view and select **Delete**, as shown in the following *Figure 10.20*. This will also give the same **Confirm Delete** message.

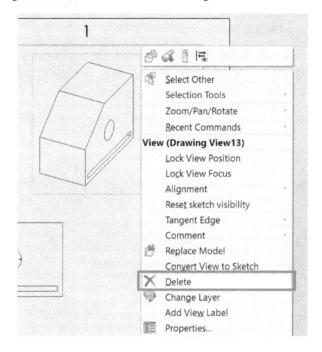

Figure 10.20 – We can delete a view using the Delete option

These are the two common methods we can use to delete a certain view from our drawing's canvas.

In this section, we have added orthographic and isometric views to our drawing file. However, note that we still have lots of empty space in our drawing sheet, which means that we can make our drawing views larger. We will address how to adjust the drawing's scale and display in the next section.

Adjusting the drawing's scale and display

With SOLIDWORKS drawings, we have the option of adjusting the size of our displayed view and its display style. In this section, we will continue working on our drawing by adjusting both its size and display style.

Adjusting the scale of our drawing

When we input the drawing views into the drawing sheet, SOLIDWORKS automatically sets a certain scale for our drawing. The scale refers to the size of the drawing as it's displayed on the drawing sheet compared to the actual size of the 3D object. Let us now explore how we can manually adjust that. To show how we can change the drawing scale, we will adjust the scale for the front view (a parent view) and the isometric view (a child view).

Changing the drawing scale for the front (parent) view

To change the drawing scale for the front parent view for our drawing, follow these steps:

1. Click on the front (parent) view.

2. In the PropertyManager, scroll down to find the scale listing. Select the **Use custom scale** option. When clicking on the drop-down menu, we will see multiple options for drawing scales, as highlighted in *Figure 10.21*:

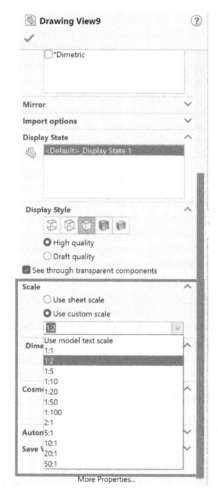

Figure 10.21 – Adjusting the scale using the PropertyManager

3. Set the scale to **1:5**. We will notice that all the drawing views become smaller, as shown in *Figure 10.22*. Note that all the views have changed in scale as well. This is because they are linked to the parent view by default.

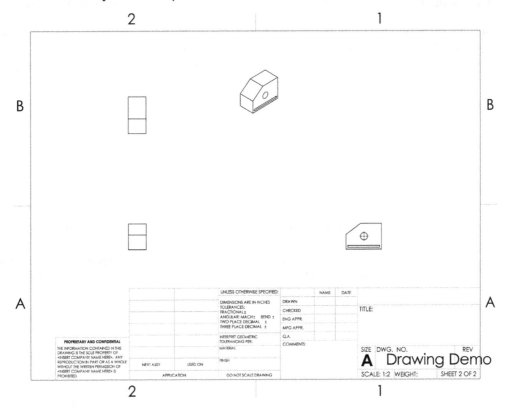

Figure 10.22 – The scale of the child views will change along with that of the parent view

4. You will now see that the new scale is very small compared to the drawing sheet. Select the front view again and change the scale to **1:2**.

There is no right or wrong scale. The best way to find the most suitable scale is to try different ratios until you are satisfied with the drawing's size and proportion considering all the required dimensions of your drawing views.

> **Note**
>
> If you would like to have a custom scale that is not listed in the drop-down menu shown in *Figure 10.21*, you can directly input it using the *x:x* or *x/x* format.

Now that we've looked at changing the scale of a parent view, we will look at changing the scale of a child view.

Changing the drawing scale of the isometric child view

To change the drawing scale of the isometric child view, we can follow the same steps that we followed for the front view. However, there is one slight difference. Follow these steps to change the scale:

1. Select the isometric view and find the **Scale** menu in the PropertyManager. Note that the scale is set to **Use parent scale**. This is why the scale of the isometric view changed when we changed the scale of the front view. To change the scale, select **Use custom scale**. Then, adjust the scale to **1:2**, as highlighted in the following *Figure 10.23*:

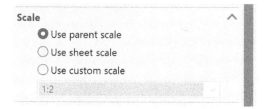

Figure 10.23 – We can set a scale for a child view that is different than that of the parent view

2. Take note that the isometric view became bigger, while all the other views remained the same size. The resulting drawing will look similar to that shown in the following *Figure 10.24*:

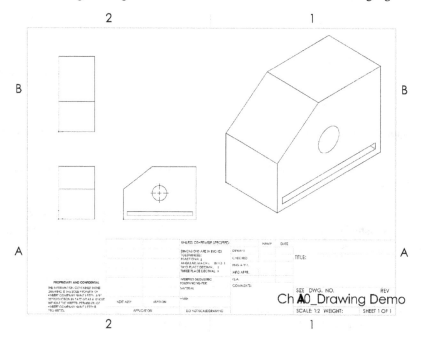

Figure 10.24 – The drawing after adjusting the scales

> **Tip**
> You can move the views around by clicking and holding the left mouse button. We usually do this to arrange our views.

Now, we have set the final scale for our drawing set. However, before we move on to adjusting the display, let's talk about the scale ratios that are provided within SOLIDWORKS.

Understanding scale ratios

In the drawing we created earlier, we had two scales: 1:2 and 1:1. Let's take a look at what the first and second numbers in these scales refer to:

- **First number**: The first number refers to the actual object size from modeling. In other words, it refers to the dimensions we used when we made the part in the first place.

- **Second number**: The second number refers to the size of the view on the drawing sheet. This relates to the final printing size of the drawing paper. Remember that when we first started our drawing file, we had to select a drawing sheet standard, which included the size of the drawing paper.

Now, let's put these two numbers together. If we were to manufacture the actual object and print the drawing sheet for our model, we would notice the following:

- The drawing sizes of the front, side, and top views are half the size of the actual object; thus, the scale is 1:2.

- The drawing size of the isometric view is the same as the actual object; thus, the scale is 1:1.

Now that we understand the scaling ratio and how to adjust our scales, we can start learning about the different displays and how to adjust them.

Different display types

Apart from the drawing view scale, we can also adjust how the drawing is displayed. There are five different displays we can use with our drawing views. These are highlighted in the table shown in *Figure 10.25*, where they are demonstrated using the isometric view we had in our previous drawing:

Name of Display	Example View
Wireframe	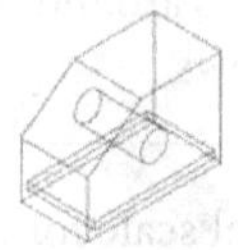
Hidden Lines Visible	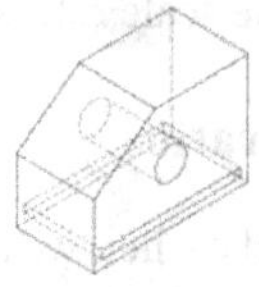
Hidden Lines Removed	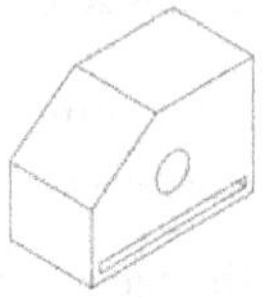
Shaded With Edges	
Shaded	

Figure 10.25 – The different display types we can use with the drawing views

To change the display of our isometric view to one of those shown in the preceding table, follow these steps:

1. Select the isometric view in the drawing sheet.

2. From the PropertyManager, find **Display Style**, as shown in the following *Figure 10.26*. The small annotated icons represent the different display styles that were shown in the preceding table. Note the **Use parent style** option here, too—we can leave that checked if we want it to match the parent display style. Selecting any of the other view styles will automatically uncheck that box:

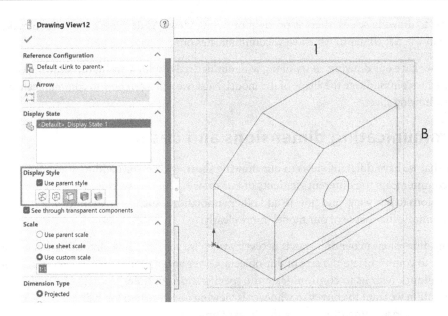

Figure 10.26 – The location of the Display Style option in the PropertyManager

3. Change the display style to shaded with edges, which is represented by the fourth icon from the left. After this, our drawing will look similar to that shown in the following *Figure 10.27*:

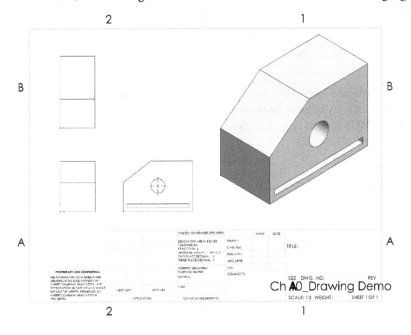

Figure 10.27 – Our drawing after changing the display style

Similar to drawing scales, there is no right or wrong view. As designers, we have to make the best decision when it comes to which view communicates our message the best.

So far, we have our drawing views, along with our desired display style, in our drawing canvas. These are used to communicate the shape of the model. Next, we will learn how to communicate dimensions in the drawing sheet.

Communicating dimensions and design

Now that we have different views in our drawing sheet, we can start adding information so that we can communicate the different elements of our drawing. In this section, we will cover how to add dimensions to our views and how to add other annotations, such as centerlines and hole callouts, to communicate the shapre of our model more clearly.

Having dimensions in our drawings is necessary when we are designing physical products. Dimensions help us to communicate the size of our objects. Other annotations, such as centerlines, notes, and hole callouts, help us to communicate the specifications of holes, centers of circles, and general information we want to convey to whoever is viewing our drawing. We will start by learning how to display numerical dimensions using the **Smart Dimension** tool.

Using the Smart Dimension tool

The **Smart Dimension** tool allows us to easily display dimensions in our drawings. We will continue working with our previous drawing and add selected dimensions to our drawing sheet. Our end result will look similar to that shown in *Figure 10.28*.

> **Note**
>
> Similar to parts and assemblies, a drawing file also has measurement unit settings that can be adjusted from the small menu in the bottom-right corner of the interface. Make sure your units are set to **MMGS** for this demonstration.

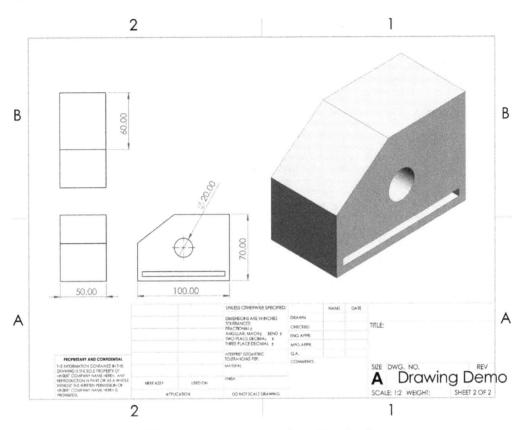

Figure 10.28 – Our finished drawing after adding the dimensions

To add dimensions, follow these steps:

1. Under the **Annotation** tab, select the **Smart Dimension** option, as highlighted in the following *Figure 10.29*:

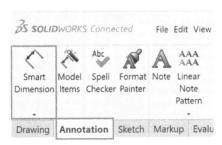

Figure 10.29 – The Smart Dimension option

2. Now, we can simply click on whatever part of the drawing we would like to dimension. This works very similarly to the **Smart Dimension** feature we used when sketching.

3. Click on the lines indicated by the arrows in the following *Figure 10.30* to add dimensions to them. The first click will show the dimension, while the other click will confirm it:

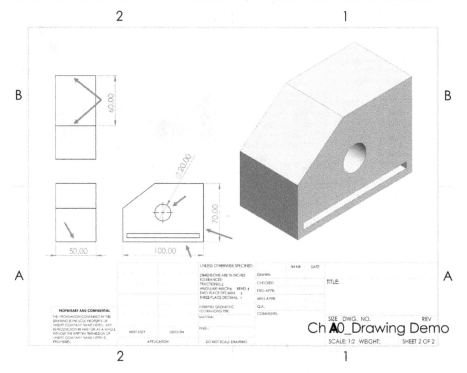

Figure 10.30 – Click areas to display the shown dimensions

This concludes the section on how we can use the **Smart Dimension** tool in drawings. In addition to dimensioning line lengths and circle diameters, we can also dimension angles or any distance between two selected points.

> **Note**
>
> Using the **Smart Dimension** tool within drawings doesn't change the dimensions of the model. By default, it will display the dimension set in the 3D model itself.

> **Tip**
>
> You can adjust the locations of the dimensions annotations by moving them around to improve the visual appeal of the drawing.

If we mistakenly input a dimension we don't want, we can delete it. To delete a dimension, we can use one of two methods:

- Right-click on the dimension and select **Delete**.

- Select the dimension and press the *Delete* key on the keyboard.

Now that we have our dimensions displayed in our drawing views, we can start inputting additional annotations to make it easier for others and ourselves to understand the drawing. In the next section, we will input centerlines, notes, and hole callouts.

Centerline, center mark, note, and hole callout annotations

In addition to dimensions, we can further clarify our drawings by adding additional annotations, such as centerlines and notes. SOLIDWORKS drawings provide an array of annotations we can use. However, we will be covering only the following in this section:

- Centerlines

- Center marks

- Notes

- Hole callouts

Centerlines

As their name suggests, centerlines highlight the center of drawing entities. They can highlight the center between two lines. In the drawing we created earlier, we will add the centerline highlighted in the following *Figure 10.31*:

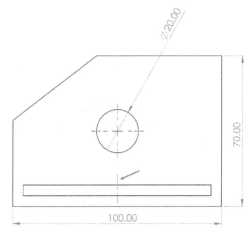

Figure 10.31 – The centerline we will add in the exercise

To add a centerline, follow these steps:

1. Under the **Annotation** tab, select **Centerline**, as shown in the following *Figure 10.32*:

Figure 10.32 – The location of the Centerline option

2. Now, we will select the two entities between which we would like to add a centerline. In our drawing, click on the two lines shown in the following *Figure 10.33*. This will automatically put the centerline between them:

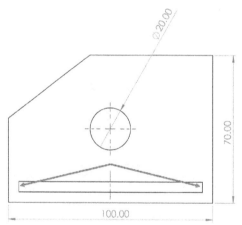

Figure 10.33 – Click on the areas indicated by arrows to add the required centerline

This concludes the section on how we can generate centerlines in SOLIDWORKS drawings. Centerlines can make it easier to interpret parts and design intents from drawings by indicating a central location between any two lines in the drawing.

> **Tip**
> We can extend the centerline as needed by dragging one end of it in a certain direction.

Next, we will address center marks.

Center marks

Center marks mark the center of circles, fillets, and slots to make them easier to identify when evaluating a drawing. The following *Figure 10.34* indicates the center mark of a circle:

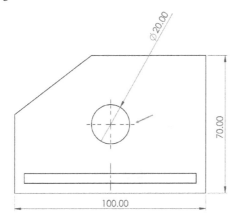

Figure 10.34 – A center mark indicating the center of a circle

To add a center mark, follow these steps:

1. Under the **Annotation** tab, select **Center Mark**, as shown in the following *Figure 10.35*:

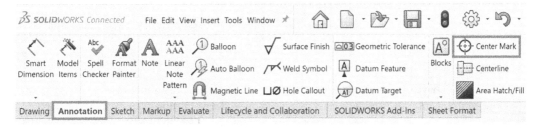

Figure 10.35 – The location of the Center Mark option

2. Now, we can directly click on circles, fillets, and slots to insert a center mark.

The **Center Mark** PropertyManager also has the option to auto-insert a center mark in our drawing views. This can save us time if we intend to add center marks to all circles, fillets, and slots. The **Auto Insert** option is highlighted in the **Center Mark** PropertyManager in the following *Figure 10.36*:

Figure 10.36 – The Auto Insert option inserts center marks automatically to applicable entities

> **Tip**
> You might have the center mark for a circle showing by default. In that case, it is set to auto-insert in the settings. You can adjust that by going to **Tool | Options | Document Properties | Detailing**. Then, in the **Auto insert on view creation** tab, you can check the options for center marks. You can turn those on and off depending on what is best for you.

Next, we will learn how to add notes to our drawings.

Notes

Notes are text indications we can add to our drawings to highlight specific aspects, functioning as open text boxes for conveying any information we want to communicate. To demonstrate how we can use notes, we will add the indicated notes to the drawing shown in *Figure 10.37*:

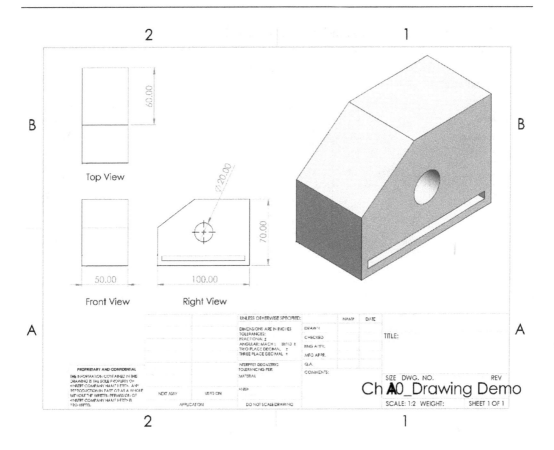

Figure 10.37 – Our drawing with notes indicating the front, right, and top views

To add a note to our drawing, follow these steps:

1. Under the **Annotation** tab, select the **Note** option, as highlighted in the following *Figure 10.38*:

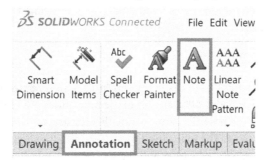

Figure 10.38 – The location of the Note option

2. Move the cursor to the drawing sheet and click under the front view. This will open a text box. Write `Front View` into it. Take note of the text format popup, as shown in the following *Figure 10.39*. This allows us to easily adjust the format of the note, including its size, color, and font:

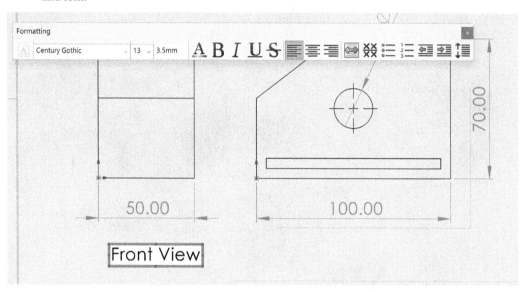

Figure 10.39 – The text format shortcut allows us to easily modify our note's format

3. Input the `Right View` and `Top View` notes in the same way.

> **Tip**
>
> While inputting a note, if we move the cursor toward lines in the drawing, SOLIDWORKS will automatically generate an arrow pointing toward that location for the note.

This concludes the section on how we can add notes to our drawing. Next, we will learn how to use the hole callout command.

Hole callout

The hole callout feature allows us to easily present information related to a particular hole in our model. This information includes the diameter of the hole, the depth of the hole, and any standards related to the creation of the hole. To demonstrate this feature, we will use it on the hole shown in the right view of our drawing. The following *Figure 10.40* also highlights the difference between the regular **Smart Dimension** we used earlier and the specialized **Hole Callout** feature:

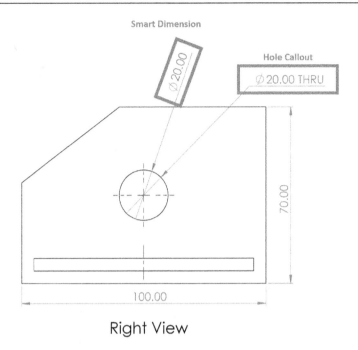

Right View

Figure 10.40 – The outcomes from the Smart Dimension versus the Hole Callout command

To add a hole callout, follow these steps:

1. Under the **Annotation** tab, select **Hole Callout**, as highlighted in the following *Figure 10.41*:

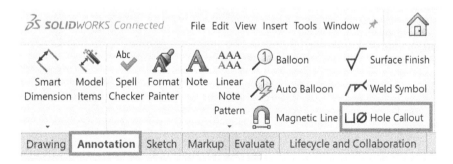

Figure 10.41 – The location of the Hole Callout option

2. Click on the hole we want to call out. In this case, it is the circle indicated in the right view, as shown in the following *Figure 10.42*. Once we click on the hole, the hole callout will appear, along with all the information linked to that particular hole:

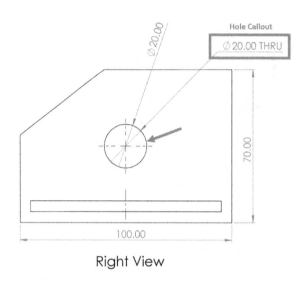

Right View

Figure 10.42 – The area we should click on to get the hole callout

3. Delete the smart dimension input to make the drawing look cleaner.

In the following example, the information that is linked to the hole details its diameter and depth. The depth of the hole is marked as **THRU**, indicating that the hole goes through all the models. If the hole had more information linked to it, such as a different hole type, this would be displayed within the callout as well. The following *Figure 10.43* contains examples of the types of hole callouts for holes with different specifications:

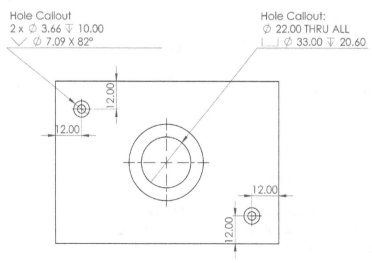

Figure 10.43 – Different types of hole callouts

This concludes the section on how we can use a hole callout. At this point, our drawing has all the required views, display types, dimensions, and annotations. Next, we will learn how to adjust the information block, which contains information that is relevant to the drawing and us as designers, such as the drawing's name, number, and the name and number of the company we work for.

Utilizing the drawing sheet's information block

In this section, we will cover how to edit the information block that's located at the bottom of our drawing sheet. This information block displays information such as the material, mass, drawer, reviewer name, and drawing number. The information block in our current drawing is highlighted with a box in the following *Figure 10.44*. We will cover how to edit the existing information and how to add new information to the block.

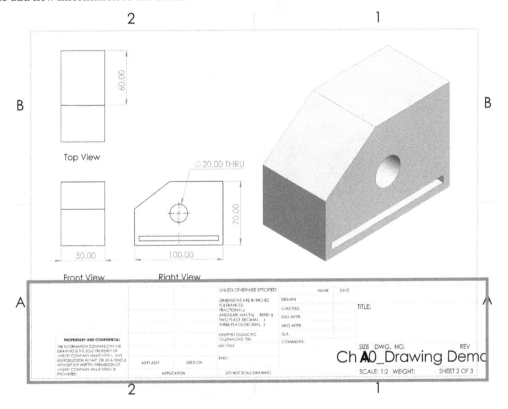

Figure 10.44 – The information block contains information related to the drawing

Now that we know what an information block is, we can start learning how to edit it.

Editing the information block

In this exercise, we will make the following edits:

- **TITLE:** Block-A

- **DWG. NO.:** 5598

- **DRAWN | NAME:** TM

To edit the information in the sheet, follow these steps:

1. Right-click anywhere in the drawing sheet. Then, select **Edit Sheet Format**, as shown in the following *Figure 10.45*. Make sure that you don't right-click on a particular drawing view:

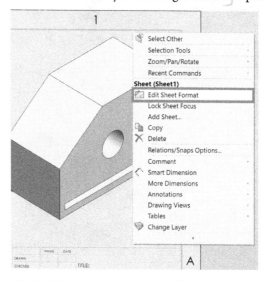

Figure 10.45 – The menu showing the Edit Sheet Format option

2. Now, the drawing lines surrounding the information block will turn blue, as shown in *Figure 10.46*. This means that we can edit the information block. The boxes for **TITLE**, **DWG. NO.**, and **DRAWN | NAME** already have text boxes. All of them are empty except that for **DWG. NO.**, which contains the name of the file by default. Most of the empty fields already have hidden text boxes that we can fill out. We can see these boxes by clicking in the middle of the empty box. The text box for **TITLE** is highlighted in *Figure 10.46* as well. To edit the **TITLE** box, we can double-click on it and input the text we want. Edit the text boxes so that they display the following information:

- **TITLE:** Block-A

- **DWG. NO.:** 5598

- **DRAWN | NAME:** TM

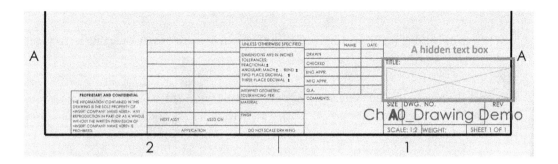

Figure 10.46 – Double-clicking on the empty boxes enables you to modify them

3. Double-check that all the inputted information is correct. Our information block should look similar to that in the following *Figure 10.47*:

Figure 10.47 – The edited information block

4. Exit the information block's edit mode by clicking on the icon highlighted in the following *Figure 10.48*. This is located in the top right-hand corner of the drawing canvas:

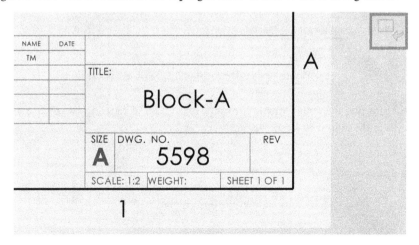

Figure 10.48 – Exit the editing mode by clicking on the indicated icon

> **Tip**
>
> When in the editing mode, we can modify the location of a text box by dragging it. You can also modify the font as well.

This concludes how we can edit information in our drawing information block. However, we've only learned how to edit existing information. Next, we will learn how to add new information boxes to the information block.

Adding new information to the information block

In addition to editing information and filling in existing text boxes, we can also add new text boxes to our information block. In this exercise, we will add the following text under **COMMENTS**: `This drawing was a practice.`. To do this, follow these steps:

1. Right-click the drawing sheet and select **Edit Sheet Format** to start editing the information sheet.

2. Note that the **COMMENTS** field doesn't have an existing text box. To add text, we need to add a normal note. We can do this by selecting the **Annotation** tab and then selecting **Note**, as highlighted in the following *Figure 10.49*. Then, we can insert the note under the **COMMENTS:** heading in the information block:

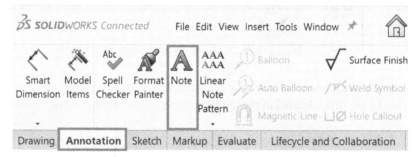

Figure 10.49 – The location of the Note option

3. After inserting the note, we can type in the text `This drawing was a practice.`, as shown in the following *Figure 10.50*:

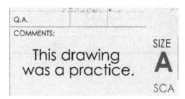

Figure 10.50 – The note typed in the COMMENTS: section

4. Exit the sheet format editing mode. The final drawing will look as shown in *Figure 10.51*:

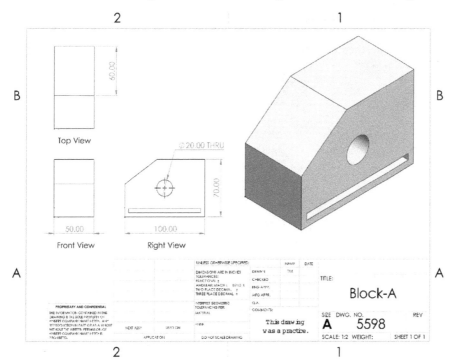

Figure 10.51 – The final look of our drawing

This concludes the section on how to add additional text to our drawing information block. At this point, our drawing is complete for this exercise. You can always add more edits and information to the information block if needed.

> **Note**
>
> It is common practice for different organizations to use custom information blocks that contain branding elements and other information used in internal communication.

To share the drawing with other individuals, especially if they don't have access to SOLIDWORKS, we will need to export the drawing as an image or PDF file so that they can view it. We will learn how to do this next.

Exporting the drawing as a PDF file or image

Now that we've completed creating a drawing with SOLIDWORKS' drawing tools, we need to export it as a PDF or image file so that we can share it with individuals who don't have access to SOLIDWORKS. This is what we'll do in this section. First, we will export the drawing as a PDF file, and then as an image.

Exporting a drawing as a PDF file

To export a drawing as a PDF file, follow these steps:

1. Click on **File**, then **Export As…**, as shown in the following *Figure 10.52*:

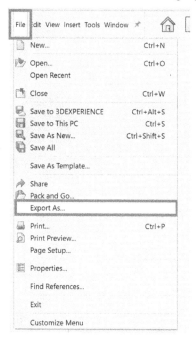

Figure 10.52 – The Export As… option

2. Under **Save as type**, open the drop-down menu and select **Adobe Portable Document Format (*.pdf)**, as highlighted in the following *Figure 10.53*:

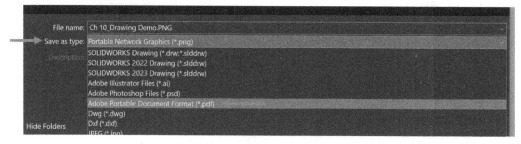

Figure 10.53 – Saving the drawing as a PDF file

3. Click on **Export**. This will export the drawing sheet as a PDF file to the designated folder.

> **Tip**
>
> You can carry out the same export process using the **Save As New…** option.

This concludes the section on how to save the drawing sheet as a PDF file. Next, we will learn how to export it as an image.

Exporting the drawing as an image

Exporting the drawing as an image is similar to exporting it as a PDF file. Follow these steps to do so:

1. Click on **File**, then **Export As…**.

2. Under **Save as type**, open the drop-down menu and select **Portable Network Graphics (*.png)**, as highlighted in the following *Figure 10.54*. We can also select other image formats if needed:

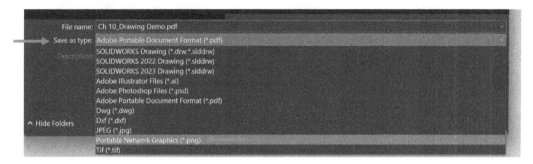

Figure 10.54 – Saving the drawing as a PNG image

3. Click on **Options…**, as shown in the following *Figure 10.55*:

Figure 10.55 – The Options… menu allows you to set the quality of the saved image

4. Adjust your options to match what is shown in *Figure 10.56*. The following is a brief explanation of the options we will adjust:

    •   **Print capture**: This exports the image as a print. When exporting the image for sharing, it is best to use the **Print capture** option. The **Screen capture** option prints whatever is shown on your drawing canvas.

    •   **DPI**: This stands for dots per inch. The higher the DPI value, the better the quality of our exported drawing print.

    •   **Paper size**: This allows you to determine the paper print size of the export:

Figure 10.56 – The settings used in this exercise

5. Click on **OK** and then on **Export** to export the image. This will save the file as a PNG in the designated folder.

This concludes the section on how to save the drawing sheet as an image. We can now share our drawings with others as images or PDF files. These formats can be accessed by larger groups of people without them needing access to special software such as SOLIDWORKS.

Note that, throughout this chapter, we have focused on generating a drawing to communicate a SOLIDWORKS part. However, the same principles apply when communicating an assembly.

Summary

Engineering drawings are what engineers and designers use to communicate their designs to other parties, such as manufacturers. SOLIDWORKS provides us with comprehensive tools that we can use to generate those drawings. In this chapter, we learned how to generate basic drawing views, including orthographic and isometric views. Then, we learned how to adjust the drawing scale and display style for a particular view. After that, we learned how to add dimensions and different annotations, such as centerlines and hole callouts. Finally, we learned how to adjust the information block and export the drawing as an image or a PDF file.

The skills we learned in this chapter will allow us to communicate our designs to external entities. Creating clear and easily interpretable drawings bridges the gap between us SOLIDWORKS users and others who don't have access to, or expertise in, the software. This is what makes the topics covered in this chapter important.

In the next chapter, we will discuss how to add a **Bill of Materials** (**BOM**) to our drawings to highlight the different parts that are used in an assembly.

Questions

1. How can we open a new drawing file?

2. What different display styles can we use with our drawing views?

3. What is the best scale to use for our drawing views?

4. What is the information block we can often find at the bottom of a standard drawing sheet?

5. Can you download the model linked to this chapter and duplicate the shown drawing (sheet specifications: **A (ANSI) Landscape**, scale 1:3)? The measurements unit system has been set to **MMGS**:

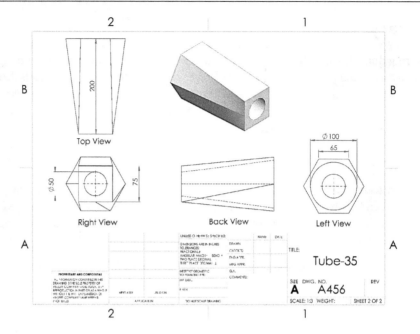

Figure 10.57 – The drawing for question 5

6.   Can you download the model linked to this exercise and duplicate the shown drawing (sheet specifications: **A4 (ANSI) Landscape**, scale 1:3)? The measurements system has been set to **MMGS**:

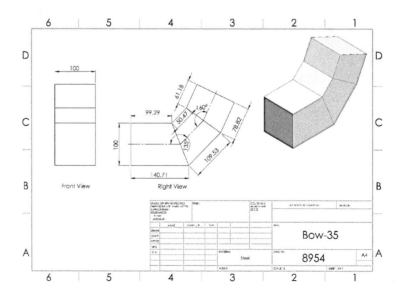

Figure 10.58 – The drawing for question 6

7. Can you create the shown model from scratch and then duplicate the shown drawing (sheet specs: **B (ANSI) Landscape**, scale: 1:1)? The measurements system has been set to **MMGS**:

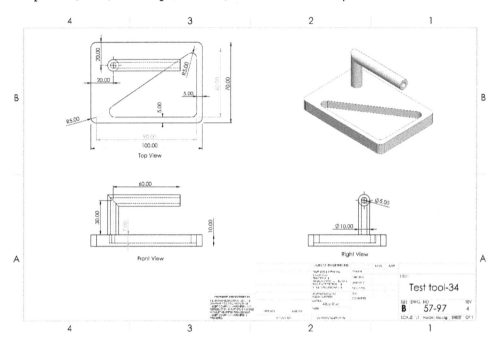

Figure 10.59 – The drawing for question 7

> **Important note**
> The answers to the preceding questions can be found at the end of this book.

Bills of Materials

Most assemblies consist of multiple parts. For example, a simple coffee table would have four legs, a top, and perhaps some screws. Other, more complex assemblies, such as engines, would contain hundreds of different parts. A **Bill of Materials** (**BOM**) can help us list and illustrate those different assembly parts. Apart from listing the parts, it can also help communicate any other information, such as cost, materials, and part numbers. This chapter will enable us to create standard BOMs. It will also enable us to modify and utilize equations in our BOMs.

The following topics will be covered in this chapter:

- Understanding BOMs
- Generating a standard BOM
- Adjusting information in the BOMs
- Utilizing equations with BOMs
- Utilizing parts callouts

By the end of the chapter, we will be able to generate a standard communicative BOM that goes with our drawings. We will also be able to fine-tune the information displayed to fully match our needs.

Technical requirements

This chapter will require access to SOLIDWORKS software.

The project files for this chapter are available at the following GitHub repository: `https://github.com/PacktPublishing/Learn-SOLIDWORKS-2025-Third-Edition`

The CiA video for this chapter can be found at `https://packt.link/iy43H`

Understanding BOMs

A BOM is an essential part of any engineering drawing representing an assembly. This is because it shows relevant information about the different parts that are present in our final product. Before we make BOMs, we need to understand what they are and their purpose. In this section, we will learn about BOMs and introduce the BOMs we will create in this chapter. Let's start.

Understanding a BOM

BOMs are an important part of engineering drawings relating to assemblies. They show more specific information about the product we are working on. For example, a typical BOM might contain the following information:

- Names of the parts in the assembly

- Part numbers

- Quantity of each part present in the assembly

- Sequential listing of each entry

However, they are not limited to this information. BOMs are customizable, depending on the established practices and the application needs. Other information that can often be found in BOMs is cost, manufacturer, materials, store locations, reference numbers, and so on. The following are two examples of assembly drawings with BOMs. Note that each table is considered a BOM. The drawing shown in *Figure 11.1* is for a mechanical cap. The BOM there includes **PART NAME**, **PartNo**, **DESCRIPTION**, **QTY.**, **Cost** (per part), and **Total Cost** (per part). It also highlights the **Total Cost** sum for all the parts and the **Highest Cost** value for one part, and also includes a subassembly and its parts, noted under **Damper Assembly**:

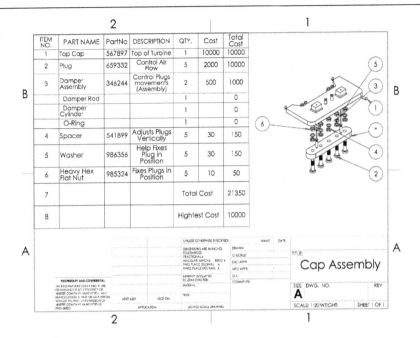

The table in the figure:

ITEM NO.	PART NAME	PartNo	DESCRIPTION	QTY.	Cost	Total Cost
1	Top Cap	567897	Top of Turbine	1	10000	10000
2	Plug	659332	Control Air Flow	5	2000	10000
3	Damper Assembly	346244	Control Plugs movements (Assembly)	2	500	1000
	Damper Rod			1		0
	Damper Cylinder			1		0
	O-Ring			1		0
4	Spacer	541899	Adjusts Plugs Vertically	5	30	150
5	Washer	986356	Help Fixes Plug in Position	5	30	150
6	Heavy Hex Flat Nut	985324	Fixes Plugs in Position	5	10	50
7				Total Cost		21350
8				Hightest Cost		10000

Figure 11.1 – A drawing with a BOM of a cap assembly

Figure 11.2 shows the BOM of a simple coffee table. The BOM includes **PART NAME, COST PER PART (USD)**, **QTY.**, and **TOTAL COST PER PART (USD)**. Note that the information in this BOM is different than the one shown previously.

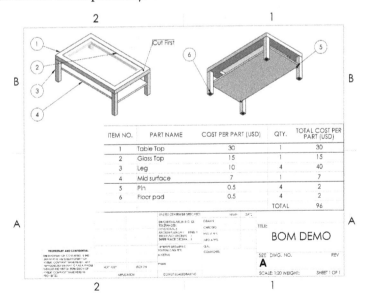

ITEM NO.	PART NAME	COST PER PART (USD)	QTY.	TOTAL COST PER PART (USD)
1	Table Top	30	1	30
2	Glass Top	15	1	15
3	Leg	10	4	40
4	Mid surface	7	1	7
5	Pin	0.5	4	2
6	Floor pad	0.5	4	2
			TOTAL	96

Figure 11.2 – An assembly drawing with a BOM for a coffee table

Throughout this chapter, we will be working to create the preceding drawing sheet and BOM from scratch. You can download all of the parts and assembly files for this chapter. Our first step will be to generate a standard BOM, which we will do next.

Generating a standard BOM

In this section, we will learn how to generate a standard BOM using the tools provided by SOLIDWORKS drawings. We will start by inserting views of our model into the drawing and then generate a BOM. A standard BOM is our starting point when generating those bills within the SOLIDWORKS drawing tools. After generating the standard bill, we will be able to modify it further so that it matches our specifications.

Inserting an assembly into a drawing sheet

A BOM often accompanies assemblies. This is because the BOMs will indicate the parts that exist in the assembly. Hence, we will start by including an assembly in our drawing sheet. Inserting an assembly works the same way as inserting a part. We can use the following steps:

1. Open a new drawing file and pick the **A (ANSI) landscape** sheet format.

2. Go to **View Palette**, browse for the assembly file we downloaded with this chapter, and select it as shown in *Figure 11.3*:

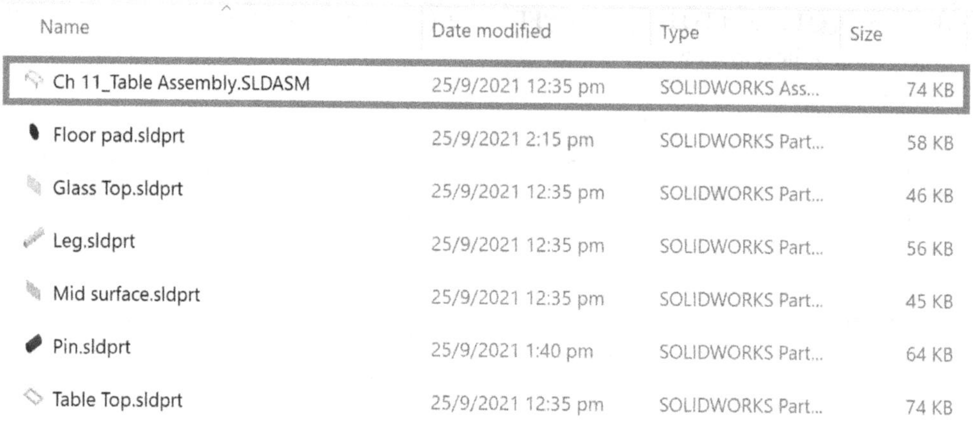

Name	Date modified	Type	Size
Ch 11_Table Assembly.SLDASM	25/9/2021 12:35 pm	SOLIDWORKS Ass...	74 KB
Floor pad.sldprt	25/9/2021 2:15 pm	SOLIDWORKS Part...	58 KB
Glass Top.sldprt	25/9/2021 12:35 pm	SOLIDWORKS Part...	46 KB
Leg.sldprt	25/9/2021 12:35 pm	SOLIDWORKS Part...	56 KB
Mid surface.sldprt	25/9/2021 12:35 pm	SOLIDWORKS Part...	45 KB
Pin.sldprt	25/9/2021 1:40 pm	SOLIDWORKS Part...	64 KB
Table Top.sldprt	25/9/2021 12:35 pm	SOLIDWORKS Part...	74 KB

Figure 11.3 – Inserting the assembly file into the drawing

3. In **View Palette**, select the two views, that is, the *Isometric view and **Bottom View**, and insert them into the sheet, as shown in *Figure 11.4*:

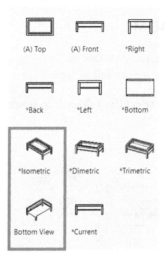

Figure 11.4 – Inserting the Isometric view and Bottom View into the drawing

4. Change the scale to 1:15 and the display to **Shaded With Edges** for both views. Our drawing sheet should look as in *Figure 11.5*:

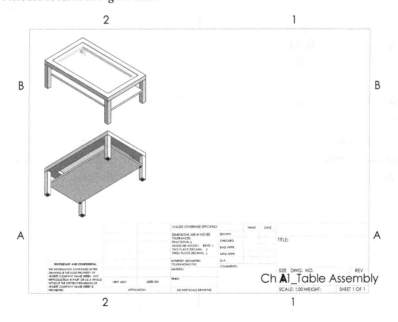

Figure 11.5 – The drawing sheet with the coffee table assembly

This concludes this section on adding an assembly to a drawing sheet. Note that adding assembly views is the same as adding parts views. Next, we will generate our standard BOMs for this assembly.

Creating a standard BOM

Now that we have inserted the assembly file into the drawing sheet, we can generate a standard BOM for the assembly. To generate that, we can follow these steps:

1. From the **Annotation** tab, select the **Tables** drop-down menu and select **Bill of Materials**, as shown in *Figure 11.6*:

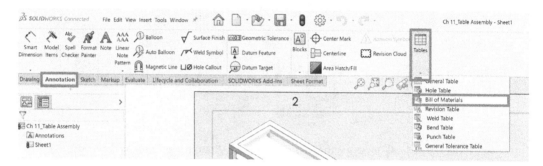

Figure 11.6 – The command to insert a BOM

> **Note**
>
> You can also insert a BOM by right-clicking on a drawing view and then going to **Tables** and selecting **Bill of Materials**.

2. Select the view for the model that you want to create the BOM. Since both the drawing views are the same, we can click on either of them.

3. This will show the **Bill of Materials** PropertyManager toward the left of the screen. As we are creating a standard default BOM, we can simply leave all of the default options, as highlighted in *Figure 11.7*. Click on the green checkmark to confirm the table:

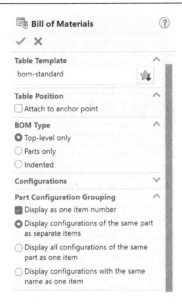

Figure 11.7 – Initial options for inserting a BOM

4. A standard BOM will be generated and displayed in our drawing sheet. We can place the BOM on the sheet to get the drawing shown in *Figure 11.8*. We can adjust the locations of the table and drawing views by dragging them.

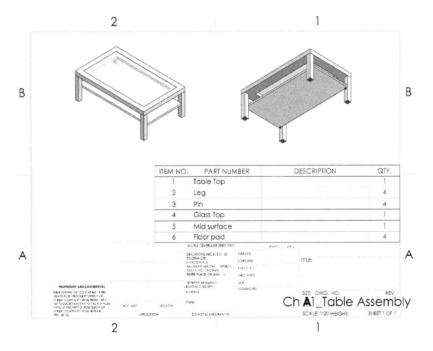

ITEM NO.	PART NUMBER	DESCRIPTION	QTY.
1	Table Top		1
2	Leg		4
3	Pin		4
4	Glass Top		1
5	Mid surface		1
6	Floor pad		4

Figure 11.8 – The initial BOM after being inserted

This concludes how to generate a standard BOM. This will be our first step whenever we generate a BOM. Next, we will be working on adjusting the information in our BOM to match the BOM that was shown in *Figure 11.2* earlier in this chapter.

Adjusting information in the BOMs

Often, the information in the standard BOM doesn't exactly match our requirements. Hence, we need to be able to adjust the information shown to match our needs and requirements. In this section, we will learn how to adjust the BOM by adjusting and adding information. By the end of this section, our drawing will look as follows. Note that the headings and information are different from how they were previously.

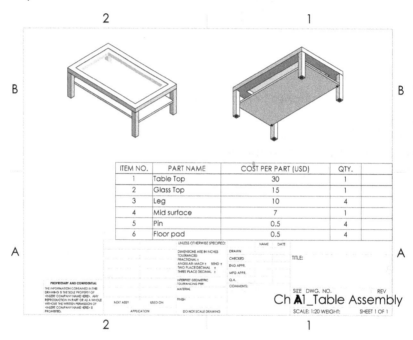

ITEM NO.	PART NAME	COST PER PART (USD)	QTY.
1	Table Top	30	1
2	Glass Top	15	1
3	Leg	10	4
4	Mid surface	7	1
5	Pin	0.5	4
6	Floor pad	0.5	4

Figure 11.9 – The drawing by the end of this section

We will start by adjusting our BOM's titles and information categories.

Adjusting listed information in the BOM

Here, we will learn how to change information that is already listed in the BOM by making the following adjustments:

- Changing a title in the BOM
- Changing a column category in the BOM

Changing a title in the BOM

Here, we will change the title of **PART NUMBER** to **PART NAME**, as highlighted in *Figure 11.10*:

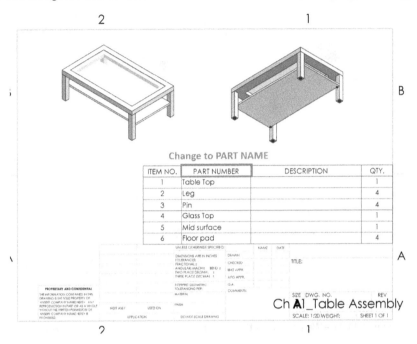

Figure 11.10 – Double-clicking on the title cell allows us to edit it

To do that, we can follow these steps:

1. Double-click on the title cell. This will open up the title for editing.

2. Change the title from PART NUMBER to PART NAME.

3. Click anywhere outside the BOM to confirm the adjusted information.

When it comes to editing information, we can think of the BOM in a similar way to tables in Microsoft Excel.

Changing a column category

When generating a BOM, usually, each column can be linked to a series of information that is gathered from the model itself. For example, the column quantity (**QTY.**) automatically links to the number of parts that are present in the assembly. This will then display the quantity without us manually inputting it. We can still add more columns that are not linked with more manual information if needed. We can also adjust the information category for a specific column. In this section, we will change the

DESCRIPTION column to **COST PER PART (USD)**, as highlighted in *Figure 11.11*. This new column will automatically be filled with the cost information linked to every part.

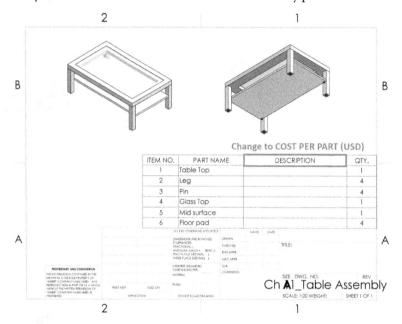

Figure 11.11 – We will adjust the DESCRIPTION column to COST PER PART (USD)

To adjust the column from **DESCRIPTION** to **COST PER PART (USD)**, we can use the following steps:

1. Hover the mouse cursor over the table, and then select the letter **C** at the top of **DESCRIPTION** to select the whole column. Then, click on the column property, as highlighted in *Figure 11.12*.

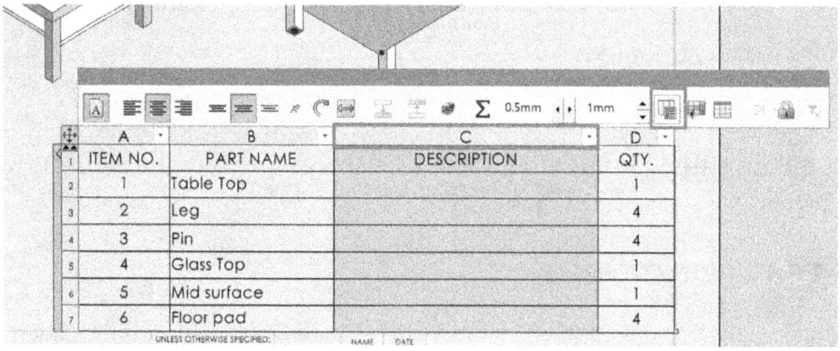

Figure 11.12 – The column property allows us to link the table to different information

2. From the shown options, make the following selections, as highlighted in *Figure 11.13*:

    - **Column type | CUSTOM PROPERTY**
    - **Property name | COST PER PART (USD)**

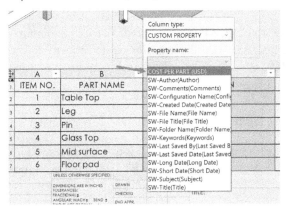

Figure 11.13 – COST PER PART (USD) as found under CUSTOM PROPERTY

3. Once we select that, the whole column will change, as well as all of the values in it. It will be filled with assigned values for the costs of each part. The BOM will look as in *Figure 11.14*:

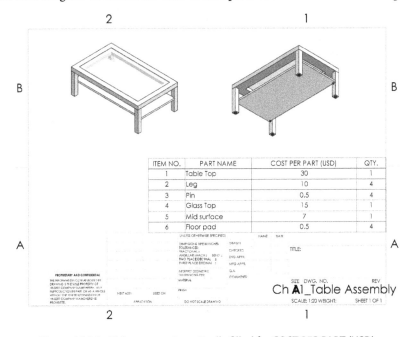

Figure 11.14 – Values are automatically filled for COST PER PART (USD)

> **Note**
>
> The values for the cost per part are not automatically generated by SOLIDWORKS. Rather, they are inputted into each part during the design process. The **CUSTOM PROPERTY** function can then auto-call those values in the BOM.

This concludes how to adjust the column category for a specific column in our BOMs. Being able to adjust categories is a necessary skill that will allow us to extract information linked to our models and put it into our BOMs. Next, we will start changing the order of listed information in our BOM by sorting it.

Sorting information in our BOMs

We can also re-sort the information in our BOMs to put it in a specific order. In our case, we will order the information based on **COST PER PART (USD)** in descending order. To do this, we can follow these steps:

1. Right-click anywhere in the table and select **Sort**, as in *Figure 11.15*.

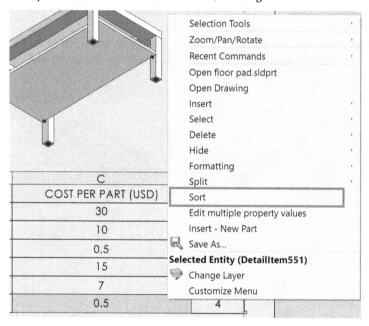

Figure 11.15 – The location of the Sort command

2. We will get the window shown in *Figure 11.16*. We can adjust the first **Sort by** option to **COST PER PART (USD)**, and then select the **Descending** option next to it. The **Method** option should be set to **Numeric**. Click **OK** after that:

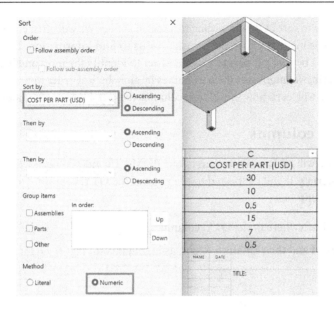

Figure 11.16 – The adapted setting for the Sort function

This will automatically re-sort the order of the whole table so that it's in descending order based on **COST PER PART (USD)**. Finally, our BOM will look as in *Figure 11.17*:

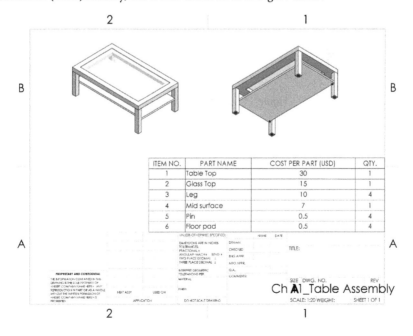

Figure 11.17 – The Sort function allowed us to sort the BOM from most to least expensive

This concludes how to sort BOMs according to a specific order. We will often end up sorting our BOMs for a variety of reasons that will make it easier for us to find information. For example, we would want to sort a BOM by cost per part to make it easier to identify the most and least costly parts. In an alternative scenario, we might sort the part names by alphabetical order to make finding a part easier by name. Next, we will learn how to add new columns to our table to accommodate new information.

Adding new columns

In this section, we will add a new column to our BOMs. The new column will be used to display the total cost of materials. It will be located to the right of the **QTY.** column. To add a new column, we can follow these steps:

1. Right-click anywhere in the **QTY.** column.

2. Select **Insert** and then **Column Right**, as highlighted in *Figure 11.18*:

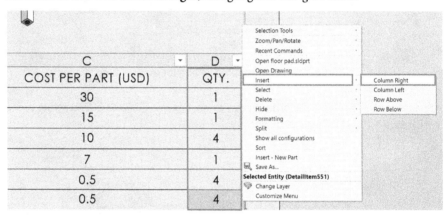

Figure 11.18 – Inserting a new column

This will insert a new empty column to the far right so that we get the BOM shown in *Figure 11.19.*

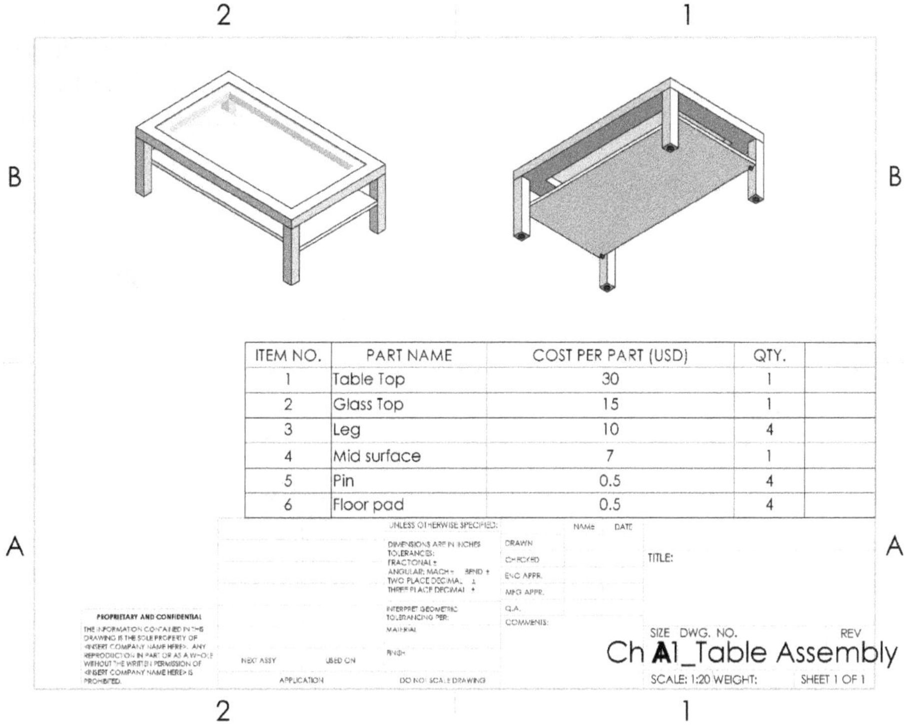
2
1
B
B
A
A
ITEM NO. | PART NAME | COST PER PART (USD) | QTY.
1 | Table Top | 30 | 1
2 | Glass Top | 15 | 1
3 | Leg | 10 | 4
4 | Mid surface | 7 | 1
5 | Pin | 0.5 | 4
6 | Floor pad | 0.5 | 4
TITLE:
SIZE DWG. NO.
REV
Ch A1_Table Assembly
SCALE: 1:20 WEIGHT:
SHEET 1 OF 1
2
1

Figure 11.19 – We can insert new columns into our BOM for more information

This concludes how to add a new column to our BOM. Note that to add a new row, we can follow the same procedure for adding a column. When we insert a new BOM, the number of columns and rows can be limiting. Hence, it will be important for us to add rows and columns to include more information. In the next section, we will fill in the new column we added using equation-based functions.

Utilizing equations with BOMs

Equations in SOLIDWORKS drawings allow us to do simple calculations within our BOMs. This will allow us to perform operations such as addition, subtraction, division, and multiplication. In this section, we will learn more about what we can do with equations. We will also use this feature for additional calculations in our BOMs.

What are equations in SOLIDWORKS drawings?

Equations allow us to perform several mathematical equations in our BOMs without leaving the SOLIDWORKS drawing interface. With equations, we can think of our BOMs as simple Excel sheets. With the equations, we can perform two types of operations, functions and mathematical operations. Let us learn what each of these includes. By the end of this section, our BOM will look as in *Figure 11.20*.

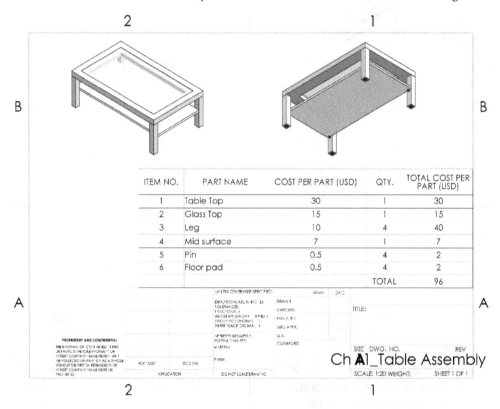

Figure 11.20 – Our BOM after applying equations to it

Functions

Functions are limited programmed calculations that we can directly use to find specific information and display it in our BOM. They include the following:

- **If**: This opens an `if` statement that will allow us to apply a condition.
- **Average**: This finds the average of different numerical values.
- **Count**: This counts the number of cells, regardless of whether the cell is empty or not.
- **Max**: This finds the maximum value within the selected cells.

- **Min**: This finds the minimum value within the selected cells.

- **Sum**: This finds the sum of values with the selected cells.

- **Total**: This totals all of the values in the column that are located above the selected cell.

Next, we will learn about mathematical operations.

Mathematical operations

In addition to functions, we can utilize different mathematical operations that are found in a basic calculator. These include the four main operations: addition, subtraction, multiplication, and division. We can use these operations by inputting them using the +, -, *, and / symbols found on the keyboard.

Now that we know about functions and mathematical operations, we will start using them in our BOM.

Inputting equations in a BOM

Here, we will demonstrate how to add equations in a BOM. We will do this by revisiting our previous BOM. We will add **TOTAL COST PER PART (USD)** and **TOTAL COST** for a full product (a coffee table, in this case). To do this, we will start by applying the mathematical operation, multiplication. Then, we will apply the *total* function.

Applying a mathematical operation

For this, we will fill the last column of our table with **TOTAL COST PER PART (USD)**, as indicated in the BOM highlighted in *Figure 11.20*. This will show the total cost of purchasing the quantity needed for each part. To calculate the value, we can multiply **COST PER PART (USD)** by **QTY.**.

Add the TOTAL COST PER PART (USD)

ITEM NO.	PART NAME	COST PER PART (USD)	QTY.	
1	Table Top	30	1	
2	Glass Top	15	1	
3	Leg	10	4	
4	Mid surface	7	1	
5	Pin	0.5	4	
6	Floor pad	0.5	4	

Figure 11.21 – We will fill the new column with calculated values for the cost

To do this, we can follow these steps:

1. Highlight the empty column at the far right by clicking the letter **E** at the top. Then, click on the equation (**Σ**) command, as highlighted in *Figure 11.22*:

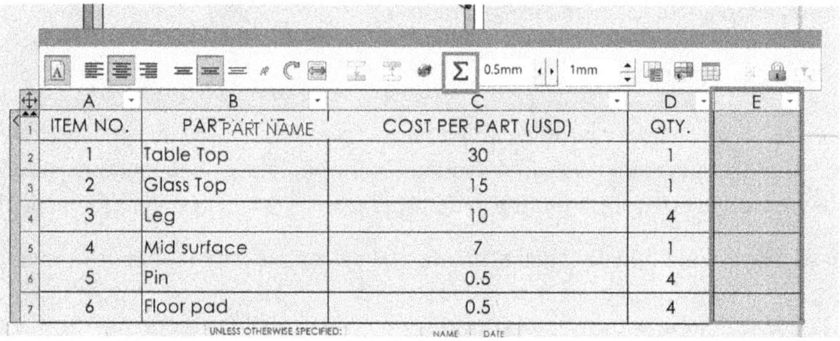

Figure 11.22 – The equation command allows us to input mathematical operations in the BOM

2. This will open the equation function, as shown in *Figure 11.23*. In the equation space, we need to input the following – COST PER PART (USD)*QTY. Instead of typing all of this in the equation space, we can use the **Columns** drop-down menu. From there, we can select **COST PER PART (USD)**. Then, we will notice that the column name was input in the equation space. After that, we can type * and select **QTY.** from the **Columns** drop-down menu:

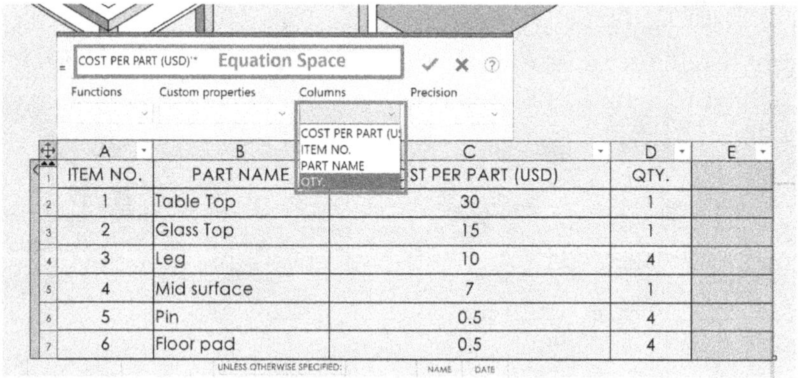

Figure 11.23 – We can input equations to calculate values in the BOM

3. Click on the green checkmark to apply the equation. We will notice that the column was filled with numbers that are equal to the number under **COST PER PART (USD)**, multiplied by **QTY.**. To make the BOM more understandable, we can add a title to the column. The title can be **TOTAL COST PER PART (USD)**. The final BOM will look as in *Figure 11.24*:

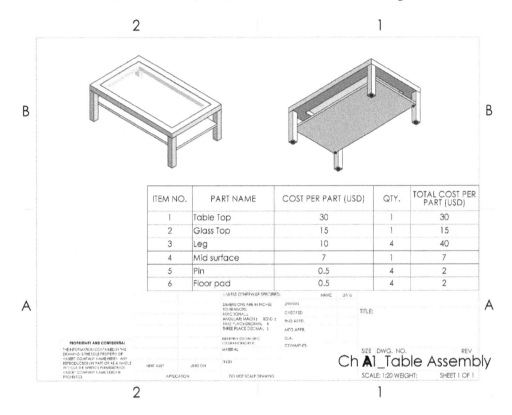

ITEM NO.	PART NAME	COST PER PART (USD)	QTY.	TOTAL COST PER PART (USD)
1	Table Top	30	1	30
2	Glass Top	15	1	15
3	Leg	10	4	40
4	Mid surface	7	1	7
5	Pin	0.5	4	2
6	Floor pad	0.5	4	2

Figure 11.24 – The TOTAL COST PER PART (USD) column is auto-calculated using a simple multiplication

This concludes using multiplication. All the other operations, such as addition and subtraction, can be applied in the same way. Next, we will learn how to apply a function.

Applying an equation function

Here, we will apply the **Total** function in our BOM. Using this function, we will add all of the values under **TOTAL COST PER PART (USD)**, which we just created. This will give us the total cost per coffee table. To do this, we can follow these steps:

1. Add a new empty row at the bottom of the table. We can do that by right-clicking any cell at the bottom and inserting a row.

2. Select the bottom cell under the **TOTAL COST PER PART (USD)** column and select equation (**Σ**).

3. In the equations panel, select **TOTAL** under the **Functions** drop-down menu, as highlighted in *Figure 11.25*. Once we select that, the word **TOTAL** will appear in the equation space.

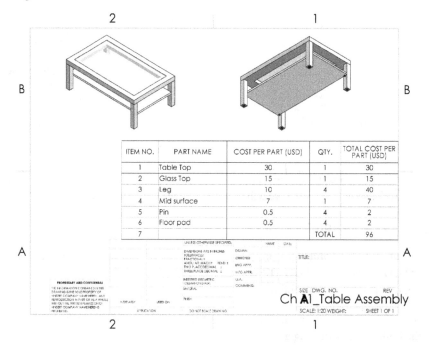

	A	B	C	D	E Σ
		PART NAME	COST PER PART (USD)	QTY.	TOTAL COST PER PART (USD)
2		Table Top	30	1	30
3	2	Glass Top	15	1	15
4	3	Leg	10	4	40
5	4	Mid surface	7	1	7
6	5	Pin	0.5	4	2
7	6	Floor pad	0.5	4	2
8	7				0

Figure 11.25 – The TOTAL function adds all the values passed to it

4. Click on the green checkmark. Then, the value of the last cell (**E8**) will change to **96**, which is the sum of all of the numbers in that column. To make the table clearer, we can add the word TOTAL in cell **D8**. The resulting BOM will be as in *Figure 11.26*. Note that when editing cell **D8**, SOLIDWORKS will ask us whether we want to break the link in the cell and continue editing it; click **Yes** to edit.

ITEM NO.	PART NAME	COST PER PART (USD)	QTY.	TOTAL COST PER PART (USD)
1	Table Top	30	1	30
2	Glass Top	15	1	15
3	Leg	10	4	40
4	Mid surface	7	1	7
5	Pin	0.5	4	2
6	Floor pad	0.5	4	2
7			TOTAL	96

Ch **A**1_Table Assembly

Figure 11.26 – Our new BOM after calculating the total cost

5. It is good practice to try cleaning our BOM as much as possible. For example, in the preceding BOM, we can remove item number **7**, as that row only has the total and does not include an item. To hide that item number, we can right-click on that cell and select **Hide Item Number**, as highlighted in *Figure 11.27*.

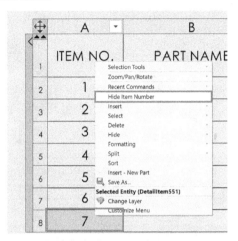

Figure 11.27 – Hiding the last item number can get us a cleaner BOM

This concludes our work with equations within BOMs. Using equations will allow us to generate new numerical information in our table that is not linked to a particular part. A common application of equations is calculating the costs of parts as we demonstrated in this demo. Next, we will learn how to add callouts to the assembly for easier referencing between our BOM and the visual display of the assembly.

Utilizing parts callouts

In this section, we will cover how to add callouts to the parts in our BOM. These can help us identify the location of the items in the drawing itself. Here, we will cover auto-balloon callouts. To create these callouts, we can do the following:

1. From the **Annotation** tab, select the **Auto Balloon** command, as highlighted in *Figure 11.28*:

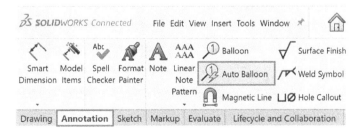

Figure 11.28 – The location of the Auto Balloon command

2. Click on the two views we have on the drawing, one after another. You will notice balloons popping up with numbers and arrows pointing toward different parts, as shown in *Figure 11.29*. Note that each number in a balloon matches the number in the BOM:

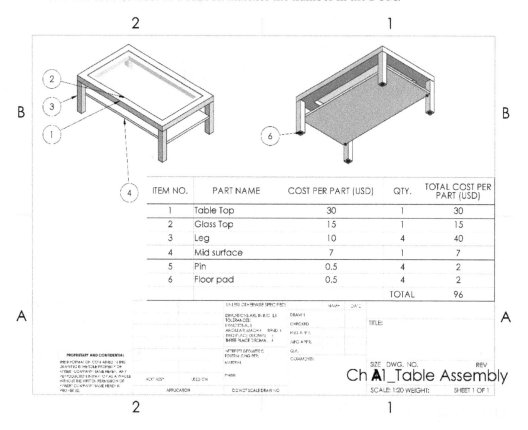

ITEM NO.	PART NAME	COST PER PART (USD)	QTY.	TOTAL COST PER PART (USD)
1	Table Top	30	1	30
2	Glass Top	15	1	15
3	Leg	10	4	40
4	Mid surface	7	1	7
5	Pin	0.5	4	2
6	Floor pad	0.5	4	2
			TOTAL	96

Figure 11.29 – Balloons indicate the item numbers matching the BOM

3. In the PropertyManager, we have multiple options to adjust the callout. For this exercise, adjust the options highlighted in *Figure 11.30*. After adjusting these settings, click on the green checkmark to apply the balloon callouts:

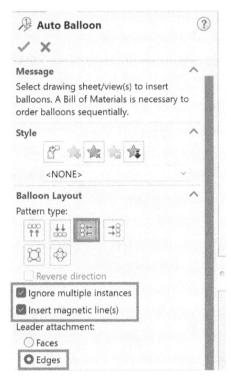

Figure 11.30 – The Auto Balloon PropertyManager

The final result of our drawing would look as in the drawing highlighted in *Figure 11.31*. Note that we can manually drag the balloons to change their position as we see fit:

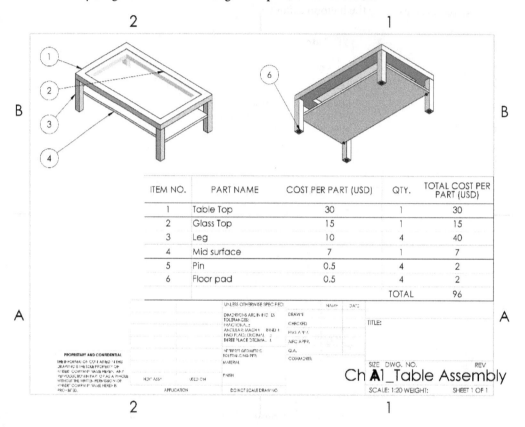

ITEM NO.	PART NAME	COST PER PART (USD)	QTY.	TOTAL COST PER PART (USD)
1	Table Top	30	1	30
2	Glass Top	15	1	15
3	Leg	10	4	40
4	Mid surface	7	1	7
5	Pin	0.5	4	2
6	Floor pad	0.5	4	2
			TOTAL	96

Figure 11.31 – The final drawing with a clean and communicative BOM

One limitation of the **Auto Balloon** command is that it only captures the visible and base parts in the drawing views. For example, item number **6**, the floor pad, is indicated on the drawing view shown on the right. If we don't have that view in the drawing, then **Auto Balloon** will not annotate that part because it is not visible in the drawing view shown on the left. Also, item number **5**, the pin, is not annotated at all even though it is visible on the right drawing view. However, the visible pins are mirrored ones, not the base pin. The base pin is the one hiding above the annotated item number **6**. We will add a balloon for item number **5**, the pin, manually when addressing the manual balloon command.

> **Tip**
>
> To ensure all parts are captured with the **Auto Balloon** command, we can change the display for our drawing display style to one where all the parts are visible, such as the **Wireframe** display style. We can then apply the **Auto Balloon** command to capture all the parts, and then change our drawing view to any other display style we want to use.

In this exercise, we used the **Auto Balloon** command to display the item number, as shown in *Figure 11.31*. However, we can also choose to display other types of information from the **Auto Balloon** PropertyManager as shown in *Figure 11.32*:

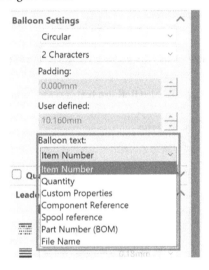

Figure 11.32 – The balloons can display different information other than the item number

This concludes how to create auto balloons for display purposes in our drawing. Note that these balloons will enable anyone who views the drawing to link the information in the BOMs to the displayed assembly.

Manual balloon command

The **Auto Balloon** command enables us to quickly display related information to more than one part at once. However, if we want to include a few balloons with specific information, such as written text notes, a missed part from the **Auto Balloon** command, or any other custom property, then manually adding a balloon can be a more efficient option. Let's try out the manual balloon function by indicating the pin (**ITEM NO. 5**) and adding a box-shaped annotation linking to the tabletop with the Cut First text. We will start by adding a balloon pointing to **ITEM NO. 5**.

To do this, follow these steps:

1. Select the **Balloon** command, as shown in *Figure 11.33*:

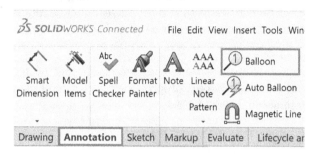

Figure 11.33 – The location of the Balloon command

2. In the PropertyManager, make sure **Balloon text** is set to **Item Number**, as shown in *Figure 11.34*.

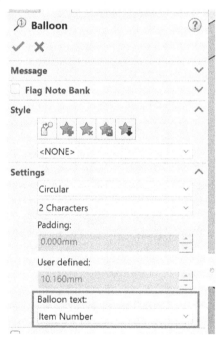

Figure 11.34 – The Balloon text option allows us to decide which information to display

3. Zoom in to the pin and click on it to position the balloon, as shown in *Figure 11.35*. This will automatically call out the item number 5 in the balloon itself.

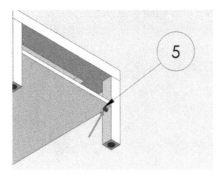

Figure 11.35 – The callout will automatically fill out the linked information

4. Press *Esc* on the keyboard on click on the green checkmark to confirm the new balloon.

This concludes how to manually add the last balloon for the last missing part. Now, we can try adding new custom text. In this exercise, we will add a box-shaped annotation linking to the tabletop with the Cut First text. To do this, we can follow those steps:

1. Select the **Balloon** command under the **Annotation** tab, as shown in *Figure 11.33*.

2. Under **Settings** in the PropertyManager, select the **Box** and **Tight Fit** options, as shown in *Figure 11.36*:

Figure 11.36 – We can set up the shape and text of the balloon per our needs

3. Under **Balloon text**, select the **Text** option and type Cut First, as shown in *Figure 11.36*.

4. Using the cursor, go back to the drawing canvas and attach the box-shaped balloon to the tabletop. We will end up with what is shown in *Figure 11.37*:

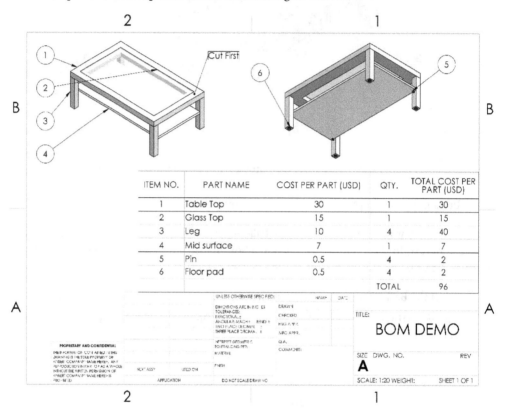

ITEM NO.	PART NAME	COST PER PART (USD)	QTY.	TOTAL COST PER PART (USD)
1	Table Top	30	1	30
2	Glass Top	15	1	15
3	Leg	10	4	40
4	Mid surface	7	1	7
5	Pin	0.5	4	2
6	Floor pad	0.5	4	2
			TOTAL	96

Figure 11.37 – The resulting drawing with a box-shaped balloon

5. Confirm the balloon annotation by clicking on the green checkmark or pressing *Esc* on the keyboard.

Other than the custom text we added in the previous example, the balloon can also link to specific parts' properties and extract information from it in the same way that custom properties work in the BOM. Thus, we can call out linked information such as the filename, quantity of each part, and component reference. At this point, we can conclude our discussion on utilizing the ballooning options to display specific communicative information. We can also conclude our exploration of BOMs within SOLIDWORKS drawings.

Summary

Most of the products we work with include more than one part put together. To easily show these parts alongside related details, such as cost, part numbers, materials, and weights, we can use a BOM. Including a BOM is a very common practice when creating drawings of assemblies. In this chapter, we learned what BOMs are and how to generate a standard one. We then learned how to adjust the information in a standard BOM by adding, removing, and regenerating information, rows, and columns. We also learned how to use equations to generate numerical values from our BOMs. Finally, we learned how to generate callouts to create visual links between the information in our BOM and the visual representation of our assembly in the drawing sheet.

Being able to create a BOM is an essential skill to create products consisting of many different parts. It is also an expected skill of a SOLIDWORKS professional.

In the next chapter, we will start learning about and using another set of advanced features that will enable us to generate more complex 3D models than what we've learned about already. These features will include draft, shell, hole wizard, features mirror, and multi-body parts.

Questions

1. What is a BOM?
2. What information can be found in BOMs?
3. What is linked information in BOMs in SOLIDWORKS?

4. Download the parts and assembly linked to this exercise and generate the following standard BOM:

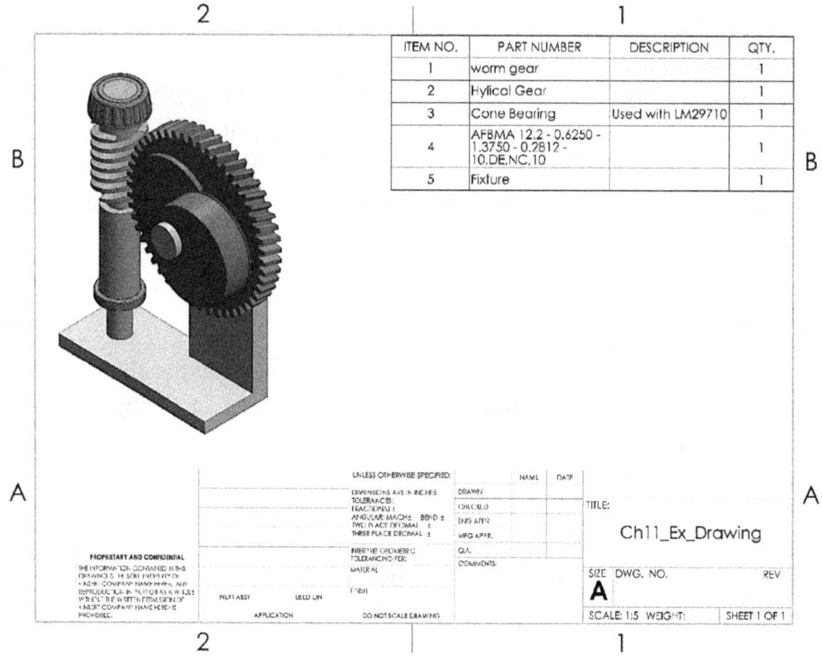

ITEM NO.	PART NUMBER	DESCRIPTION	QTY.
1	worm gear		1
2	Hylical Gear		1
3	Cone Bearing	Used with LM29710	1
4	AFBMA 12.2 - 0.6250 - 1.3750 - 0.2812 - 10.DE,NC,10		1
5	Fixture		1

Figure 11.38 – The drawing for question 4

5. Modify the BOM from the previous question by adding and filling up the cost column (note – cost numbers are not automatically generated; input them manually, as shown in the following screenshot). Also, change the **PART NUMBER** and **DESCRIPTION** columns to **PartNo** and **Vendor**. Your BOM would look similar to the one shown here:

ITEM NO.	PartNo	Vendor	/Cost (USD)	QTY.
1	DP-6739	International Gears Limited	800	1
2	DP-5946	International Gears Limited	1200	1
3	DP-9548	International Bearing Limited	50	1
4	DP-3464	International Bearing Limited	60	1
5	DP-4638	Almattar Machine Shop	150	1

Figure 11.39 – The BOM for question 5

6. Use equations to calculate the total cost for all the parts. You will end up with the following BOM shown in *Figure 11.40*:

ITEM NO.	PartNo	Vendor	/Cost (USD)	QTY.
1	DP-6739	International Gears Limited	800	1
2	DP-5946	International Gears Limited	1200	1
3	DP-9548	International Bearing Limited	50	1
4	DP-3464	International Bearing Limited	60	1
5	DP-4638	Almattar Machine Shop	150	1
		Total	2260	

Figure 11.40 – The BOM for question 6

7. Use the **Auto Balloon** command to link the item numbers in the drawing views. Your result will look similar to the following *Figure 11.41*:

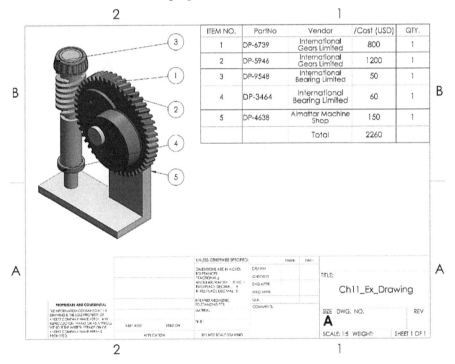

Figure 11.41 – The drawing and BOM for question 7

Important note
The answers to the preceding questions can be found at the end of this book.

Get This Book's PDF Version and Exclusive Extras

UNLOCK NOW

Scan the QR code (or go to `packtpub.com/unlock`). Search for this book by name, confirm the edition, and then follow the steps on the page.

Note: Keep your invoice handy. Purchases made directly from Packt don't require an invoice.

Part 6:
Advanced Mechanical Core Features – Professional Level

The Professional level is the second level of proficiency for SOLIDWORKS users in core mechanical design applications. This part covers all the features expected for this level. Those include features for building 3D models, including drafts, shells, hole wizards, mirroring, and ribs. Also, you will learn about features that can optimize your 3D modeling approach, such as multi-body parts, equations, configurations, and design tables.

This part has the following chapters:

- *Chapter 12, Advanced SOLIDWORKS Mechanical Core Features*
- *Chapter 13, Equations, Configurations, and Design Tables*

12

Advanced SOLIDWORKS Mechanical Core Features

In this chapter, we will cover more advanced and less commonly used features in SOLIDWORKS mechanical modeling. These features include the **draft** feature, the **shell** feature, the **Hole Wizard**, **features mirroring**, and the **rib** feature. We will use these features to build 3D models, while also covering the concept of **multi-body parts**. These features will greatly enhance your SOLIDWORKS skills to an advanced level by further simplifying complex model creation and manipulation. They are also essential for passing the SOLIDWORKS professional certification exam.

The following topics will be covered in this chapter:

- Understanding and applying the draft feature
- Understanding and applying the shell feature
- Understanding and utilizing the Hole Wizard
- Understanding and applying features mirroring
- Understanding and applying the rib feature
- Understanding and utilizing multi-body parts
- Understanding and applying the linear, circular, and fill feature patterns

By the end of this chapter, you will be able to generate more complex models with features matching international standards with holes. Also, the draft, shell, mirror, rib, patterns, and multi-body parts features will provide us with more means to meet specific design requirements.

Technical requirements

This chapter requires you to have access to the SOLIDWORKS software. The project files for this chapter are available at the following GitHub repository: `https://github.com/PacktPublishing/Learn-SOLIDWORKS-2025-Third-Edition`

The CiA video for this chapter can be found at `https://packt.link/mexVN`

Understanding and applying the draft feature

Drafting in engineering refers to adding a taper or angle to sharp edges or steps in parts, typically transitioning them into chamfered surfaces. Drafting primarily comes from the casting and plastic injection molding industry to make it easier to release parts out of molds. As SOLIDWORKS professionals, we will be expected to apply drafts to a variety of applications where necessary. In this section, we will explore what drafting is and how to use the draft feature.

What are drafts?

Drafts are commonly applied to parts that are made with injection molding or casting. It is a slight tilt between two different surfaces at different levels. In practice, drafts help make parts fit better with the mold and make the parts easier to remove from the mold compared to without it. Also, drafts help increase the success rate of the mold taking effect.

The following diagrams highlight the effect of the draft feature. The first diagram in *Figure 12.1* shows the model at hand, while *Figure 12.2* highlights the cross-section area where we are showing the effect of the draft:

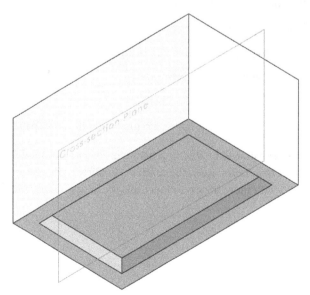

Figure 12.1 – An example of a draft

The second diagram in *Figure 12.2* shows the cross-section of the model with a draft and without a draft. Note that the draft looks similar to a chamfer:

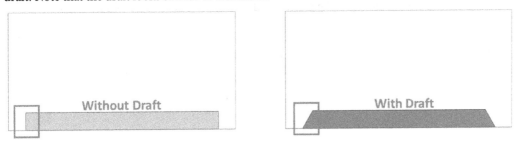

Figure 12.2 – Examples of a design with and without draft

Now that we know what drafts are, we can apply them to create a particular 3D model.

Applying drafts

In this section, we will cover how to apply the draft feature. We will apply a draft to the model shown in *Figure 12.3*. You can download this model from the package provided for this chapter.

Alternatively, you can create the model from scratch using the drawing in *Figure 12.3*. All of the lengths are in millimeters. Note that the draft is measured by the angle shown in the *Draft DETAIL D* view:

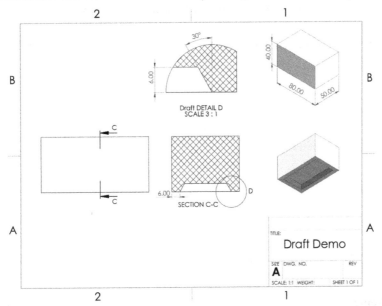

Figure 12.3 – The 3D model we will build in this exercise

To create the draft, do the following:

1. First, download and open the model for this exercise. The model looks like *Figure 12.4*:

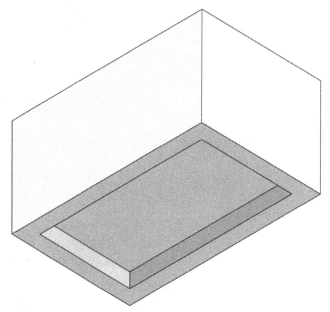

Figure 12.4 – The model to be downloaded for this section

2. Select the **Features** command category, then select the **Draft** command, as highlighted in *Figure 12.5*:

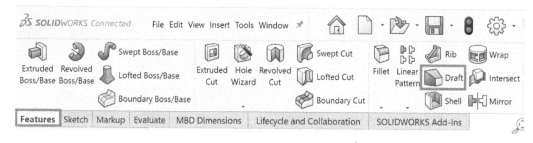

Figure 12.5 – The location of the Draft command

3. Adjust the PropertyManager options and selections so that they match those shown in *Figure 12.6*. Here is a brief description of some of the options:

    *   **Neutral plane**: Here, we can select a surface that will be adjacent to the draft. This neutral surface will remain unchanged during the creation of the draft. Hence, the other adjacent surface will be affected.

- **Faces to Draft**: Here, we can select all of the faces to be drafted. We can select more than one surface as we see fit. In this exercise, we are selecting all four inside faces, as demonstrated in *Figure 12.6*.

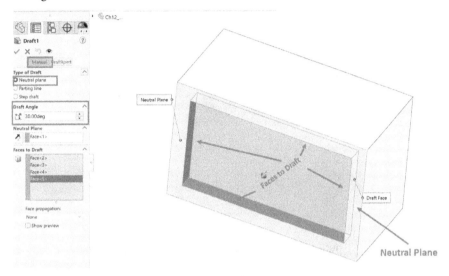

Figure 12.6 – The settings for the draft feature

4. Click on the green checkmark to apply the draft. The resultant draft will be as shown in *Figure 12.7*. In the applied draft, note the following:

   - The position of the draft angle. In some drawings, we might be given the complementary angle in a drawing instead of the draft angle. In that case, the draft angle = 90 - the complementary angle.

   - The neutral surface and its surface area remained unchanged compared to all of the other surfaces related to the draft.

Figure 12.7 – The selected neutral surface remains unchanged once the draft has been applied

Note that, in this exercise, we used the **neutral plane** draft type. Other types of drafts are available, including the **parting line** and **step draft**, which can be created in a similar manner. *Figure 12.8* highlights an example of a parting line draft application.

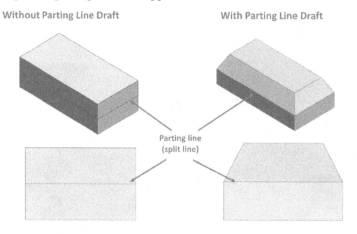

Figure 12.8 – A demo application of the parting line draft

Figure 12.9 highlights an example of a step draft application:

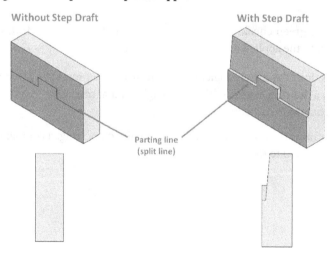

Figure 12.9 – A demo application of a step draft

Both draft options can be found in the draft feature's PropertyManager. This concludes our coverage of the draft feature. We learned what the feature is and how to generate and define it.

Next, we will cover another feature known as the shell feature.

Understanding and applying the shell feature

In this section, we will cover the **shell feature**. As its name suggests, the shell feature enables us to create a shell out of an existing shape without much effort. In our everyday lives, we interact with shells in multiple products, such as cans, laptops, and phone exteriors. In this section, we will explore what the shell feature is and how to apply it.

What is a shell?

The **shell feature** enables us to make a shell of an existing model. It makes it easy to create objects such as cans and containers. *Figure 12.10* shows a box with the effect of the **Shell** command implemented:

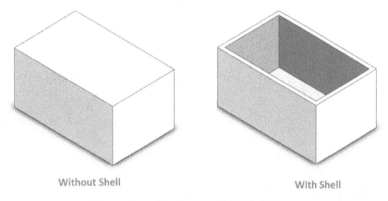

Without Shell   With Shell

Figure 12.10 – The impact of the shell feature

Now that we know what the shell feature does, we can apply it to create a specific model, which we will do next.

Applying a shell

In this section, we will cover how to apply the shell feature. We will apply the **Shell** command to create the model highlighted in *Figure 12.11*. You can download this model from the package provided for this chapter. Alternatively, you can create the model from scratch. Note that this is a continuation of the previous model. The thickness of the wall is 3 mm, as highlighted in the review cloud in the diagram:

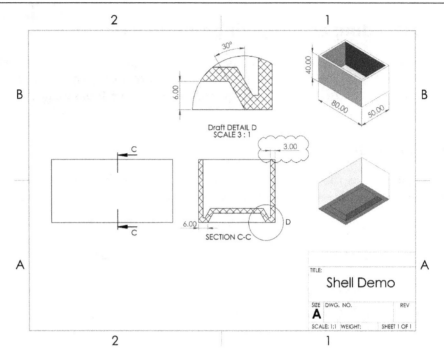

Figure 12.11 – The 3D model we will build in this exercise

To create the highlighted shell, follow these steps:

1. First, download and open the model for this exercise, which looks similar to the one shown in *Figure 12.12*. This is the same model we created in the previous exercise on drafts.

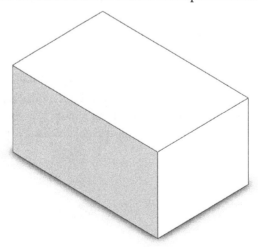

Figure 12.12 – The model that is included with the downloads for this section

2. Select the **Features** option category, then select the **Shell** tool, as highlighted in *Figure 12.13*:

Figure 12.13 – The location of the Shell command

3. Adjust the PropertyManager options and selections so that they match the parameters highlighted in *Figure 12.14*:

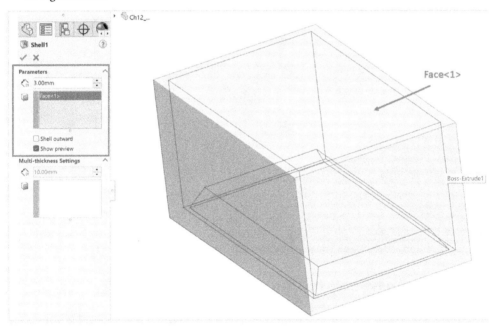

Figure 12.14 – The selected face parameters will be removed after applying the shell

Here is a brief description of some of the options:

- **Faces to Remove**: This is the field below the thickness of the draft. The faces we select here will be removed after we apply the shell. If we leave this field empty, the model will still be shelled from the inside; however, the outer surface will remain unchanged.

- **Shell outward**: This will flip over the shell direction. In other words, the current model will be removed and replaced by a shell starting from its outer surface. You can give it a try to figure out the effect.

- **Multi-thickness Settings**: This allows us to have different wall thicknesses on different sides. We will look at this in more detail in this section.

4.   Click on the green checkmark to apply the shell. Our model will look like *Figure 12.15*:

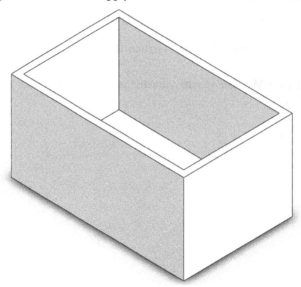

Figure 12.15 – The resulting 3D model after applying the shell feature

This concludes this section on applying a unified shell using the shell feature. Next, we will cover a special condition where we can specify different thicknesses for different sides.

Multi-thickness shell settings

In the shell's PropertyManager, we have the option of making a shell with different thicknesses for different walls. To explore this option, we will go back to our model and modify it. This modification is highlighted by the review clouds in *Figure 12.16*:

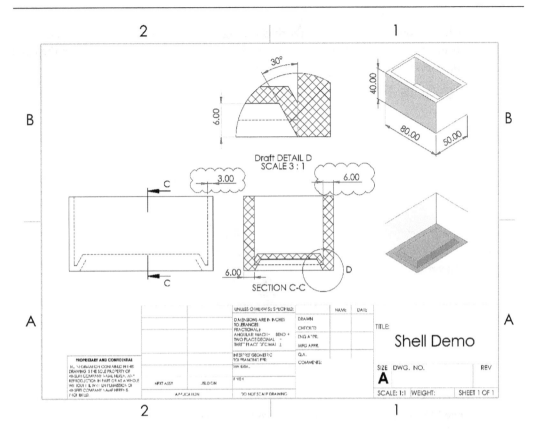

Figure 12.16 – The multi-thickness shell we will apply

Note that the model is still shelled in the same way; however, two of the walls have different thicknesses than the others.

To apply this modification, we can do the following:

1. Right-click on the **Shell** feature from the design tree and select **Edit Feature**.

2. Via the PropertyManager, we will adjust the fields under **Multi-thickness Settings** by doing the following steps. *Figure 12.17* highlights **Multi-thickness Settings**:

   - Under the **Multi-thickness Settings** face selection, select one of the faces we want to make thicker. Then, adjust the thickness from 3 mm to 6 mm.

- Select the other face and adjust the thickness again from 3 mm to 6 mm. Note that we have to adjust the thickness of each side separately. The rest of the faces will keep the default thickness we set at the top of the PropertyManager, which is 3 mm in our case.

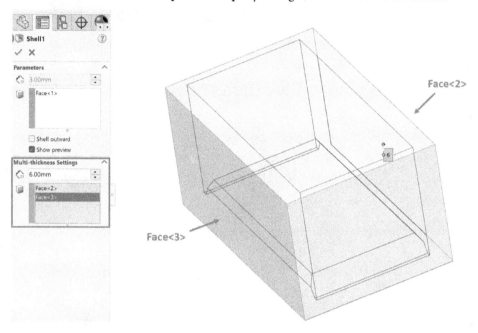

Figure 12.17 – Multi-thickness Settings allows us to apply different thicknesses to the walls

3. Click on the green checkmark to apply these **Multi-thickness Settings**. The result of the model will look like *Figure 12.18*. Note that two of the side walls are thicker than the others:

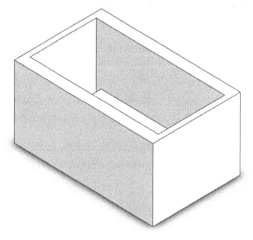

Figure 12.18 – The resulting shell with multiple wall thicknesses

This concludes this exercise on using the shell feature. We covered the shell feature and how to apply it, as well as how to apply both shell thickness and a multi-thickness shell. In the next section, we will cover the Hole Wizard, which will enable us to create different types of holes.

Understanding and utilizing the Hole Wizard

Holes are very common features in most products. If we look at any product, we will likely see screws that hold different parts together. In essence, these are different holes. Usually, these holes are made according to common international standards. The Hole Wizard allows us to create holes as per those standards. In this section, we will explore the Hole Wizard and how to utilize it to create holes.

What is the Hole Wizard and why use it?

The **Hole Wizard** in SOLIDWORKS enables us to create holes in our model that match different international standards for holes. This includes drilling and threading the holes as well. The Hole Wizard makes it easy to make those holes by selecting the hole standard and type and placing the hole directly on the part.

To identify a hole in the SOLIDWORKS Hole Wizard, we must have the following information:

- **Hole type**: This includes the following:

  - **Overall shape**: Nine shapes are supported by the Hole Wizard; these include counterbore holes, countersink holes, tapered tap holes, slots, and others. We have a graphical presentation of each hole shape in the SOLIDWORKS interface.

  - **Standard**: This includes internationally recognized standards for defining holes. The most commonly used standards are the ones from the **International Standard Organization** (**ISO**) and the **American National Standard Institute** (**ANSI**). The Hole Wizard also includes more standards, such as those from the **British Standard Institute** (**BSI**), the **Japanese Industrial Standards** (**JIS**), **Korean Standards** (**KS**), and many others. These standards mainly differ in hole sizes, hole fittings, and referencing.

  - **Type**: The type of the hole depends on the preceding two options. Each combination of the overall shape and standard will have a different set of types to choose from.

- **Hole specifications**: This includes the following:

  - **Size**: This allows us to specify the diameter of the hole, as per the standard and type we pick. Note that each standard references sizing differently in terms of naming.

  - **Fit**: For selected hole shapes, we will be able to specify whether we would like to have a normal, loose, or close fit for our hole.

  - **Custom sizing**: This allows us to further customize the size of the hole from the standard size if needed.

- **End condition**: Similar to the end conditions for the common features we used previously, this allows us to specify the depth of the hole. This will also allow us to specify the depth of the thread if the hole differentiates that.

We have just learned what the Hole Wizard is and what specifications we need to know to identify and call out a hole. Next, we will learn how to put all of that into practice by using the Hole Wizard to create holes.

Utilizing the Hole Wizard

In this section, we will use the Hole Wizard to create multiple holes in a box, as shown in *Figure 12.19*. You can download the basic box from the models for this chapter. Alternatively, you can create it from scratch. Note that each of the holes is identified with all of the information we need to create it. All of the dimensions in the diagram are in millimeters.

The following diagram highlights the two holes that we will be creating:

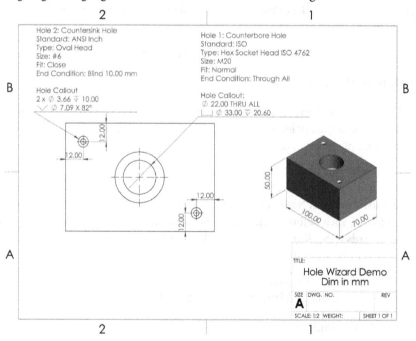

Figure 12.19 – The 3D model we will build in this exercise

This figure shows multiple holes that have been made with different standards. When working with a realistic project, we will mostly use one standard for the whole product. However, as this is an exercise for demonstration and learning purposes, we are using different standards.

To create the model shown in *Figure 12.19*, we will create *Hole 1* and then *Hole 2*. Follow these steps to do so:

1. First, download and open the model linked to this section.

2. Select the **Hole Wizard** command, as shown in *Figure 12.20*:

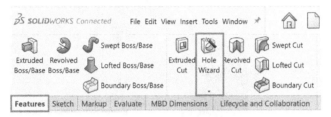

Figure 12.20 – The location of the Hole Wizard command

3. **Making Hole 1**: At this point, we will be making Hole 1, which will be in three stages which are choosing the hole specifications, choosing the hole's position, and confirming the hole's position:

    I.   **Hole specifications**: We can fill out the information shown in *Figure 12.19*. The settings will be as shown in *Figure 12.21*. After deciding on the hole specifications, we can choose the hole's position. The hole's shape is only shown with figures; however, if we hover our cursor over the icons, the names will appear:

Figure 12.21 – The hole specifications can be applied to match the industry standards

II.   **Hole position**: To start working on the positioning of the hole, we can select the **Positions** tab in the PropertyManager. This will prompt us to select which surface we want the hole on. Select the bluish top surface, as highlighted in *Figure 12.22*:

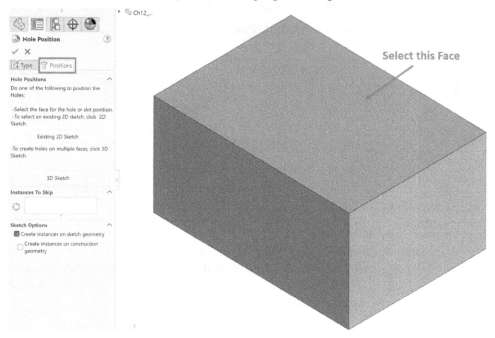

Figure 12.22 – Holes can be positioned on an existing surface

Then, we can place a dot, which will be the center of the hole. We will position the hole in the center of the shape, which coincides with the origin. Hence, we can place the dot so that it coincides with the origin, as highlighted in *Figure 12.23*. Note that we can see a preview of the hole:

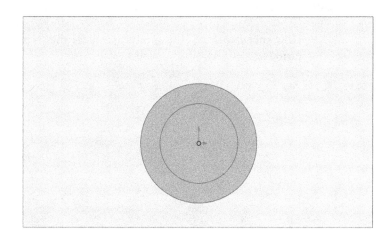

Figure 12.23 – A hole can be positioned using sketching commands

---

Note

If you don't select any surface for the Hole Wizard, a 3D sketch will be started to locate the hole. You can also click on the **3D Sketch** option shown in *Figure 12.22* if you intend to use a 3D sketch for positioning. You can pre-select a surface to use for positioning the hole in the Hole Wizard.

III.   **Confirmation**: If we are happy with the hole preview, we can click on the green checkmark to implement the hole. The hole will be added, as shown in *Figure 12.24*:

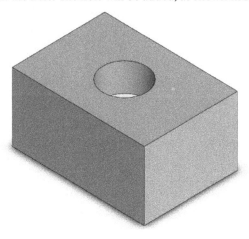

Figure 12.24 – The hole that matches the specified standards will be applied after confirmation

4. **Making Hole 2**: To make the other holes, we will follow the same procedure we did for the first one. However, the specifications, positions, and the number of holes are different:

    I. **Hole specifications**: The specifications for the second hole can be set as shown in *Figure 12.25*:

Figure 12.25 – The specifications of the second hole's types

II. **Hole position**: Now that we have two holes, we can place two dots for each hole. Positioning a hole follows the same sketching commands we used previously. The only exception is that any *point* we place will be interpreted as the center of a hole. Hence, we can position our holes using the **Smart Dimension** command to match what's shown in *Figure 12.26*:

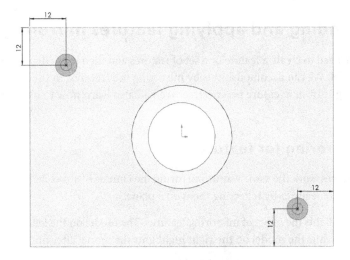

Figure 12.26 – Dimensions can be used to locate a hole

III. **Confirmation**: Click on the green checkmark to apply the two holes. The final shape will look like *Figure 12.27*:

Figure 12.27 – The final 3D model after confirming the holes

This concludes how to utilize the Hole Wizard. In this section, we learned what the Hole Wizard is and how to identify details for the Hole Wizard. Then, we learned how to use the Hole Wizard's functions in SOLIDWORKS to generate a hole based on different industry standards and identifications. Next, we will cover how to mirror any feature we apply in SOLIDWORKS.

Understanding and applying features mirroring

Sometimes, we need to create a feature or a set of features and then try to duplicate them on the other side of the model. We can accomplish this by mirroring the features. In this section, we will discuss what **mirroring** is from a feature perspective. We will also learn how to use this feature to mirror other features.

What is mirroring for features?

Mirroring features work the same way as mirroring the entities of a sketch. It enables us to duplicate a feature or a set of features by reflecting them on a plane.

Figure 12.28 highlights the effect of mirroring features. The model on the left highlights a model with a set of features, while the model on the right highlights the model after mirroring selected features:

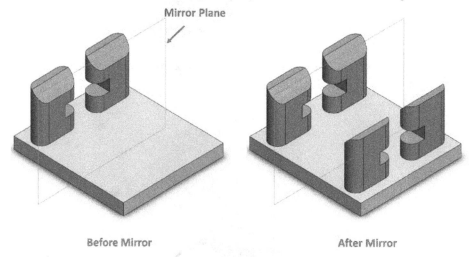

Figure 12.28 – The impact of the mirror feature

Now that we know what is meant by mirroring features, we can start learning how to apply the feature to 3D model creation.

> **Note**
> The mirror plane could be an existing plane or an existing surface. Also, it can be a plane that we create for mirroring purposes.

Utilizing the Mirror command to mirror features

In this section, we will learn how to use the **Mirror** command to mirror features. We will create the model shown in *Figure 12.29*. Note that the pillars in the model are mirrors of each other:

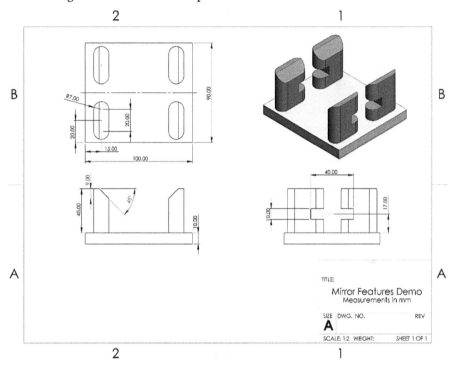

Figure 12.29 – The 3D model we will build in this exercise

To use the **Mirror** command, follow these steps:

1. Download and open the part linked to this section. The model looks like *Figure 12.30*. Alternatively, you can create the model from scratch using the information provided in *Figure 12.29*:

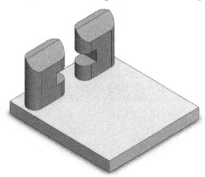

Figure 12.30 – Our starting model, which you can download with this section

2. From the **Features** tab, select the **Mirror** command, as shown in *Figure 12.31*:

Figure 12.31 – The location of the Mirror command

3. The mirror PropertyManager will be shown on the left, as demonstrated in *Figure 12.32*. We can make the following selections:

    -   **Mirror Face/Plane**, in this case, is the same as the default **Right Plane**. Hence, we can select **Right Plane** from the design tree, as shown in *Figure 12.32*:

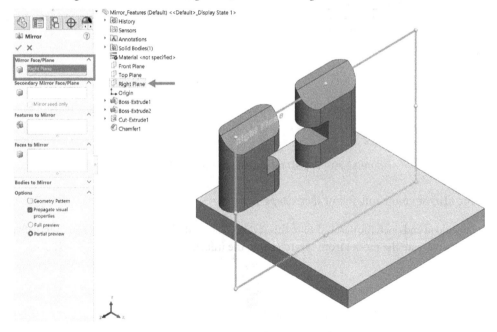

Figure 12.32 – A plane is used to reflect the features

Tip

You can preselect the mirror plane before clicking the **Mirror** command.

- For **Features to Mirror**, we want to mirror a total of three features, which are **Boss-Extrude**, **Cut-Extrude**, and **Chamfer**. We can select all of these features from the design tree, as shown in *Figure 12.33*. Alternatively, we can select the features directly by selecting them from the 3D model shown in the canvas:

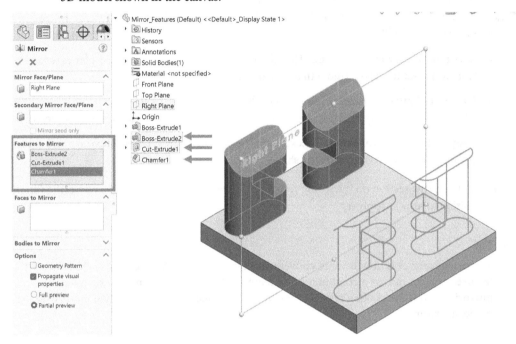

Figure 12.33 – We can select features to mirror from the design tree

4. After confirming the preview, click on the green checkmark to apply the mirror. The final model will look like *Figure 12.34*:

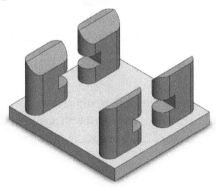

Figure 12.34 – The final 3D model after mirroring the specified features

In this exercise, **Mirror Face/Plane** happened to be the same as the default **Right Plane**. However, it can be any planar surface from the model itself. It can also be a new plane that we generate ourselves using reference geometries. We can also mirror any number of features in one go. Note that, similar to mirroring sketches, any modifications we apply to the original features will be reflected in the mirrored features. Before we conclude our discussion of the **Mirror** command, let's explain three notable options shown in *Figure 12.33*:

- **Secondary Mirror Face/Plane**: This allows us to select another mirror plane, resulting in mirroring features across more than one plane at once.

- **Geometry Pattern**: Checking this option will mirror a geometrical replica of the original shape while disregarding any logical arrangements that were used, such as feature end conditions, if any.

- **Propagate visual properties**: Checking this option will copy the visual textures and colors of the original shape to the mirrored shape.

This concludes this section on mirroring features. We learned what the features mirroring function is, in addition to how to apply it.

> **Tip**
>
> Mirroring features is mostly more convenient than mirroring sketches. This is because, with features, we can mirror a combination of features that inherently include sketches. Hence, mirroring end features often results in less modeling time and less time when modifying the model afterward.

Next, we will cover another feature, known as the rib feature.

Understanding and applying the rib feature

Ribs are reinforcement structures that are used to help fix two sides together. In this section, we will learn what ribs are and how to create them using the SOLIDWORKS rib feature. We will also learn how SOLIDWORKS interprets the creation of ribs using the rib feature.

Understanding ribs

Ribs are reinforcements that enhance the support, stiffness, and overall integrity of a given structure. They can be welded into metal frameworks, embedded with plastic toys, featured in architectural designs, and many other applications. *Figure 12.35* shows two models, one without ribs and one with ribs:

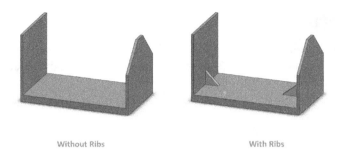

Without Ribs   With Ribs

Figure 12.35 – 3D models with and without ribs

Note that we can create ribs out of other features, such as extruded boss and extruded cuts. However, the rib feature provides us with an easier method to both build and define a rib. Now that we know what ribs are and what the rib feature does, we will start learning how to apply it to a SOLIDWORKS model.

Applying the Rib command

In this section, we will learn how to use the **Rib** command to generate ribs in our models. We will create the model shown in *Figure 12.36*. The base model we will use can be downloaded from this book's GitHub repository. Note that the ribs are highlighted in detail views *A* and *B*, which we will generate in this exercise:

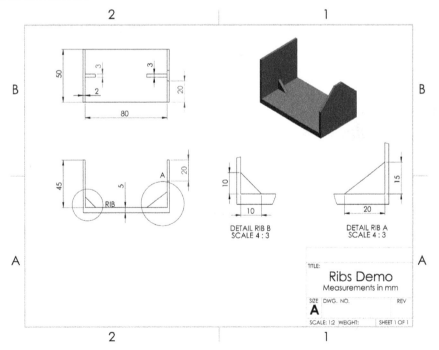

Figure 12.36 – The 3D model we will build in this exercise

To use the rib feature for this exercise, we can follow these steps:

1. First, download and open the SOLIDWORKS part linked to this section concerning ribs. The model will look like *Figure 12.37*. Alternatively, you can create the model from scratch using the information provided in *Figure 12.36*.

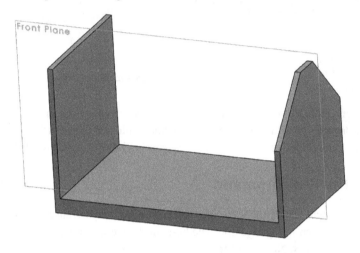

Figure 12.37 – You can download the following model with this section

2. Select **Front Plane** from the design tree. The front plane is shown in *Figure 12.37*.

3. In the **Features** tab, select the **Rib** command, as highlighted in *Figure 12.38*:

Figure 12.38 – The location of the Rib command

4. Sketch a line that outlines the outer boundary of rib A, as shown in *Figure 12.39*. This sketch matches the *DETAIL RIB A* view on the drawing. Note that *Steps 2 to 4* are interchangeable; we can choose the feature first, then select the plane and sketch. We can also sketch, and then select the **Rib Feature** command:

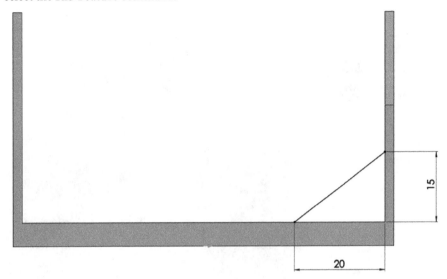

Figure 12.39 – The rib can be built with a single line sketch

5. Exit the sketch mode.

6. After we exit the sketch mode, we will notice the **Rib** command's PropertyManager window on the left, as shown in *Figure 12.40*. For rib A, the parameters and the preview can be set as shown in *Figure 12.40*. These parameters will allow us to define our rib in terms of the following:

    *   **Thickness**: The width of the rib.

    *   **Extrusion direction**: The direction of the rib concerning the guiding sketch. Here, we can choose to have all of the width extending toward either direction or to the midway point.

- **Draft**: We can also specify whether we require a draft with the rib and what the draft angle is.

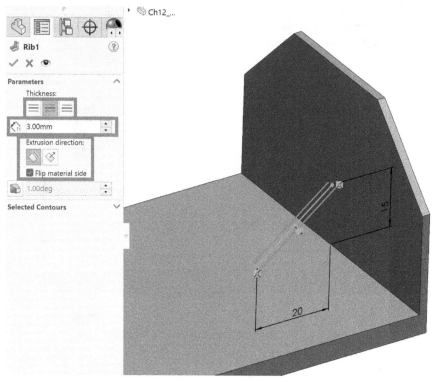

Figure 12.40 – The Rib command's PropertyManager window and its settings

7. Click the green checkmark to apply the rib. The model will look like *Figure 12.41*:

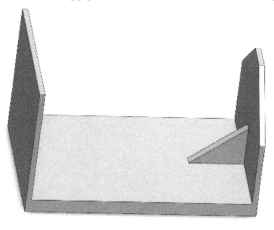

Figure 12.41 – The resulting rib after its application

8. Follow *Steps 2* to *6* again to create rib B. Make sure that you use the dimensions shown in the *DETAIL RIB B* diagram highlighted in *Figure 12.36*. The final model will look like *Figure 12.42*:

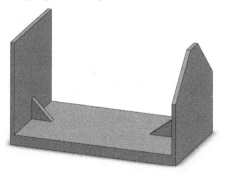

Figure 12.42 – The final 3D model after applying both ribs

We likely have identical rib dimensions in products that we interact with within our day-to-day lives. However, we created ribs with different dimensions to practice how to use the tools.

Drafted rib

One of the key options we can utilize with ribs is adding a draft to them. This option can be used by enabling the **Draft outward** option shown in the feature's PropertyManager window, as highlighted in *Figure 12.43*:

Figure 12.43 – The rib feature allows us to make a drafted rib

A drafted rib will look the same as the drafts we covered in the *Understanding and applying the draft feature* section.

This concludes how to use the **Rib** command. We covered what the rib feature is and how to apply it. All of the features we covered earlier in this chapter can be used to construct models directly. Next, we will cover multi-body parts, which is a method for creating parts rather than a feature.

Understanding and utilizing multi-body parts

In all of the applications we have explored in this book, each part file we made consisted of one body. We used assembly files to combine the different parts. In this section, we will explore a different approach with multi-body parts. We will cover what multi-body parts are, how they are created, and what the advantages of multi-body parts are.

Defining multi-body parts and their advantages

Multi-body parts are models made within a SOLIDWORKS part file that contains more than one separate body. Hence, they are called multi-body parts. *Figure 12.44* shows the contents of one SOLIDWORKS part file. However, the diagram on the left consists of one solid body, while the one on the right consists of two solid bodies. Note that the difference between these two diagrams is that the right-hand one has an extrusion cut that separates the large triangle (one solid body) into two triangles (two solid bodies):

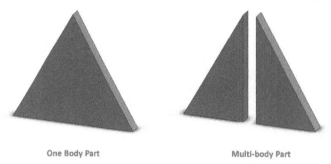

One Body Part   Multi-body Part

Figure 12.44 – A single-body part and a multi-body part

However, we should not confuse multi-body parts with assemblies. The different bodies in multi-body parts are not dynamic, as is the case with the different parts in an assembly. This makes multi-body parts appropriate for certain applications that involve static interactions, such as frames. This is due to the following advantages:

- The frame and other static elements will all be contained in one file, making them easier to access and modify.

- The work process is faster as we won't need to use more than one SOLIDWORKS file. Also, we won't need to create mates to ensure the different parts fit together, as is the case with assemblies.

However, assemblies also have other advantages over multi-body parts. These include the following:

- We can showcase the moving dynamics within a product.

- There's more flexibility in reusing parts in different assemblies. Also, we have more flexibility in exchanging an independent part.

- Creating separate drawings for each component is more convenient as the parts are set up in different files.

- Having parts in different files makes it easier to have separate part names and numbers for archiving or inventory references.

There is no right or wrong answer to what approach to choose between assemblies and multi-body parts. As designers and practitioners, we will have to make the choice, weighing up the advantages of both approaches. To do that, we have to be familiar with both approaches. Next, we will create a multi-body frame to put what we just learned into practice.

Generating and dealing with a multi-body part

In this section, we will learn how to generate a part with multiple bodies. To demonstrate this, we will create the frame shown in *Figure 12.45*:

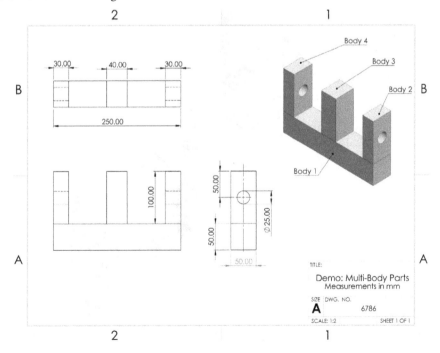

Figure 12.45 – The 3D model we will build in this exercise

Note that each element of the frame is indicated with a different number in the drawing, and the frame consists of four different bodies. To model this frame, we will follow these steps:

1. Create *body 1* as per the dimensions shown in the preceding diagram using the extruded boss feature. It should look like *Figure 12.46*:

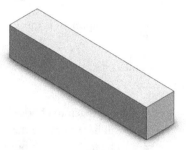

Figure 12.46 – A one-body part of body 1

2. Create the three additional bodies with one more extruded boss feature. To do this, we will follow steps similar to those that we have followed previously. However, before applying the extruded boss feature, we will uncheck the **Merge result** box, as highlighted in *Figure 12.47*.

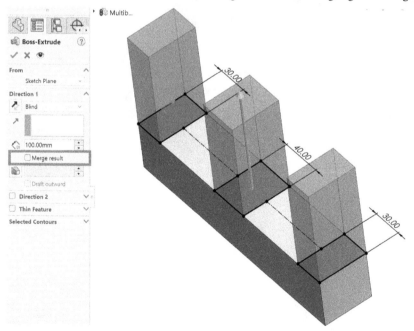

Figure 12.47 – Unchecking the Merge result option will build the extrusions as separate bodies

3. Click on the green checkmark to apply the extrusion. After doing that, we will see two differences compared to our usual extrusion. They are as follows:

- We will notice a line separating the different frames, as indicated in *Figure 12.48*. These lines indicate a separation between the different bodies on our part:

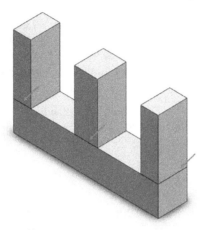

Figure 12.48 – The lines indicating separate bodies

- In the design tree shown in *Figure 12.49*, we will notice a tab with the title **Solid Bodies**, as shown in the highlighted design tree diagram. If we expand this, we will see a list of all of the different bodies we have on our part; clicking on any of them will highlight the body in the canvas. From this list, we can selectively hide, delete, or change the display or assign materials to a specific body by right-clicking on the listing and choosing from the different options.

Figure 12.49 – The design tree will show different bodies

This concludes one of the common ways to generate a multi-body part. We can also intentionally create separate bodies in the canvas, which will automatically result in multi-body parts. Also, whenever we apply features such as an extruded cut, which would result in physically separating bodies, we will have a multi-body part.

> **Note**
>
> One important aspect to note is that SOLIDWORKS, by default, will tend to merge bodies as that is a more common practice. Hence, any feature we apply that physically connects separate bodies will merge them unless we specify otherwise by unchecking the **Merge result** option.

Two important and useful elements concerning multi-body parts are the **feature scope** and being able to save bodies in different SOLIDWORKS part files. We will cover these two aspects next and apply them to our model.

Feature scope applications

The **feature scope** refers to the extent to which a feature is applied. For example, in a multi-body part, we can apply a feature such as an extruded cut and specify which body can be included in the cut and which body should not be included. In our exercise, notice in the drawing provided that there is a hole that goes through bodies 2 and 4 and skips body 3.

To utilize the feature scope, we can follow the same steps that we followed when applying a normal extruded cut. However, we will see the options under the **Feature Scope** tile in our cut extrude PropertyManager. We can see some options highlighted in the following screenshot with both the sketch and the other options for the extruded cut feature. Under the **Feature Scope** options, we can select the **Selected bodies** option and uncheck **Auto-select**.

When using the feature scope, follow the steps for applying a normal extruded cut. In the cut extrude PropertyManager, you'll find options under the **Feature Scope** tile. Select **Selected bodies** and uncheck **Auto-select**, as highlighted in *Figure 12.50*.

Then, we can manually select bodies 2 and 4, as highlighted in *Figure 12.50*:

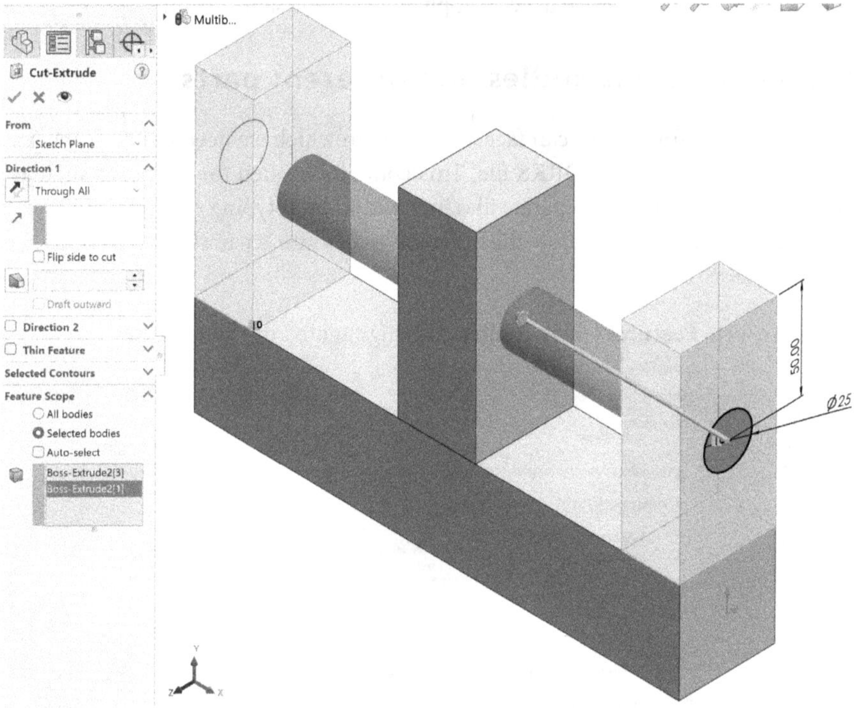

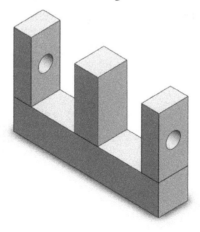

Figure 12.50 – Feature Scope allows us to select which body a feature will affect

As usual, click on the green checkmark to apply the extruded cut feature. Note that the resultant hole is only applied to the selected bodies, as shown in *Figure 12.51*:

Figure 12.51 – The final multibody 3D model

Being able to scope features enables us to apply our design intent faster and more efficiently as it reduces the number of features we need to apply to reach the same result.

Separating different bodies into different parts

Now that we have the frames, we can face a situation in which we need to have each frame element or body in a separate SOLIDWORKS file. This could be needed for purposes such as generating separate drawings, inputting the separate files into a rapid prototyping machine, or other applications. SOLIDWORKS enables us to separate the different bodies into separate SOLIDWORKS part files. To do this, follow these steps:

1.   Go to **Insert | Features | Save Bodies…**, as highlighted in *Figure 12.52*:

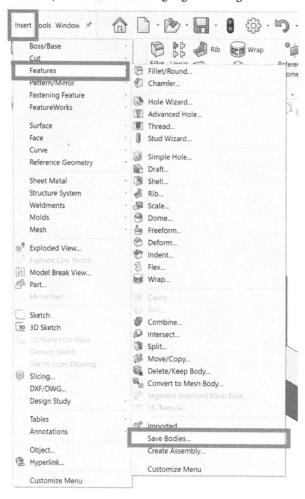

Figure 12.52 – Different bodies can be saved in different SOLIDWORKS part files

2. This will prompt you to select which bodies you want to save separately. You can select the bodies directly from the canvas or the list by checking the box under the save icon. Once you approve the **Save Bodies** command, a new file will be generated for each body. By default, the new files will be located in the same folder as the original file. These command options are shown in *Figure 12.53*:

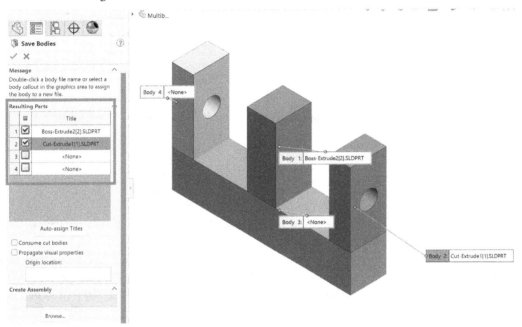

Figure 12.53 – The Save Bodies command allows us to pick which specific bodies we want to save

After applying the **Save Bodies** command, we will notice that this command is listed in the design tree, as highlighted in the following screenshot. The separate files will only reflect the shape from before that feature's listing. Hence, applying more features after using the **Save Bodies** command will not update the already saved bodies. We can drag the features we want to be reflected above the **Save Bodies** command in the design tree, as highlighted in *Figure 12.54*:

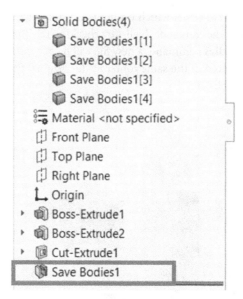

Figure 12.54 – The Save Bodies command will appear in the design tree after its application

This concludes our coverage of multi-body parts. In this section, we learned what multi-body parts are, how they differ from assemblies and their advantages, how to create a multi-body part, how to use the features scope function, and how to save different bodies into separate SOLIDWORKS part files.

> **Note**
>
> In this section, we have explored building multi-body parts by not merging bodies together. However, we can also split an existing solid body into multiple ones using the **Split** command. You can find the **Split** command by going to **Insert** in the top toolbar, then **Molds**, and then **Split…**.

Knowing how to utilize multi-body parts to our advantage will enable us to optimize the software when targeting different applications, such as a static furniture design or a beam structure design.

Understanding and applying linear, circular, and fill feature patterns

Feature patterns allow us to duplicate features quickly according to a certain pattern. In this section, we will learn about the linear, circular, and fill patterns. In addition, we will learn how to apply them using the available SOLIDWORKS tools.

Understanding feature patterns

Feature patterns enable us to swiftly replicate features following a predefined arrangement. They are similar to sketching patterns, which we covered in *Chapter 4*. However, with feature patterns, we can build patterns of features and bodies rather than patterns of sketch entities. In this section, we will cover three types of patterns, as follows:

- **Linear pattern**: This duplicates features or bodies in a linear fashion.

- **Circular pattern**: This duplicates features or bodies in a circular fashion.

- **Fill pattern**: This duplicates features within a set boundary by following a specific pattern. Possible patterns include perforation, circular, square, and polygon.

The following table in *Figure 12.55* highlights the difference between the three types of patterns:

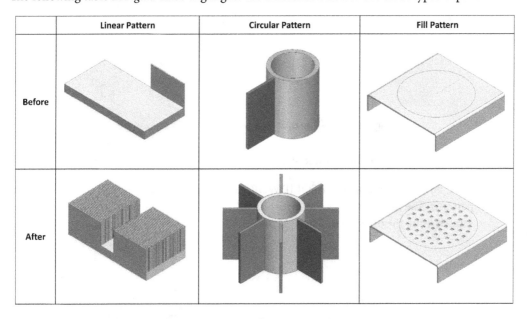

Figure 12.55 – The different types of patterns

Now that we know what to expect of each type of pattern, we can start applying them in the software. Let's get started with linear patterns.

Applying a linear pattern

In this section, we will learn how to use the **Linear pattern** command to create pattern features. We will create the heat sink shown in *Figure 12.56*. The base model we will use can be downloaded from this chapter's GitHub repository. In the following diagram, note the repeated fins, which we will generate using linear patterns.

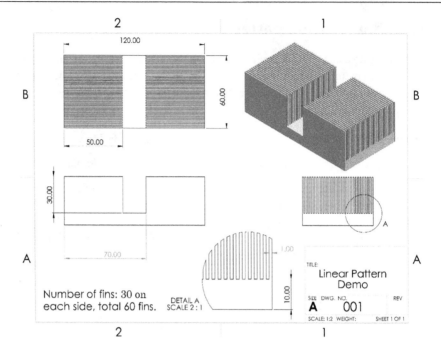

Figure 12.56 – The heat sink we will build in this exercise

To create the linear pattern, follow these steps:

1. First, download and open the SOLIDWORKS part linked to this section concerning feature patterns. The model is shown in *Figure 12.57*. Alternatively, you can create the model from scratch using the drawing in *Figure 12.56*.

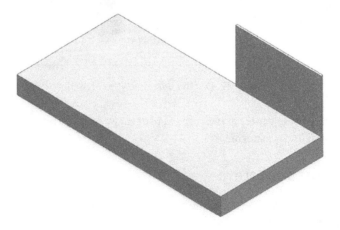

Figure 12.57 – The 3D model that's available for this exercise

2.   Select the **Linear Pattern** command, as shown in *Figure 12.58*:

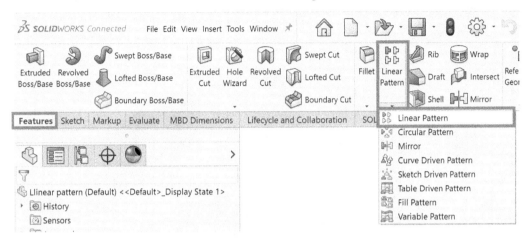

Figure 12.58 – The location of the Linear Pattern command

3.   Check the **Features and Faces** box and select the fin, as shown in *Figure 12.59*. You can use the canvas design tree to select features as well.

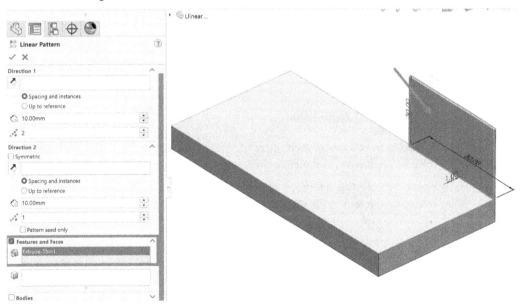

Figure 12.59 – The Features and Faces section of the linear pattern

4. For **Direction 1**, select the smaller side edge, as highlighted in the following screenshot. A preview will then show the projected pattern as highlighted in *Figure 12.60*:

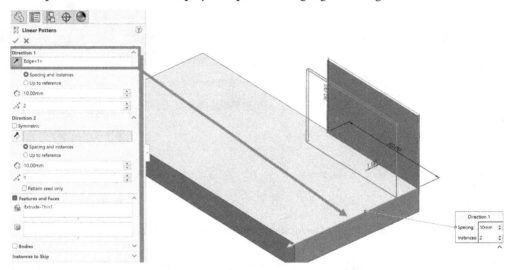

Figure 12.60 – The selection for Direction 1 using an edge

> **Hint**
>
> To indicate the direction of the pattern, we can also select lines from sketches, planes, planar surfaces, axes, and temporary axes.

5. For the end condition, do the following:

   I.   Select **Up to reference**, then select the end surface of the heat sink base for the first reference geometry, as shown in the following screenshot.

   II. Select the **Selected reference** option, then select the indicated face of the fin for the second seed reference, as shown in the following screenshot.

III.   Define the pattern by the number of instances and input 3 0 for that, as shown in *Figure 12.61*:

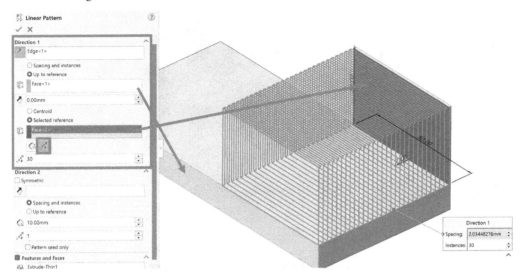

Figure 12.61 – Defining Direction 1 for the linear pattern

With that, we have defined the pattern for one side of the heat sink, as you will see from the preview on your screen. To do the other side of the heat sink, we can work more on the pattern feature under **Direction 2**. So, let's start defining the other side of the heat sink in the same way.

6.   Repeat *Steps 4* to *5* for **Direction 2**, as follows:

I.   Select the long edge of the base for the direction, as shown in the following screenshot.

II.   Select **Up to reference**, then select the side end surface of the heat sink base, as shown in the following screenshot.

III.   Select the **Selected reference** option, then select the indicated thin side face of the fin for the seed reference, as shown in the following screenshot.

IV.   Define the pattern by the number of instances and input 2 for that, as shown in *Figure 12.62*:

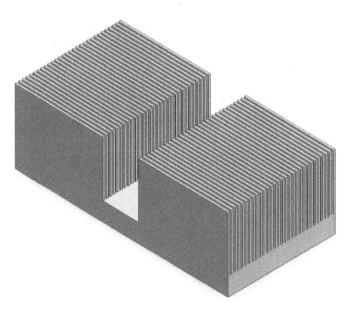

Figure 12.62 – Defining Direction 2 for the linear pattern

7.   Click on the green checkmark to apply the pattern. The resulting 3D model will look like *Figure 12.63*:

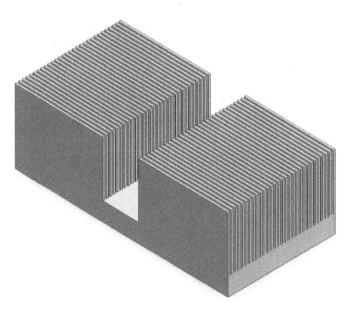

Figure 12.63 – The final 3D model after applying the linear pattern

With that, we have applied the linear pattern. Now, let's define two more important options we did not use in this exercise – **Bodies** and **Instances to Skip**. We can find these options in the **Linear Pattern** PropertyManager window, as shown in *Figure 12.64*:

Figure 12.64 – The different options available for a linear pattern

Let's look at these options in more detail:

- **Bodies**: This allows us to build a pattern of bodies instead of features. This option can be used when we're working with multi-body parts.

- **Instances to Skip**: This allows us to exclude some of the duplicates that are generated by the pattern manually. This works similarly to the sketch patterns we covered in *Chapter 4*.

With that, we can conclude our discussion on linear patterns. Next, we will take a closer look at circular patterns.

Applying a circular pattern

In this section, we will learn how to use the **Circular Pattern** command to pattern features. Note that setting up a circular pattern follows a similar procedure to setting up a linear pattern. We will create a simple water wheel design, as shown in *Figure 12.65*. Note the repeated blades that we will generate using the circular pattern:

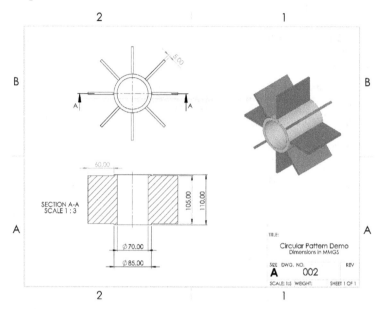

Figure 12.65 – The water wheel we will create in this exercise

To complete the exercise, follow these steps:

1. First, download and open the SOLIDWORKS part linked to this section regarding feature patterns. The model is shown in *Figure 12.66*. Alternatively, you can create the model from scratch using *Figure 12.65*.

Figure 12.66 – The 3D model for this exercise

2. Select the **Circular Pattern** command, as shown in *Figure 12.67*:

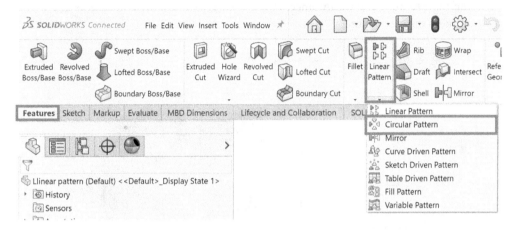

Figure 12.67 – The location of the Circular Pattern command

3. Check the **Features and Faces** box and select the blade, as shown in *Figure 12.68*:

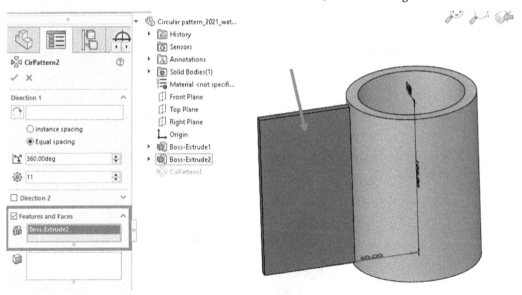

Figure 12.68 – Features section for the linear pattern

> **Hint**
> You can use the canvas design tree to select features as well.

4. For **Direction 1**, select the circular face, as shown here. Also, set the pattern to **Equal spacing**, the angle to 360, and the instances to 8. Note that the number of instances in *Figure 12.69* includes the base feature:

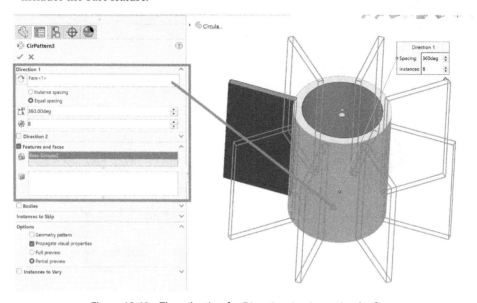

Figure 12.69 – The selection for Direction 1 using a circular face

> **Tip**
> For the direction, we can use faces, circular edges, axes, and temporary axes.

5. Click on the green checkmark to apply the pattern. The resulting 3D model will look like *Figure 12.70*:

Figure 12.70 – The final 3D model after applying the linear pattern

With that, we can conclude our application of the circular pattern. Other options, such as **Bodies** and **Instances to Skip**, have the same functionality as their counterparts for linear patterns. Next, we will discuss the fill pattern feature.

Applying a fill pattern

In this section, we will learn how to use the **Fill Pattern** command to pattern features. We will create the simple ventilation grilles shown in *Figure 12.71*. Note the repeated square holes on the top surface. We will use the fill pattern feature to generate those. Fill patterns are commonly used for grills that are used in sound systems, in ventilation for electronics, and for weight reduction purposes:

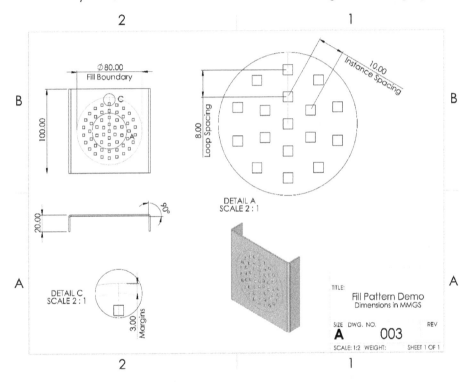

Figure 12.71 – The 3D model we will create in this exercise

Before applying the fill pattern, we need to define a boundary and a direction to be applied. Thus, we will be defining these before applying the pattern.

To complete this exercise, follow these steps:

1. First, download and open the SOLIDWORKS part linked to this section regarding feature patterns. The model is shown in *Figure 12.72*. Alternatively, you can create the model from scratch by using *Figure 12.71*.

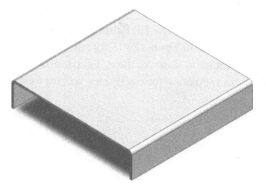

Figure 12.72 – The base model for this exercise

2. Sketch the 80 mm circular boundary on the top surface using a solid line. Then, sketch a construction line for the direction, as shown in *Figure 12.73*:

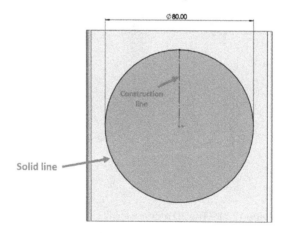

Figure 12.73 – The fill pattern requires a boundary and direction to apply

> **Tip**
>
> The boundary and direction do not have to be sketched separately. The existing linear surface can be used as a boundary, and existing edges can be used for direction. However, making new sketches allows us to build custom boundaries.

3. Select the **Fill Pattern** command, as shown in *Figure 12.74*:

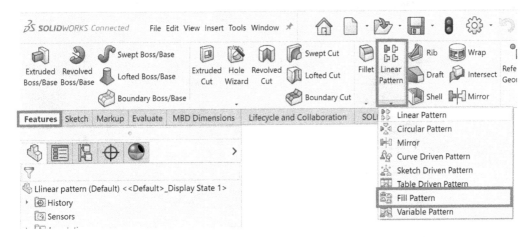

Figure 12.74 – The location of the Fill Pattern command

4. Under **Fill Boundary**, select the circular sketch, as shown in *Figure 12.75*:

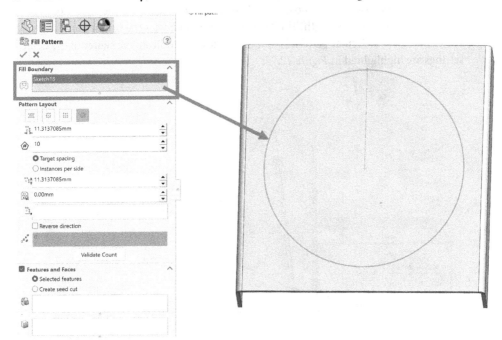

Figure 12.75 – A fill boundary is required for the fill pattern

5. For **Pattern Layout**, select the **Circular** pattern and select the construction line for the direction, as shown in *Figure 12.76*:

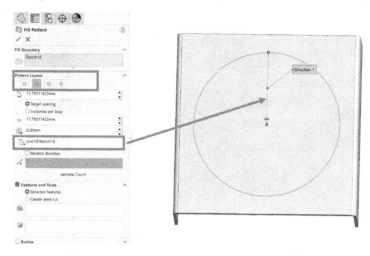

Figure 12.76 – Selecting the layout and direction

6. Check the **Features and Faces** box, select **Create seed cut**, and pick the square option. We can specify 3 mm as the side length. The **Vertex** selection indicates the origin point of the pattern. For example, for a circular pattern layout, the vertex will indicate the center of the circle. These settings are highlighted in *Figure 12.77*:

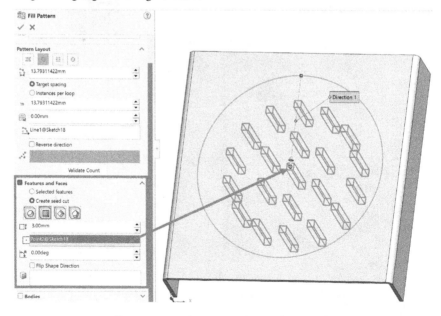

Figure 12.77 – The settings for the feature cut

> **Note**
>
> The **Create seed cut** option allows us to create common cuts that are associated with fill patterns. However, we can create fill patterns for any other shape as well.

7. Now, we can fine-tune the pattern. Make sure that the **Target spacing** option is selected. Then, set **Loop Spacing** to 8 mm, **Instance Spacing** to 10 mm, and **Margins** to 3 mm. **Margins** refer to the clear space between the boundary and the patterned instance. The onscreen preview will change as we adjust those parameters. Also, note that the **Instance Count** value will change according to the number of repeated instances. *Figure 12.78* highlights these settings:

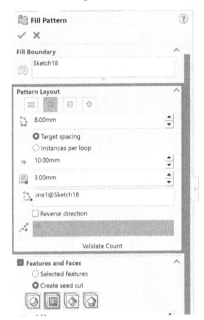 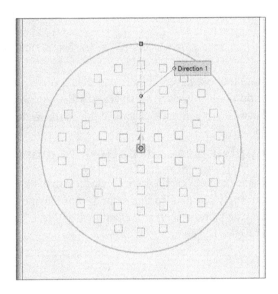

Figure 12.78 – The pattern layout instance's settings

> **Tip**
>
> It is common to keep adjusting the instance settings in a trial-and-error fashion until we get the desired result.

8. Click on the green checkmark to apply the fill pattern. The resulting 3D model will look like *Figure 12.79*:

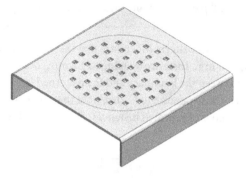

Figure 12.79 – The final 3D model after applying the fill pattern

With that, we have implemented the fill pattern feature. In this exercise, we applied the fill pattern within one boundary. However, it is also possible to apply a fill pattern that covers more than one boundary at a time. We'll look at this in the next section.

Applying the fill pattern to more than one boundary

In the previous exercise, we applied a simple fill pattern to one boundary area. However, the feature can apply one pattern that extends more than one boundary. This allows us to create harmonious-looking patterns with an elegant look and feel. To highlight this, let's look at the pattern highlighted in *Figure 12.80*:

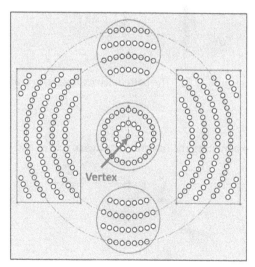

Figure 12.80 – Mutiboundary circular fill pattern with a central vertex

This fill pattern follows a circular pattern originating from the rectangular piece's center, as indicated by the *Vertex*. Here, we can see the formation of the circular fill pattern with the selected vertex as its center.

> **Note**
> The different boundaries can be in one sketch, or they can be in more than one sketch.

The vertex can also be located outside the areas of the boundaries as we see fit for our design. For example, the circular fill boundary shown in *Figure 12.81* has the same specifications as the preceding one, with the vertex located differently. Here, the overall circular alignment of the pattern is preserved. However, the center of the base circle is located differently:

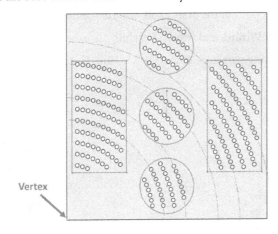

Figure 12.81 – Mutiboundary circular fill pattern with a corner vertex

With that, we have finished looking at the major types of feature patterns. We covered the linear, circular, and fill patterns. The linear and circular feature patterns are very similar to the patterns available for sketching, while the fill pattern allows us to apply patterns within selected boundaries easily. All these feature boundaries allow us to build patterns for specific features or whole bodies.

Summary

In this chapter, we learned about a variety of relatively advanced features for building more complex models. We covered the draft, shell, and rib features for creating specific geometries faster. We also learned about using the Hole Wizard to create industry-standard holes and covered how to mirror features to save us time that would otherwise be spent remaking features. We also learned about multibody parts, their advantages, and how to utilize them. At the end of this chapter, we learned how to apply linear, circular, and fill patterns for features and bodies, which allow us to duplicate features or bodies in specific formations.

Knowing about the topics that were covered in this chapter is what separates professional users of SOLIDWORKS from amateurs. Mastering this chapter's topics will help you save time and create complex shapes faster while capturing more specific design intents.

In the next chapter, we will cover equations, configurations, and design tables. These skills will allow us to create more connected models and allow us to have multiple variations of a part within one part file.

Questions

Answer the following questions to test your knowledge of this chapter. The following questions will reinforce the main topics we learned in this chapter. However, it is also good practice to pick random objects and model them in SOLIDWORKS to improve your skills:

1. Describe the functions of the draft, shell, and rib features.

2. What is the Hole Wizard, and why is it useful?

3. Create the following part in SOLIDWORKS. Hint: Use the draft and shell features:

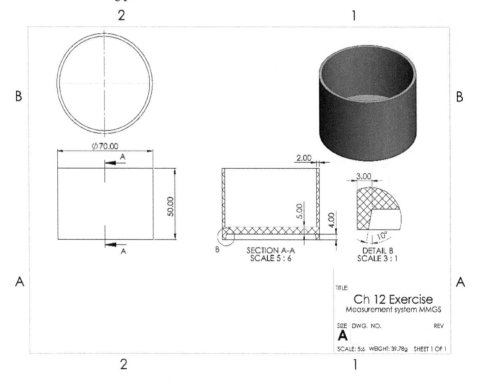

Figure 12.82 – The drawing for Question 3

4. Create the following part in SOLIDWORKS. Hint: Use the rib feature and the Hole Wizard:

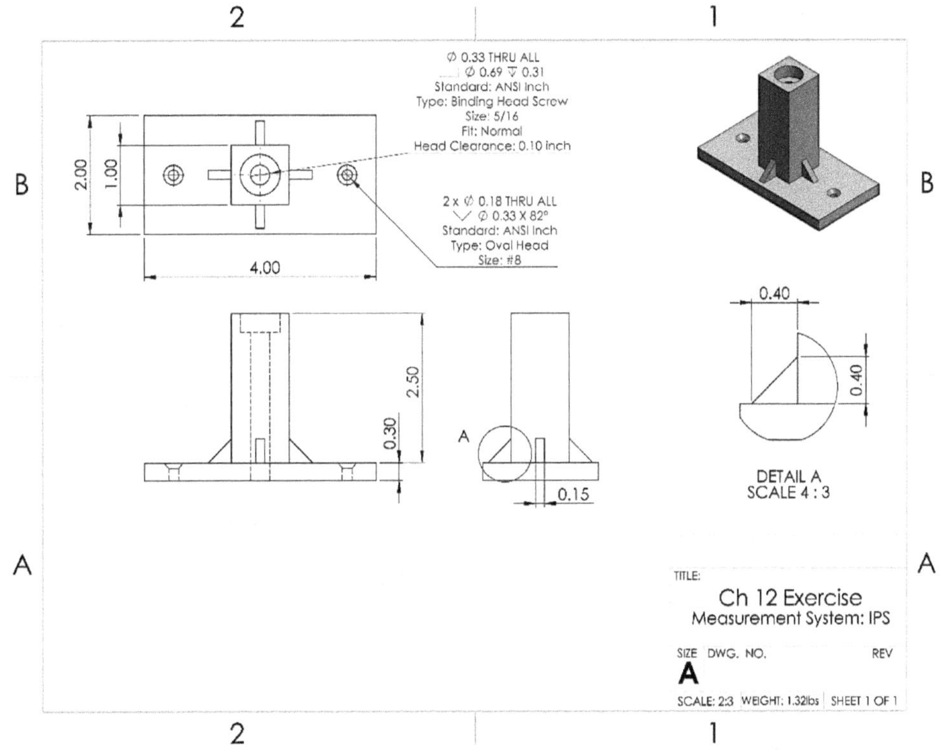

Figure 12.83 – The drawing for Question 4

5. Create the following part in SOLIDWORKS. Hint: Use the rib and draft features and the Hole Wizard. You may also use reference geometries and the swept boss. Due to the amount of information that this part contains, the drawing has been split into two diagrams. Both diagrams are for this one question:

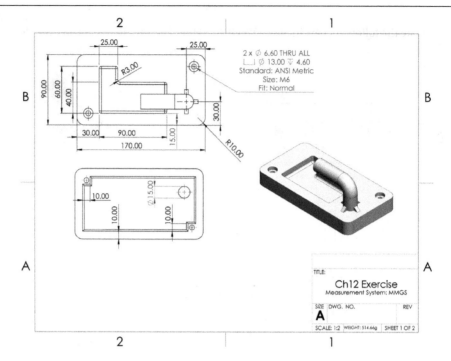

Figure 12.84 – The first drawing for Question 5

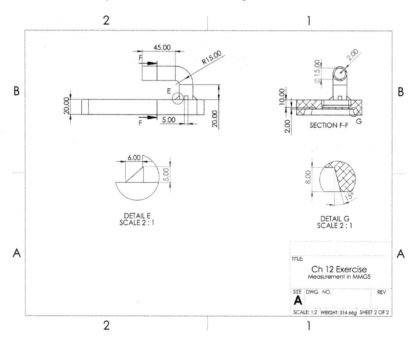

Figure 12.85 – The second drawing for Question 5

6. What are multi-body parts, and what are some advantages of using them?

7. Create the following frame in a multi-body format. Use the annotations in the drawing for the different bodies. Hint: Use features scope to extrude cut the slot shown on bodies 2 and 4:

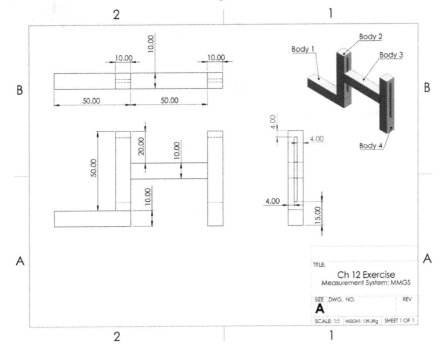

Figure 12.86 – The drawing for Question 7

8. Which of the following models represents a suitable application of a circular pattern?

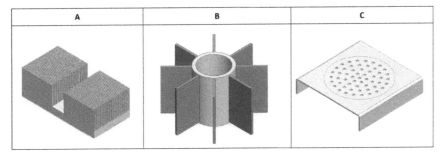

Figure 12.87 – multiple patterns for Question 8

Important note

The answers to the preceding questions can be found at the end of this book.

Get This Book's PDF Version and Exclusive Extras

UNLOCK NOW

Scan the QR code (or go to packtpub.com/unlock). Search for this book by name, confirm the edition, and then follow the steps on the page.

Note: Keep your invoice handy. Purchases made directly from Packt don't require an invoice.

13

Equations, Configurations, and Design Tables

In this chapter, we will cover how to use equations, configurations, and design tables. These functionalities are not used to directly add or remove materials or features, such as extruded bosses, ribs, and sweeps. Rather, we will cover how to link different lengths with equations and how to generate multiple versions of a part using configurations and design tables. Mastering these functionalities will enable us to generate more interlinked models that are more robust and easier to modify. Also, it will enable us to generate multiple model versions for testing and evaluation.

The following topics will be covered in this chapter:

- Understanding and applying equations in parts
- Understanding and utilizing configurations
- Understanding and utilizing design tables

By the end of this chapter, you will be able to link different dimensions with equations and create multiple variations of a part within one SOLIDWORKS file. This will enable you to both optimize and accelerate your design process. In this chapter, we will focus on applications within part files. However, similar functions are also available within assemblies.

Technical requirements

This chapter will require access to SOLIDWORKS and Microsoft Excel software on the same computer.

The CiA video for this chapter can be found at `https://packt.link/AYKYD`

Understanding and applying equations in parts

When creating 3D models, we often use a variety of dimensions to define sketch entities, such as squares and arcs. We also use dimensions to define features such as an extruded boss, an extruded cut, a revolved boss, and a revolved cut. In many applications, these dimensions are not isolated from each other. Rather, they are connected with mathematical relations. For example, it could be that the length of a rectangle should be 75% of its width or the height of a cylinder should be double the diameter.

In this part, we will learn how to set up these relations with equations. First, we will explain what equations are, and then we will apply equations in the modeling of a SOLIDWORKS part.

Understanding equations

Equations within parts allow us to both define and link different dimensions together within the parts. By defining variables and equations, we will be able to build a more interconnected 3D model. This will give us certain advantages, such as the following:

- It will be easier to access and adjust specific defined dimensions. This is because we will be able to access those dimensions from the **Equations** panel rather than by looking at them on the design tree.

- It will be easier to modify connected dimensions, as equations will enable us to modify one dimension and have all other linked dimensions updated accordingly. This is in contrast to modifying each dimension separately.

To theoretically demonstrate equations in the SOLIDWORKS part context, we will look at the rectangular cuboid scenario highlighted in *Figure 13.1*. Let's assume we were required to model a rectangular cuboid with the following specifications:

- A width equal to 5 millimeters

- A depth that is double the width (which equals 10 millimeters)

- A height that is 3 millimeters longer than the width (which equals 8 millimeters)

According to the given specifications, we can define the rectangular cuboid dimensions in two different ways. The first is what we did previously using **numerical values**, and the other is using **equations**. The following diagram shows both methods; the rectangular cuboid on the left highlights directly inputting numerical values, while the one on the right highlights inputting equations:

Numerical Values   Equations

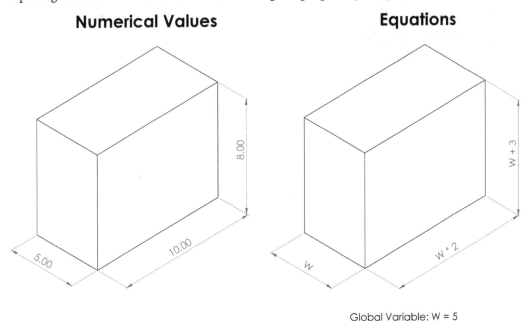

Global Variable: W = 5

Figure 13.1 – A rectangular cuboid built using two methods

The initial result of both methods gives the exact same cuboid. However, the way the dimensions are defined is different. Let's assume that, after creating this cuboid, we were required to adjust the width from 5 mm to 8 mm. With equations, all we will need to do is to change the W global variable from 5 to 8, and then all of the other dimensions for depth and height will change accordingly. However, if we input numerical values, we will have to change all three dimensions individually. This is why mastering equations is key when building more connected models.

Now that we know what equations are in a part modeling context, their advantages, and how they work, we can start learning how to use equations to enhance our 3D modeling process.

Applying equations in parts

To demonstrate how to use equations when modeling parts, we will create the following simple rectangular cuboid with the variables shown in the drawing in *Figure 13.2*:

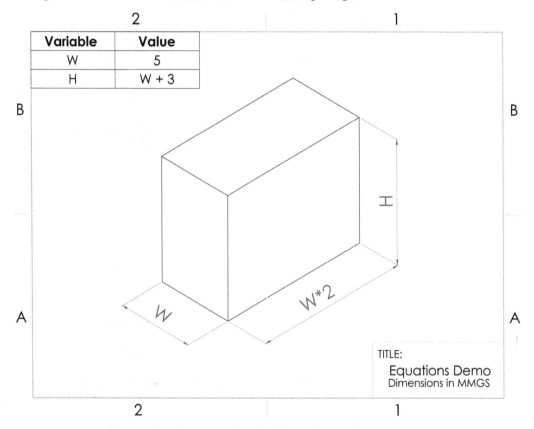

Variable	Value
W	5
H	W + 3

Figure 13.2 – The rectangular cuboid we will create in this exercise

To start, we will define our global variables. To do this, follow these steps:

1. Open the Equations Manager by right-clicking on the part name on the top of the design tree, and then choose **Hidden Tree Items | Equations | Manage Equations...**, as highlighted in *Figure 13.3*:

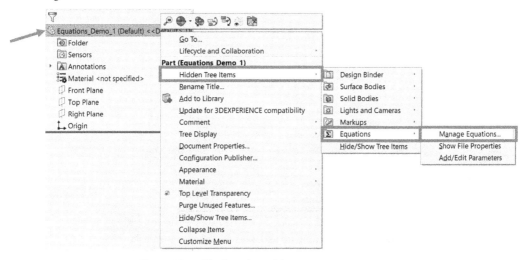

Figure 13.3 – The location of the equations manager

Tip

You can also go to **Equations** by going to **Tools** and then **Equations**.

---

2. Input the W and H variables under **Global Variables** and input the values, as highlighted in the following screenshot. Click **OK** after defining the variables:

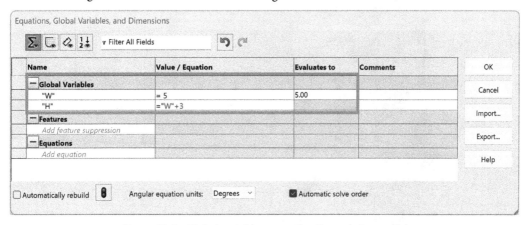

Figure 13.4 – Global variables are active through the model

3. Verify that the variables were defined by checking the new folder in the design tree under **Equations**, which will show all of our variables. *Figure 13.5* highlights the new Equations folder:

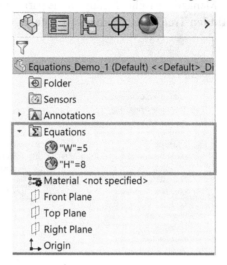

Figure 13.5 – The defined variables are listed in the design tree

4. Start creating the rectangular cuboid by inputting the variables instead of the lengths. When dimensioning the width, we can input = "W" instead of the number 5, as shown in *Figure 13.6*. Note that the dimension will then show as Σ **5.00** to indicate the involvement of the function of the equation:

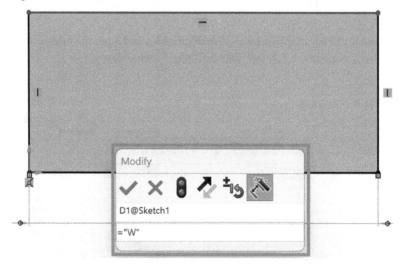

Figure 13.6 – Inputting the variable in the smart dimension field instead of the dimension

5. Dimension the depth by inputting $= "W" * 2$, as shown in *Figure 13.7*. Similar to the width, the depth dimension will be displayed as Σ **10** to indicate the involvement of the equation function:

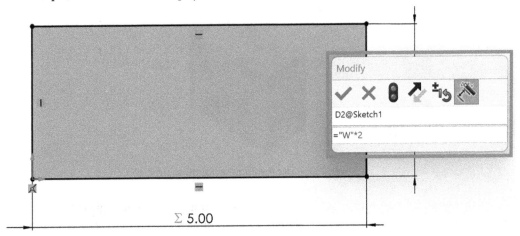

Figure 13.7 – Inputting the multiplication with a variable in place of the dimension value

6. Apply the **Extruded Boss** feature; we can follow the same technique for this. When specifying the dimension, we can input $= "H"$, as shown in *Figure 13.8*. Recall that we have previously assigned the value of 8 to the H variable.

Figure 13.8 – Inputting the defined variable in the PropertyManager feature instead of the value

After applying the **Extruded Boss** feature, we will have the complete rectangular cuboid, as shown in *Figure 13.9*:

Figure 13.9 – The resulting rectangular cuboid

Note that we can define our variables using any term we want. It can be a single letter or a word. A good practice is to use a term that would make it easier for us to recall while creating the model.

Now that we have learned how to use equations, let's see how this will help us make changes easily.

Modifying dimensions with equations

To illustrate how to modify dimensions, let's change the width (W) from 5 millimeters to 8 millimeters. To do this, we can follow these steps:

1. Open the Equations Manager, which we will be able to find by right-clicking on the Equations folder in the design tree.

2. Change the value of the W global variable from 5 to 8, as shown in the following screenshot, and then click **OK**:

Name	Value / Equation	Evaluates to
− **Global Variables**		
"W"	= 5 Change to 8	5.00
"H"	= "W" + 3	8.00
Add global variable		

Figure 13.10 – Changing the variable's value in the Equations Manager

After implementing this change, we will notice that all of the dimensions linked to the W variable have changed as well. In this, the height, H, changed from 8 to 11, and the depth changed from 10 to 16. Not only do equations allow us to change our dimensions faster but they also allow us to keep the design intent while doing so.

Before we cover more in the next topic, let's examine a couple of more notes when dealing with equations. One is regarding the Equations Manager and the other is regarding design intents with equations.

Equations within the Equations Manager

If we go back to the Equations Manager, we will notice a tab at the bottom titled **Equations**, as highlighted in *Figure 13.11*:

Equations, Global Variables, and Dimensions

Name	Value / Equation	Evaluates to	Comments	OK
− Global Variables				Cancel
"W"	= 8	8.00		
"H"	= "W" + 3	11.00		Import...
Add global variable				
− Features				Export...
Add feature suppression				
− Equations				Help
"D1@Sketch1"	= "W"	8mm		
"D2@Sketch1"	= "W" * 2	16mm		
"D1@Boss-Extrude1"	= "H"	11mm		
Add equation				

☐ Automatically rebuild   🔒   Angular equation units: Degrees ⌄   ☑ Automatic solve order

Figure 13.11 – Used equations are listed in the Equations Manager

This shows all of the equation applications we have in the model. It also gives us quick access to all of them in case we want to apply any change without needing to look up the actual feature in the design tree. The first column shows the names of the dimensions. For example, **D1@Sketch1** refers to *dimension 1* from *Sketch1*, which is listed in the design tree.

You may think that it is not convenient for us to recognize these codes (for example, **D1@Sketch1**), especially when we build models that contain many different measurements and sketches. To make this process easier, we can change these names as we are inputting the dimensions. The following screenshot shows the dialog box we get when we enter a specific dimension, highlighting where we can change the name of that specific dimension. The highlighted box in *Figure 13.12* will enable us to change the first part of the name, **D1**. To change the second part, **@Sketch1**, we can rename the sketch entry found in the design tree:

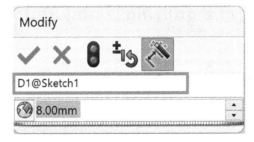

Figure 13.12 – We can change the name of the dimension as we are inputting it

Now, let's elaborate more about design intent when using equations.

Design intent with equations

Whenever we 3D model anything in SOLIDWORKS, we have to keep in mind the design intent we are aiming for. For example, if we are to sketch a rectangle with a width of 5 mm and a length of 10 mm, one question is whether we should link the two dimensions with an equation.

To answer this, we have to ask ourselves what is important. If we intend to always have the length of the rectangle double the width, then applying an equation stating that would be the better practice. However, if we intend to have the length as 10 mm regardless of the width's value, then entering a direct numerical value would be the better practice. Referring to *Figure 13.13*, the rectangle on the left shows the dimensions input if our priority is to keep the length as double the width, while the rectangle on the right shows the dimensions input if our priority is to keep the length as a constant value:

Figure 13.13 – Equations can help us preserve design intent

The rectangle example is a very simple one. However, the same principle is applicable to more complex parts or multiple parts linked together.

This concludes the topic of equations. We learned what equations are, how to apply and modify them, and, finally, important considerations with design intent and equations. Now, we can move on to exploring what configurations are within SOLIDWORKS parts modeling.

Understanding and utilizing configurations

Often, when creating a product, we will create multiple versions of it, with each of the versions having a slight variation from the others. SOLIDWORKS provides special tools for such configurations. In this part, we will learn all about configurations and how to use them to create different variations of a certain 3D model.

What are configurations?

Configurations are different variations of a particular product or object. These variations would have small differences when compared to each other. For example, the four drawings in *Figure 13.14* show different configurations of the same object:

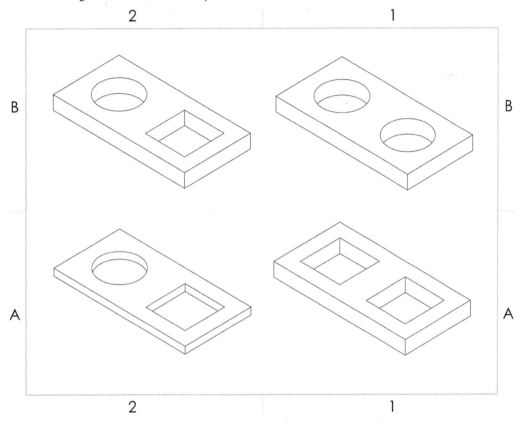

Figure 13.14 – Different configurations of the same object

Note that the four configurations do not have major differences from each other. Because of this similarity, it is an advantage for us to be able to create all of the different configurations in one SOLIDWORKS file rather than having a separate file for each configuration.

Now that we have a better idea of configurations, let's start applying them in SOLIDWORKS.

Applying configurations

Whenever dealing with configurations, we can start by creating a base model, and then we can create configurations of it. To highlight the application of configurations, we will create the model and the configurations shown in *Figure 13.15*. To make the exercise easier to follow, the dimensions in bold refer to dimensions that are different from one configuration to another:

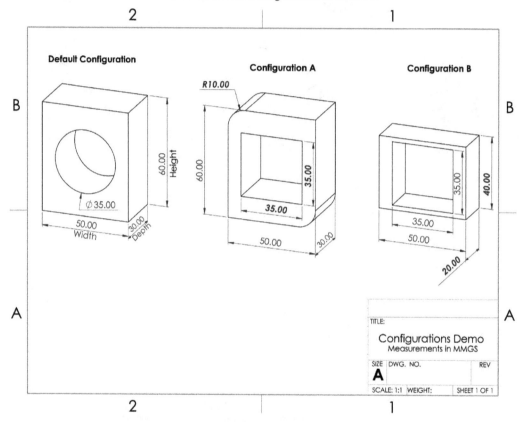

Figure 13.15 – The 3D model we are creating in this exercise

To complete this exercise, we will follow these steps:

1. Create the model titled **Default Configuration**, as shown in *Figure 13.15*. You can follow the procedure highlighted in *Figure 13.16* for your first extrusion. Note that the circle is at the center of the rectangular face:

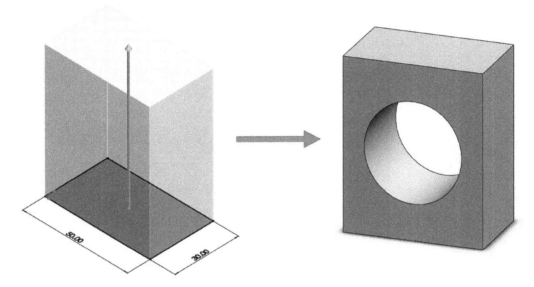

Figure 13.16 – The steps for creating the default configuration

2. To start creating the first configuration, we have to add it by going to the ConfigurationManager at the top of the design tree, and then *right-clicking* on **Configurations** and selecting **Add Configuration...**, as shown in *Figure 13.17*:

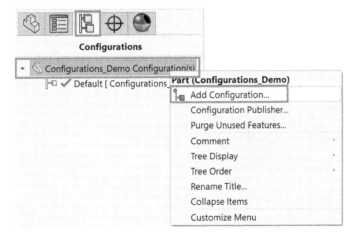

Figure 13.17 – The location of the Add Configuration… command

3. After doing that, we will be prompted to enter a name in the **Configuration name** field as well as a small description in the **Description** field, as highlighted in *Figure 13.18*. You are free to choose a name and description that would help you identify the configuration. Click on the green check mark to introduce the **Square hole with fillets** configuration:

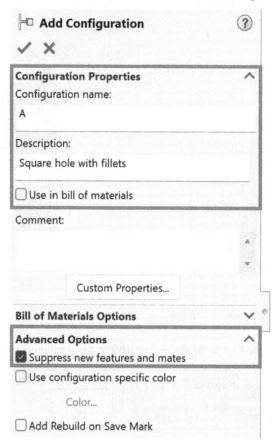

Figure 13.18 – Naming and describing the configuration can help us identify it later on

Important note

The indicated **Suppress new features and mates** advanced option will have all new features applied to other configurations suppressed in this one.

4. Note the different shading on the **Configurations** list in the ConfigurationManager, as highlighted in *Figure 13.19*. The active configuration will be solid. Double-check that you are working in the **A** configuration. Then, go back to the design tree by clicking on the icon indicated with a square in *Figure 13.19*:

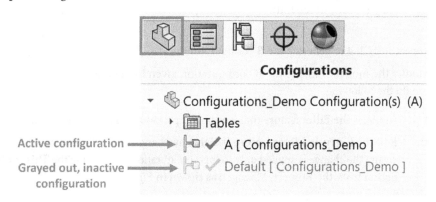

Figure 13.19 – Available configurations are listed in the ConfigurationManager

5. Now, we can modify our model to match the **A** configuration given in the initial drawing shown in *Figure 13.15*. To do that, we will do the following:

    I.   Suppress the **Extruded Cut** feature by right-clicking on the feature from the design tree and selecting **Suppress**.

    II. Apply the new extruded cut and the **Fillet** features to create **Configuration A**, as shown in *Figure 13.15*. Note that the square is at the center of the rectangular face. We should have the model shown in *Figure 13.20*, which presents **Configuration A**:

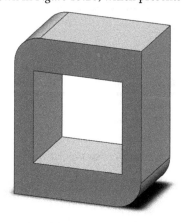

Figure 13.20 – The 3D model for Configuration A

> **Tip**
>
> After creating new configurations, a good practice is to go back to the ConfigurationManager and double-check the status of other configurations. In this case, in the ConfigurationManager, double-click on the **Default** configuration to check the difference and ensure that we created the new configuration successfully.

6. Repeat *steps 2–4* to generate the **B** configuration.

7. Modify the model to match the **B** configuration given in the initial drawing. To do that, we can do the following:

    I.   Suppress the **Fillet** feature that we applied previously from the design tree.

    II. Edit the **Extruded Boss** feature (height) by changing the dimension to 4 0 mm. Also, ensure the change is applied only to this configuration by selecting the **This configuration** option from the PropertyManager, as shown in *Figure 13.21*:

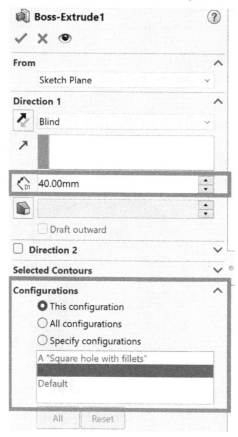

Figure 13.21 – Specifying the related configuration when changing an existing feature's dimension

8. Modify the depth by modifying the dimension in the initial sketch from 30.00 mm to 20.00 mm. Make sure this modification only applies to this configuration by selecting **This configuration**, as shown in *Figure 13.22*:

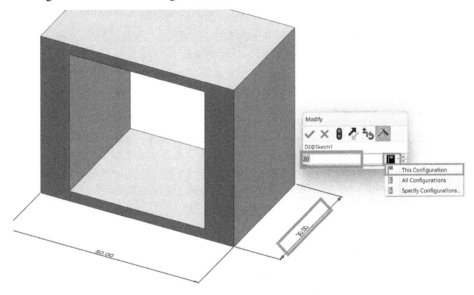

Figure 13.22 – Specifying the applicable configuration when adjusting a sketch dimension

9. Go back to the ConfigurationManager and double-click on each of the configurations to double-check that they were applied correctly. At this point, our three configurations should be as they were in the initial drawing shown in *Figure 13.15*.

Important note

When editing a dimension within an existing feature or a sketch, SOLIDWORKS allows us to pick which configuration this change should apply to. For example, in the exercise, we applied all of our edits to only the active configuration by selecting the **This Configuration** option. We can also choose to apply the edits to all configurations or specify which configurations we want the edits to apply to.

This concludes this exercise in creating different configurations for a specific model. Remember that it is a good practice to go back to the ConfigurationManager and double-check that all of our configurations are accurate by double-clicking on each of them. Here are some important takeaways from this section:

- We can suppress and unsuppress features to set them to different configurations.

- When modifying dimensions on existing features or sketches, we have to select the scope in terms of which configuration(s) the adjustment will apply to.

Configurations can be particularly useful in scenarios where multiple variations of a product are required, such as creating different sizes of a component or producing customized versions of a standard part. Additionally, they are beneficial when testing different design iterations to optimize performance or functionality without having to create multiple separate models.

Now, we can start working with design tables, which is another method that will enable us to create many different configurations quicker using an Excel table.

Understanding and utilizing design tables

Through configurations, we were able to create different variations of a particular 3D model. Design tables will also allow us to create different variations of a 3D model. However, unlike directly setting up configurations, design tables will enable us to generate more than one variation at the same time instead of generating them one after the other. In this section, we will cover how to set up design tables and some scenarios in which design tables will give us an advantage over directly setting up configurations.

What are design tables?

Setting up a **design table** is one method that will enable us to create multiple variations of a specific 3D model at once. Design tables make it easy and efficient to set up different dimensions for lengths and angles. They also allow easy manipulation for suppressing certain features in specific model variations or configurations. The drawing shown in *Figure 13.23* is a simple application of a design table with the multiple configurations it generated. Note that, with design tables, we do not enter the dimensions for the different configurations manually one by one.

Rather, we enter a table highlighting how the different configurations will differ, and SOLIDWORKS will then generate all of the configurations at once. In *Figure 12.23*, all the configurations are different, based on the **Length**, **Width**, and **Thickness** parameters:

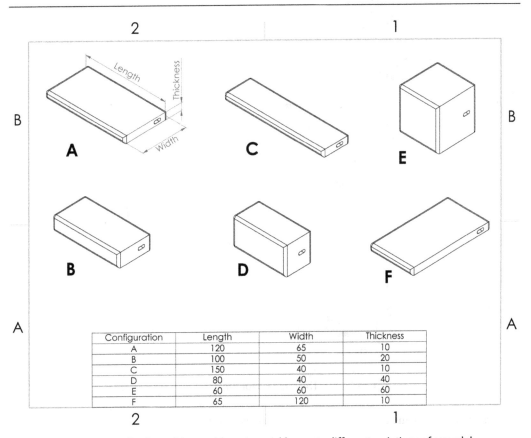

Configuration	Length	Width	Thickness
A	120	65	10
B	100	50	20
C	150	40	10
D	80	40	40
E	60	60	60
F	65	120	10

Figure 13.23 – Design tables enable us to quickly create different variations of a model

When working with design tables, a good practice is to start by creating a base model that includes all of the dimensions and features we will vary on other configurations. It is also a good practice to make custom names for all of the dimensions and features we want to vary for easier identification. Now that we have an idea of what design tables are, we can start applying them in creating a 3D model.

Setting up a design table

When dealing with design tables, we can start by creating a base model. After that, we can use a design table to create multiple configurations out of it. This will be the process we follow in this exercise. We will create the model shown in *Figure 13.24*. SOLIDWORKS can use Microsoft Excel to generate design tables, so you must have Microsoft Excel installed on your computer to follow this exercise.

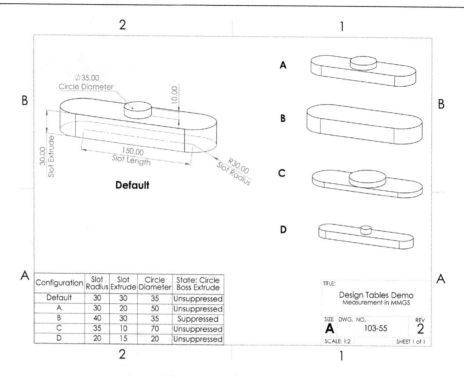

Configuration	Slot Radius	Slot Extrude	Circle Diameter	State: Circle Boss Extrude
Default	30	30	35	Unsuppressed
A	30	20	50	Unsuppressed
B	40	30	35	Suppressed
C	35	10	70	Unsuppressed
D	20	15	20	Unsuppressed

Figure 13.24 – The drawing for the design table exercise

To create the following model utilizing design tables, follow these steps:

1. Create the base model, as shown in *Figure 13.24*. When doing that, you can use the following good practices. These will minimize confusion when generating and using a design table:

    - Give names to your dimensions as you are inputting them. *Figure 13.26* shows inputting the name for Slot Radius.

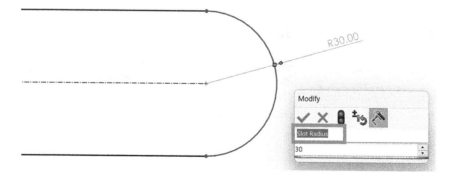

Figure 13.25 – Naming the dimensions can help identify them easier

- Rename your features and sketches listed in the design tree to make them easier to identify. The screenshot shown in *Figure 13.26* highlights a sample of renamed features and sketches for use in this exercise. Note the names are as shown in *Figure 13.24*:

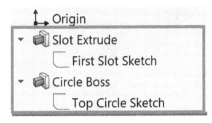

Figure 13.26 – Naming the features can make them easier to identify

2. To add a design table, go to **Insert**, then **Tables**, and select **Excel Design Table...**, as highlighted in *Figure 13.27*:

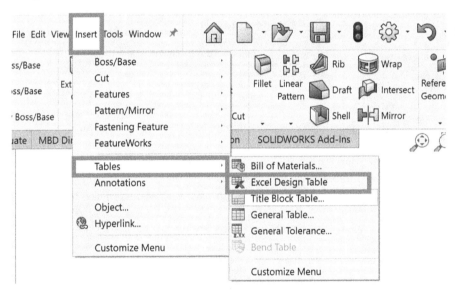

Figure 13.27 – Inserting a design table

3. Set **Source** as **Auto-create**, as shown in *Figure 13.28*. **Edit Control** is set by default to **Allow model edits to update the design table**; we will leave that selected. We will discuss this option later in this section. Click on the *green check mark* to initiate the design table:

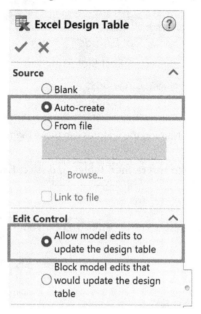

Figure 13.28 – The initial options for generating a design table

4. Select the dimensions we want to vary, as shown in *Figure 13.29*. Note that this selection only includes dimensions related to sketches and features, not whether a feature is applied or suppressed. Also, note how naming the dimensions and features makes them easier to identify:

Figure 13.29 – We can select the dimensions to be inserted into the design tree

5. Fill in the design table, as shown in *Figure 13.30*. The first column in the table shows the name of the configuration, while the others show dimensions unique to that particular configuration. Note that empty cells will automatically take the values of the active default configuration:

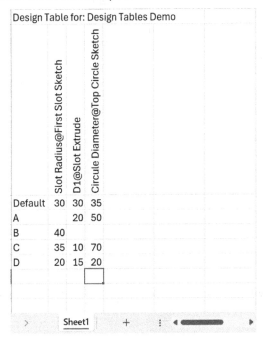

Design Table for: Design Tables Demo

	Slot Radius@First Slot Sketch	D1@Slot Extrude	Circule Diameter@Top Circle Sketch
Default	30	30	35
A		20	50
B	40		
C	35	10	70
D	20	15	20

Sheet1   +

Figure 13.30 – We can input all our variations directly in the design table

6. The table only has dimensions so far. However, the **B** configuration has the **Circle Boss** feature suppressed. To add the state of the feature to the design table, select the first empty title cell, and then double-click on the feature on the design tree, as highlighted in *Figure 13.31*. This will automatically add the state of the feature to the design table.

7. To adjust the status of the feature, we can type the letter U for *unsuppressed* or the letter S for *suppressed*. Alternatively, we can type the full words (unsuppressed or suppressed) or 0 and 1, which stand for *unsuppressed* and *suppressed*, respectively. Adjust the feature status, as shown in *Figure 13.31*:

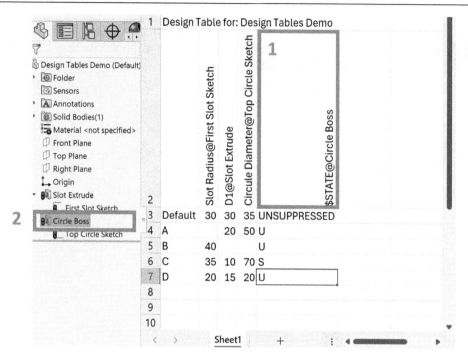

		Slot Radius@First Slot Sketch	D1@Slot Extrude	Circule Diameter@Top Circle Sketch	$STATE@Circle Boss
3	Default	30	30	35	UNSUPPRESSED
4	A		20	50	U
5	B	40			U
6	C	35	10	70	S
7	D	20	15	20	U

Figure 13.31 – We can indicate the state of a feature in the design tree

> **Tip**
>
> We can also add any dimension to the design table by directly clicking on it from the sketch on the canvas, similar to adding the feature status.

8. Click anywhere on the canvas outside of the table to confirm the table. We will see the message shown in *Figure 13.32*, confirming the new configurations:

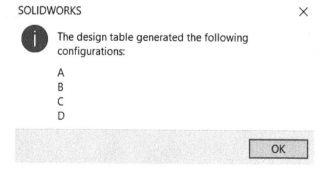

Figure 13.32 – A conformation message after generating the design table

After generating the different configurations, it is a good practice to double-check them. So, go to the ConfigurationManager, check all of the configurations that we just generated, and note how they are different from each other.

This concludes our coverage of how to generate a design table to create multiple configurations. However, now that we know how to initiate a design table, we will also learn how to edit it.

Editing a design table

There are two ways in which we can edit or update our design table. The first one is through the design table itself and the other is through directly modifying the dimensions in the model. We will examine both ways.

Editing directly from the design table

To edit the design table, we can find it in the ConfigurationManager. Then, right-click on the table and select **Edit Table**, as shown in *Figure 13.33*:

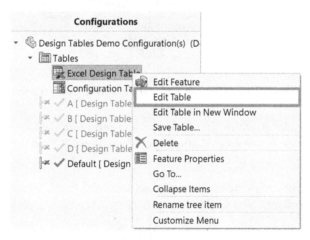

Figure 13.33 – The Edit Table command for design tables

After selecting the **Edit Table** command, the table will open for us to modify as we see fit. After modification, we can simply click anywhere on the canvas for all of the modifications to be applied.

Editing the design table by modifying the model

Another way of editing the design table is by directly editing the sketches and features in the model. This will update the corresponding cells in the design tree. However, this will only happen if we select the **Allow model edits to update the design table** option for **Edit Control**, as shown in *Figure 13.34*. Recall that we selected this option when we were creating the table. We can adjust this option by right-clicking on the design table from the ConfigurationManager and then selecting **Edit Feature**:

Figure 13.34 – Manual 3D model edits can automatically update the design table

Note that applying a design table does not prevent us from adding additional configurations, as we explored in a previous section when discussing configurations. We can use both methods together as we see fit.

Design tables are particularly valuable when it comes to creating multiple variations of a design quickly and efficiently. For instance, they are ideal for generating different sizes of components or producing various configurations of a mechanical part with distinct features. By using design tables, you can streamline the process of parameter-driven design studies, allowing you to swiftly switch between configurations and test performance under diverse conditions. This significantly accelerates your design process and enhances your productivity.

This concludes this section about design tables. We covered what design tables are and how to apply and edit them. Design tables are a very efficient tool that enables us to generate multiple configurations at once and is essential for SOLIDWORKS professionals.

Summary

In this chapter, we learned skills that will enable us to create more robust and agile models. We covered equations that will enable us to create more connected models to help us deliver our design intents. We also learned how to create different configurations of a specific model. We learned how to do that by directly and manually adding and adjusting configuration, or by using design tables to accomplish a similar objective.

The new skills in this chapter will enable us to generate more connected models. They will also allow us to generate many different variations of a model in a single SOLIDWORKS file. These will enable us to more efficiently conduct variation testing and quicker adjustments, which were the goals of this chapter.

In the next chapter, we will cover advanced mates within assemblies, which will help us create assemblies with parts that have more complex interactions with each other.

Questions

The following questions will help to emphasize the main points we have learned in this chapter. However, in terms of practical exercises, do not limit yourself to what we provide you with here. Try modeling random objects around you or come up with your own innovative mode to increase your fluency using the software:

1. What are the equations when modeling parts? Why do we use them?
2. What are the configurations for a specific part?
3. What are design tables?

4. Create the model shown in the following screenshot, utilizing global variables and equations:

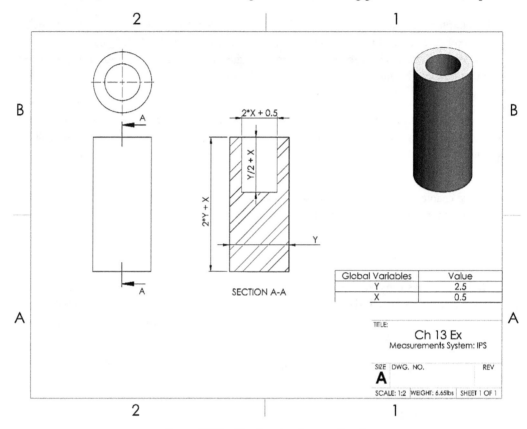

Figure 13.35 – The drawing for question 4

5. Create the base model shown, including the different configurations it highlights. Note that the two squares and the two circles in the **A** and **B** configurations are mirrors of each other:

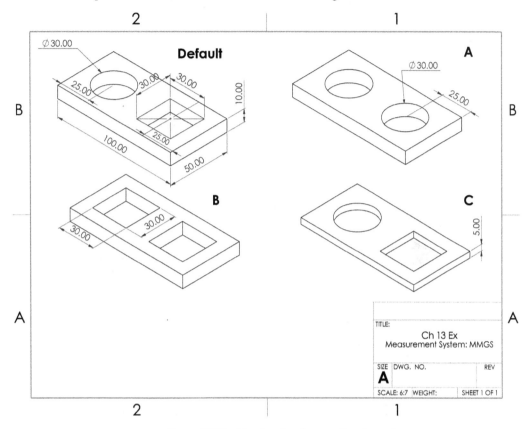

Figure 13.36 – The drawing for question 5

6. Create the following model and the different configurations shown using design tables. Use the MMGS measurement system when completing this exercise:

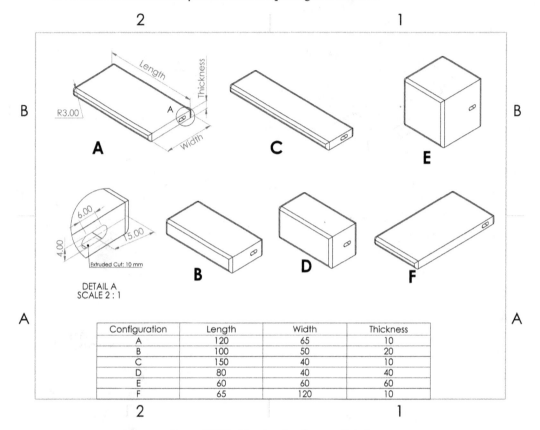

Configuration	Length	Width	Thickness
A	120	65	10
B	100	50	20
C	150	40	10
D	80	40	40
E	60	60	60
F	65	120	10

Figure 13.37 – The drawing for question 6

7. Create the following model and the different highlighted configurations. The model is displayed in two different screenshots to cover all of the requirements. Hint – use the mass values to double-check your model's accuracy:

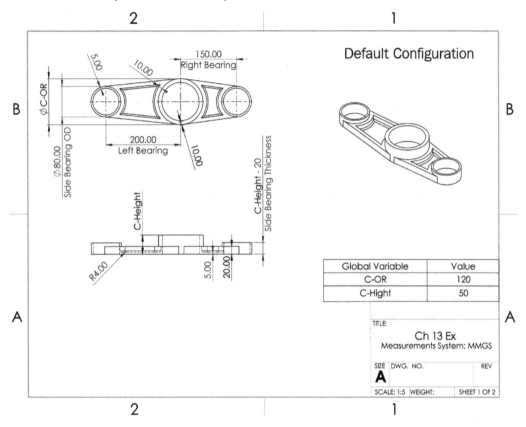

Figure 13.38 – The first drawing for question 7

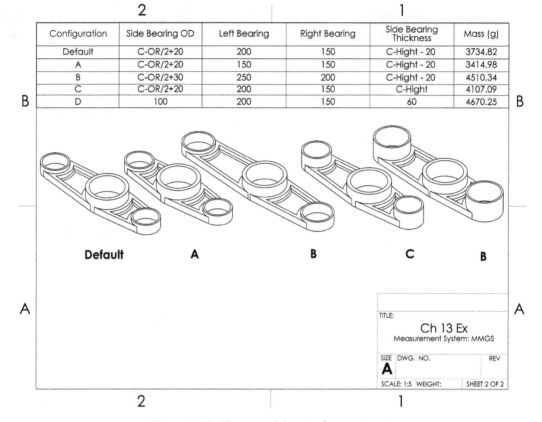

Configuration	Side Bearing OD	Left Bearing	Right Bearing	Side Bearing Thickness	Mass (g)
Default	C-OR/2+20	200	150	C-Hight - 20	3734.82
A	C-OR/2+20	150	150	C-Hight - 20	3414.98
B	C-OR/2+30	250	200	C-Hight - 20	4510.34
C	C-OR/2+20	200	150	C-Hight	4107.09
D	100	200	150	60	4670.25

TITLE:
Ch 13 Ex
Measurement System: MMGS

SIZE DWG. NO.   REV
A

SCALE: 1:5 WEIGHT: SHEET 2 OF 2

Figure 13.39 – The second drawing for question 7

> **Important note**
> The answers to the preceding questions can be found at the end of this book.

Part 7: Advanced Assemblies – Professional Level

This part will take your assembly skills to a higher level than the one we covered at the Associate level. Here, you will learn about advanced mates and other assembly functions, including interference and collection detection, assembly features, design tables, and configurations for assemblies. This part will also introduce the SOLIDWORKS connect cloud services.

This part has the following chapters and projects:

- *Chapter 14, SOLIDWORKS Assemblies and Advanced Mates*
- *Chapter 15, Advanced SOLIDWORKS Assembly Competencies*
- *Project 2 – 3D Modeling an RC Helicopter Model*
- *Chapter 16, Introduction to SOLIDWORKS Cloud Services*

SOLIDWORKS Assemblies Advanced Mates

Mates within SOLIDWORKS assemblies fall into three major types – **standard**, **advanced**, and **mechanical**. This chapter will cover using advanced mates within the SOLIDWORKS assembly's environment. These include profile center, symmetric, width, path mate, linear/linear coupler, distance range, and angle range. For each of the mates, we will learn what they do and how to use them. These advanced mates will enable us to generate assemblies with more complex part-to-part interactions than when only using standard mates.

The following topics will be covered in this chapter:

- Understanding and using the profile center mate

- Understanding and using the symmetric and width mates

- Understanding and using the distance range and angle range mates

- Understanding and using the path mate and linear/linear coupler mates

By the end of this chapter, you will be able to generate more complex interactions within different parts of an assembly. Those include both dynamic and static interactions.

Technical requirements

This chapter will require access to SOLIDWORKS software.

The project files for this chapter are available at the following GitHub repository: `https://github.com/PacktPublishing/Learn-SOLIDWORKS-2025-Third-Edition`

The CiA video for this chapter can be found at `https://packt.link/ln2us`

Understanding and using the profile center mate

The **profile center** advanced mate allows us to create a centered relationship between two planar profiles in one step. This can save us time as well as deliver a precise centered design intent in one go. However, the mate also has its own limitations. The profile center mate is an essential part of our assembly toolkit. In this section, we will learn about the profile center mate and how to apply it.

Defining the profile center advanced mate

The profile center advanced mate helps us to center two surfaces in relation to each other in an assembly. *Figure 14.1* highlights the effect of the profile center mate. The assembly on the left is before applying the mate, while the one on the right is after applying the mate:

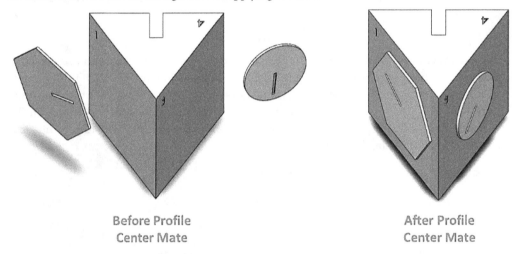

Before Profile
Center Mate

After Profile
Center Mate

Figure 14.1 – The profile center mate can position a part at the center of another

Note that SOLIDWORKS interprets the profile of the shape by its outer outline only. Also, we can only use the profile center mate with limited profiles that include circles and polygons. Polygons can include triangles, rectangles, hexagons, and so on. Any other irregular profiles cannot be used with the profile center mate. Now that we know what the profile center advanced mate does, we can start applying the mate to different parts, which we will do next.

Applying the profile center mate

In this section, we will apply the profile center advanced mate to generate the assembly shown in *Figure 14.2*. All of the parts needed for this assembly are available for you to download with this chapter. Note that in the screenshot, both the circular and hexagonal parts are in the center of the faces, indicated by the numbers *1* and *3*:

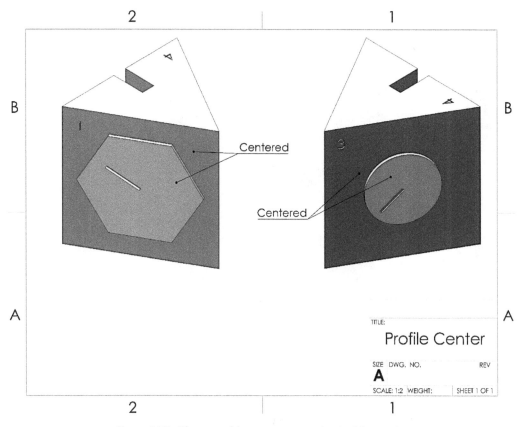

Figure 14.2 – The assembly we are generating in this exercise

To generate the assembly, we can perform the following steps:

1. Download all of the parts linked to this part and input them in an assembly file. Have the triangular prism as the fixed base part.

2. We will first start by generating the mate between the hexagonal and side *1* in the triangular prism. We will first select two profiles from each part, as indicated in the screenshot. Then, click on **Mate**:

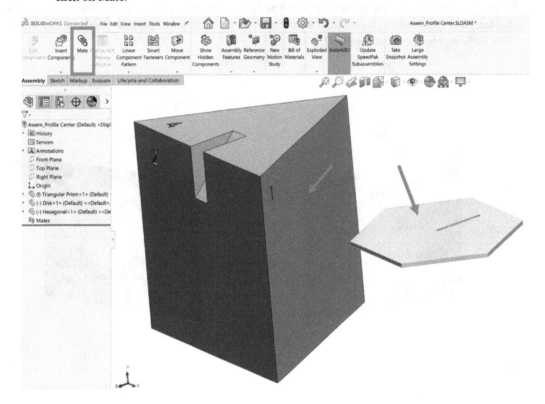

Figure 14.3 – The first selections for the mate

3. This will apply the standard mate, **Coincident**, by default. Change it by selecting the **Advanced** tab option and then the **Profile Center** mate, as shown in *Figure 14.4*.

4. This will drive the parts together for a preview with the hexagon centered in face *1*. Options for **Orientation** and **Mate alignment** can be used to adjust the positioning of the centered part as needed. Both options are highlighted in *Figure 14.4*.

This will show us a preview of the mate. Note that the hexagon became stationed at the center of side *1* of the triangular prism. Note that there are more options under the mate that we can use to orient the parts to each other:

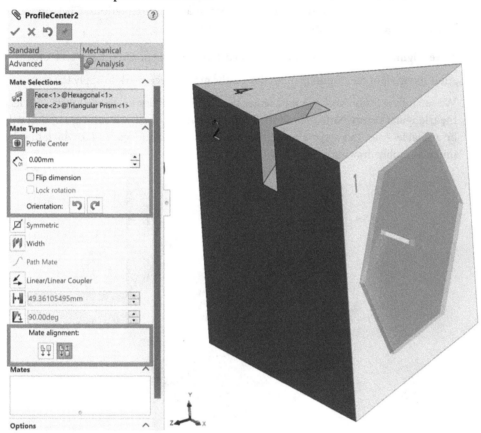

Figure 14.4 – The PropertyManager and setting for the profile center mate

Here is an explanation of all of the options available on the mate's PropertyManager:

- **Offset distance (D1)**: The distance field allows us to have a certain distance to separate the two profiles. You may enter a random number in the field to see a preview of the effect.

- **Flip dimension**: This relates to the distance field explained in the previous point. It will flip the distance measurement to the other side. You can try checking the box to see the effect of the option. In this exercise, the offset distance is zero, so the option will not make a difference.

- **Lock rotation**: This option only applies when the matted face is circular. Having this option checked will stop the circular profile from rotating.

- **Orientation**: This option allows us to change the orientation of the centered profile. This option is activated where there is a countable number of orientations in which our profiles can be centered in relation to each other. In this exercise of a hexagonal and a square profile, the profiles can be centered in four different rotational orientations. Click on the arrows until you get the orientation position you require.

5. **Mate alignment**: This will flip the two mated surfaces to opposite sides of each other.

6. After fixing all of the requirements, click on the green check mark to apply the mate.

7. We can follow *Steps 2* to *5* again to center the circular profiles for the disk part with face *3* of the triangular prism. Note that, in the PropertyManager, the **Lock rotation** option will now be available. After this step, our assembly will look similar to *Figure 14.5*, which matches the initial assembly drawing.

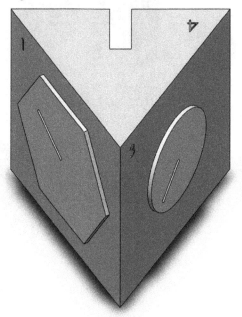

Figure 14.5 – The final result of the assembly

One aspect to note is that the profile center mate will only work on circular or polygonal profiles. In the preceding exercise, faces *2* and *4* do not have a regular polygonal profile because the profiles are not continuous. As a result, we will not be able to use the profile center command with those sides. To address these restrictions, we can instead use a reference sketch with the profile center mate. For example, we can create a new sketch on face *4* of a regular triangle and then use that to represent face *4* in applying the profile center mate.

> **Tip**
> Sketches can be used as a reference for mate selection in the same way that planar surfaces work.

This concludes this section on the advanced mate profile center. We covered its definition and how to use it. Next, we will cover two advanced mates, symmetric and width.

Understanding and using the width and symmetric mates

Symmetric and **width** are two advanced mates that we can use within SOLIDWORKS assemblies. They allow us to more flexibly control part movements, both in terms of symmetry and width adjustments. In this section, we will cover the width and symmetric mates and how to use them in our assemblies.

Defining the width advanced mate

The **width** advanced mate involves adjusting surfaces relative to each other within the width dimension. A common application of the width mate is in mechanical joints and mechanical slots. For example, take note of the mechanical joint shown in *Figure 14.6*. The width advanced mate will help adjust the location of the inner joint in relation to the outer joint.

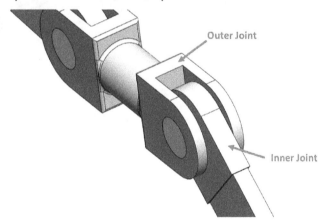

Figure 14.6 – The width mate allows us to adjust the location of the inner joint in relation to the outer one

Using the width mate, we can achieve one of the following:

- Allow the inner joint to move freely within the space defined by the outer joint boundary.
- Center the inner joint within the space defined by the outer joint boundaries.
- Fix the inner joint at a specific distance or percentage within the outer joint boundary.

Now that we know what the width relation does, we can start applying it next.

Applying the width advanced mate

In this section, we will learn how to apply the width advanced mate. We will do that by fixing the width of the inner joint to the outer joint, as highlighted in *Figure 14.7*. All of the parts needed for this exercise are available for you to download in this section.

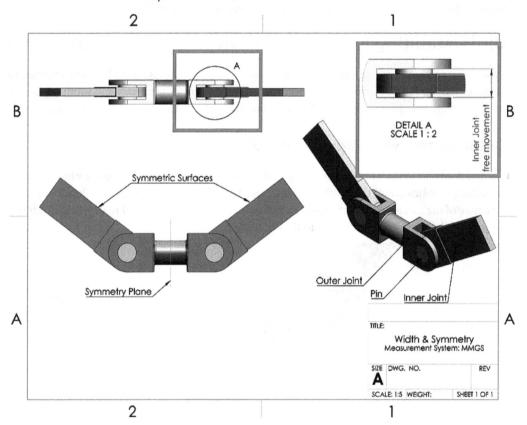

Figure 14.7 – The assembly we will build in this exercise

To apply this mate, follow these steps:

1. Download all of the parts and the assembly file linked with this part. Then, open the assembly file as a starting point.

2. Select the **Width** mate by selecting the **Mate** command and then the **Advanced** tab option. Then, select the **Width** mate, as highlighted in *Figure 14.8*.

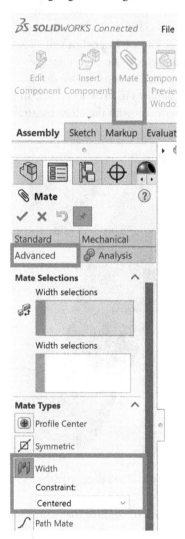

Figure 14.8 – The location of the width mate

3. Once we select the mate, we will get more options to set it up. We can follow the selections highlighted in *Figure 14.9*, which are as follows:

- **Width selections**: For the first **Width selection**, we can select the outer boundaries for our width constraints. In this case, they are the inner surfaces of our outer joint.

- **Width selections**: For the second **Width selection** , we can select the inner boundaries for our width constraints. In this case, they are the outer surfaces of our inner joint.

- **Constraint**: This allows us to determine the relationship between our **Width selections** settings and our **Tab selections** settings. There are four options that we can put in the context of this example. **Centered** will make the inner joint centered in the space available between the outer joint boundaries. **Free** will allow the inner joint to move both sides within the boundaries of the outer joint. **Dimension** will allow us to set the inner joint to a certain distance from the boundaries of the outer joint. **Percent** is similar to **Dimension**; however, instead of setting a distance dimension, we can set a percentage. In this exercise, we are required to set the constraints to **Free**, as per the initial drawing. However, while you are here, you can also experiment with the other constraints:

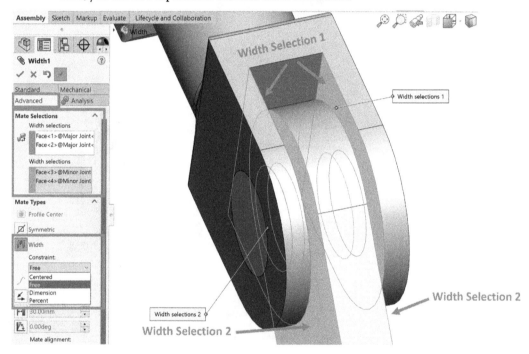

Figure 14.9 – The selection of the width and tab

4. Click on the green check mark to confirm the mate. Once you confirm the mate, it is a good practice to check its effect. So, try moving the inner joint to different sides to see the effect of the new width advanced mate. You will notice that the inner joint will move to the sides only within the space available before it hits the outer joint.

5. Repeat *Steps 2* to *4* for the other joints.

> **Note**
> We can also switch **Width selections** with the second **Width selections settings**. This will give you the same result.

This concludes this exercise on the width advanced mate. We covered what the width mate is and how to use it. Now, we will move on to another advanced mate, which is symmetric.

Defining the symmetric advanced mate

The **symmetric** mate allows us to set two different surfaces to have a symmetric dynamic relation to each other around a symmetric plane. A key thing to note with the symmetric advanced mate is that it builds a relationship between the selected surfaces, not the parts themselves. Hence, it establishes symmetric surfaces rather than symmetric parts. *Figure 14.10* highlights the functionality of the symmetric mate. The symmetric surfaces are symmetric around the symmetry plane. Hence, if we rotate any of the surfaces, clockwise or anticlockwise, the other surface will rotate as well. *Figure 14.10* also highlights the components involved in constructing symmetric surfaces:

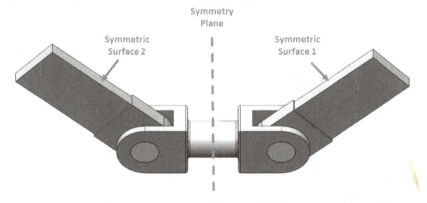

Figure 14.10 – The symmetric mates allow us to build movement symmetry

If we look at the assembly from a different angle, as shown in *Figure 14.11*, we can see the two parts (inner joints) are not fully symmetrical, as indicated by the circled space discrepancy in *Figure 14.11*.

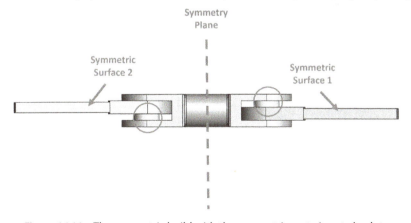

Figure 14.11 – The symmetric build with the symmetric mate is not absolute

We can also apply the symmetric mate to edges or points. It's important to note that this symmetrical relation is directional, based on our mate selections. For example, if we select two surfaces, the symmetry will be applied to movements that are normal to those surfaces. The same principle applies when selecting edges for symmetry. Now that we know what the symmetric mate does, we can start applying it to our assembly.

Applying the symmetric advanced mate

In this section, we will learn how to apply the symmetric advanced mate. We will do that by applying it to the indicated surfaces around the shown symmetry plane in *Figure 14.12*. In this exercise, we are continuing our work on the assembly that we started with the width mate.

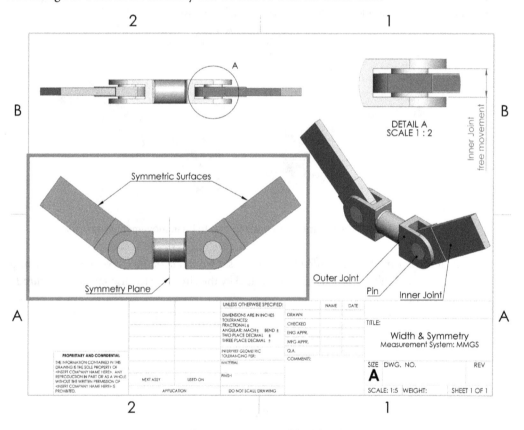

Figure 14.12 – The symmetry we will build in this exercise

To set up the symmetric mate, we need two requirements: a symmetry plane and surfaces to make them symmetrical. To apply the symmetry mate, follow these steps:

1. Check if an existing plane can match the required symmetry plane highlighted in *Figure 14.12*. In this case, there are none. Hence, we can create a new plane using the reference geometry function. The symmetry plane should be in the middle of the outer joint, as highlighted in *Figure 14.12*. *Figure 14.13* highlights the settings for our new reference geometry plane.

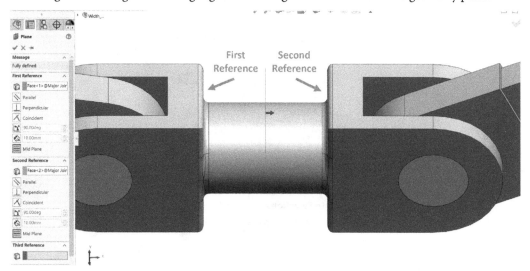

Figure 14.13 – Defining a new reference plane might be needed for our symmetry

---

Note

SOLIDWORKS automatically switches to mid-plane when two parallel planar surfaces are selected as references.

2. Select the **Symmetric** mate by selecting the **Mate** command, the **Advanced** tab option, and then the **Symmetric** mate, as highlighted in *Figure 14.14*.

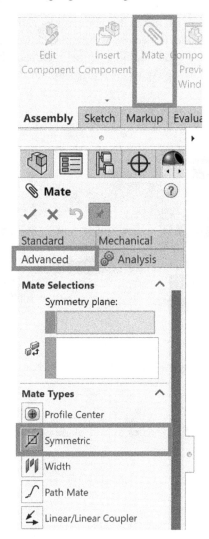

Figure 14.14 – The location of the symmetric mate

3. Now, we will be prompted to set up the mate. We can set it up as highlighted in *Figure 14.15*. Here is more explanation about the selection:

    • **Symmetry plane**: Here, we can select the plane we created in *Step 1*. If the plane is not visible in the canvas, we can find it and select it by expanding the design tree available in the canvas, as indicated in *Figure 14.15*.

- **Entities to mate:** This is the field under **Symmetry plane**. Here, we can select the two surfaces we want to mate. In this case, they are the two surfaces highlighted in *Figure 14.15*.

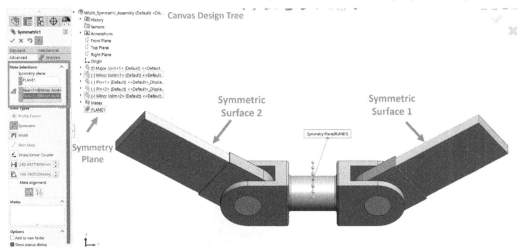

Figure 14.15 – The selection and PropertyManager for our symmetric mate

4. Click on the green check mark to apply the mate. Once you confirm the mate, try moving the different parts to see its effect.

This concludes how to use the symmetric mate. We covered what it is and how to apply it by making two surfaces dynamically symmetrical about a plane. Next, we will explore the distance range and angle range mates.

Understanding and using the distance range and angle range mates

In this section, we will explore the advanced mates, **distance range** and **angle range**, also known as **limit mates**. With these mates, we will be able to limit the movement of a specific component to be within a defined range rather than a set distance or an angle. We will first define what these mates are and then we will use them by following a practical exercise.

Defining the distance range and angle range

Distance range and **angle range** mates allow us to set up a restricted range of movement, including both angular and linear movements. These mates are commonly used in scenarios such as joints, where we want to limit the angular movement of an arm. Other examples include gas springs and distance-based switches, where we need to restrict linear movement to a specific range. *Figure 14.16* highlights an assembly and demonstrates what we can achieve with the distance range and angle range advanced mates. By setting specific angle and distance ranges, we can ensure that the lever and switch move freely only within the designated range.

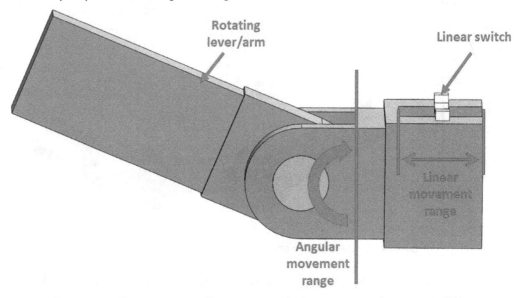

Figure 14.16 – The ranges mate allows us to restrain the movement of our parts to limits

After learning about what the distance range and angle range mates are, we will cover how to apply them in an assembly.

Applying the distance range mate

In this section, we will learn how to apply the distance range advanced mate. We will do that by setting a linear movement range of 0.00 to 50.00 mm for the switch, as indicated in *Figure 14.17*. All of the part files, as well as the initial assembly file, are available for you to download with this chapter.

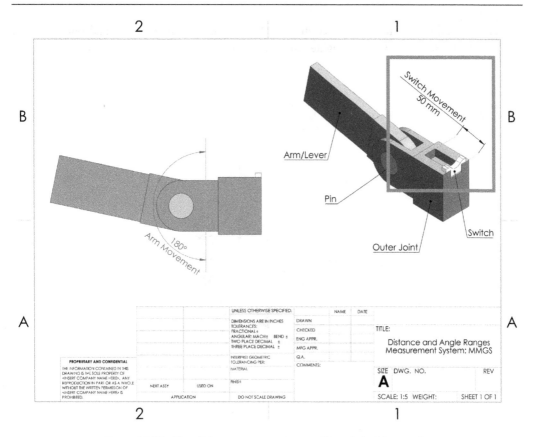

Figure 14.17 – The distance range mate we will apply in this exercise

To set up the distance range mate, we are only required to have two surfaces, edges, or vertexes for two different parts. In this example, we can follow these steps to apply it:

1. Download the linked SOLIDWORKS files and open the provided assembly, as we will use it as a starting point.

2. Select the **Distance** mate by selecting the **Mate** command, the **Advanced** tab option, and then the **Distance** mate, as highlighted in *Figure 14.18*.

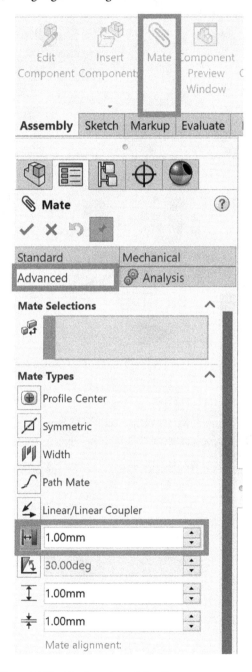

Figure 14.18 – The location of the distance range mate

3. Once we select the mate, we will be prompted to set it up, as shown in *Figure 14.19*.

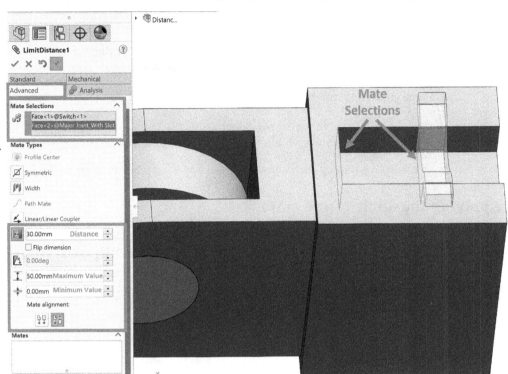

Figure 14.19 – The selections and PropertyManager for our mates

Here is an overview of the settings indicated on the PropertyManager:

- **Mate Selections**: Here, we should select the entitles we want to mate. This can include surfaces, edges, or vertexes. In this exercise, we can select the two faces indicated in *Figure 14.19* between the major joint and the switch.

- **Distance**: This acts as an overview of what a specific distance would look like. In *Figure 14.19*, the distance is set to 30.00mm, which is the actual distance shown on the canvas. Also, the distance listed here will be the initial distance applied to the parts after confirming the mate.

- **Flip dimension**: Checking and unchecking this option will flip the direction in which the distance is measured.

- **Maximum Value**: Here, we can input the maximum limit of the distance range. We can fill this with 50.00mm, as highlighted in *Figure 14.19*. This field can be filled with both positive and negative values.

- **Minimum Value**: This is the opposite of the maximum distance field in which we can set our minimum allowable value for linear movement. This can also be a negative value. In this exercise, our minimum distance is 0.00mm.

- **Mate alignment**: This adjusts how the two parts are aligned with each other. We can always try the different alignments to check which one matches our requirements.

4.   Click on the green check mark to confirm the mate.

Once you confirm the mate, try moving the switch horizontally to double-check the functionality of the mate. The switch should only move within the slot built on the major joint part.

Now that we have finished setting up our distance range, we can set up the angle range for the arm next.

Applying the angle range mate

In this part, we will learn how to apply the angle range advanced mate. We will do that by limiting the range of angular movement of our lever/arm to be between 0 and 180 degrees, as indicated in *Figure 14.20*. In this exercise, we are continuing our work on the assembly that we did in the last distance range mate part.

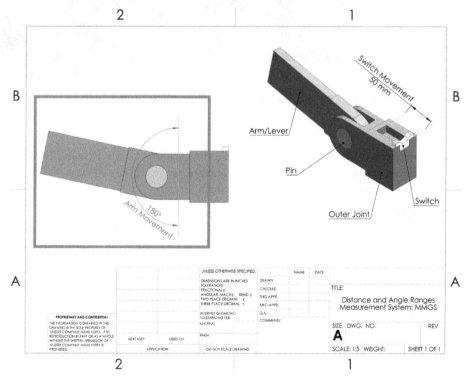

Figure 14.20 – The angle range we will apply in this exercise

Setting the angle range is similar to setting up a distance range. To set up an angle range, we will require two surfaces or edges for two different parts. In this example, follow these steps to apply it:

1. Select the **Angle** mate by selecting the **Mate** command, the **Advanced** tab option, and then the **Angle** mate, as highlighted in *Figure 14.21*.

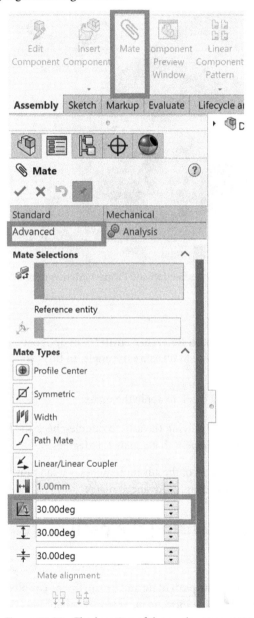

Figure 14.21 – The location of the angle range mate

2. Set up the PropertyManager, as shown in *Figure 14.22*. This setup is similar to the distance range mate. Configure the angle to be between 0 and 180 degrees using **Mate Selections**.

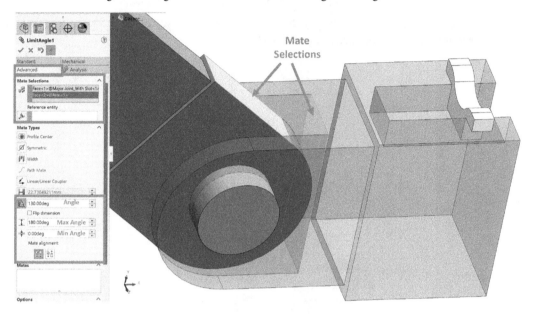

Figure 14.22 – The selection and PropertyManager for our angle range

> **Note**
>
> The additional field in this mate, **Reference entity**, is an optional entry in which we can select an edge, plane, or face as a base to measure the angle. In this exercise, we can leave it empty.

3. Click on the green check mark to apply the mate.

Once you confirm the mate, try moving the arm to double-check the functionality of the mate. The arm should have a range of movement that equals 180 degrees.

Now that we are familiar with most of the advanced mates, we can move to the last set, which includes the path mate and linear/linear coupler advanced mates.

Understanding and using the path mate and linear/linear coupler mates

In this section, we will explore the **path mate** and **linear/linear coupler** advanced mates. Both mates will allow us to introduce dynamic movements into our assemblies. We will learn what these mates are, what they do, and how to apply them.

Defining the path mate

The advanced mate's path mate allows us to restrict the movement of a specific part to follow a designated path. For example, we can have tiles follow a path, as shown in *Figure 14.23*.

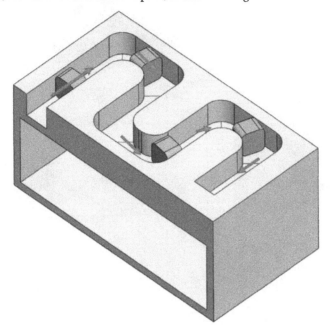

Figure 14.23 – The impact of the path mate

To apply the path mate relationship, we are required to have a path to follow and a point/vertex to follow that path. Next, we will apply the path mate command to an assembly.

Applying the path mate

In this section, we will learn how to apply the advanced mate's path mate. We will do that by limiting the movement of a tile to stick to a specifically designated path, as highlighted by the arrows in *Figure 14.23*. All of the parts and assembly files needed for this exercise are available for download with this chapter.

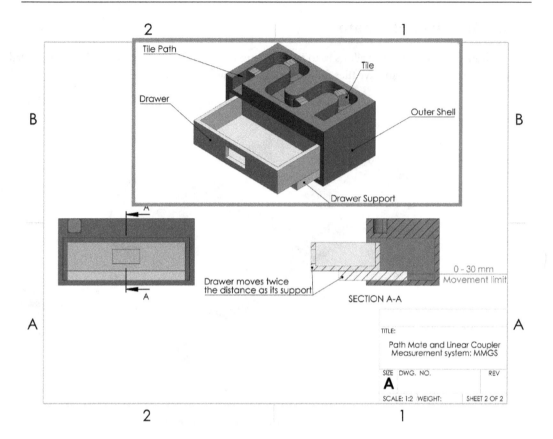

Figure 14.24 – The path mate we will apply in this exercise

To apply the path mate, we will be required to set up a path to follow. For this, we have the option of selecting an existing path that includes ones from existing sketches or existing edges. Another option is to create a new sketch to represent our path. However, when creating or selecting a path, note that a point within the moving part will have to coincide with the path at all times. For this exercise, follow these steps:

1. Create a new sketch, as shown in *Figure 14.25*. The line for the path goes through the middle of the carved path in the outer shell. After creating the sketch, exit the sketch mode.

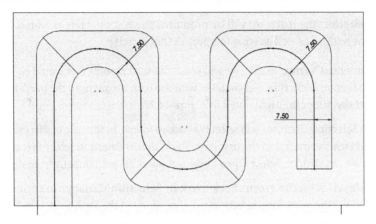

Figure 14.25 – To use the path mate, we must have a path

2. Select the **Path Mate** option by selecting the **Mate** command, the **Advanced** tab option, and then the **Path Mate** option, as highlighted in *Figure 14.26*.

Figure 14.26 – The location of the path mate

3. Once we select the mate, we will be prompted to set it up. Here is a brief description of the different settings as well as what to select in this exercise:

- **Component Vertex**: In this, we will select the vertex/point that would be following the path. Note that this selection will coincide with the path throughout the part's movement. We can select the midpoint highlighted in *Figure 14.27*.

- **Path Selection**: Here, we will select the path to follow. In the case of this exercise, we can select the sketch we created in the first step. To do that, we need to select **SelectionManager**, and then we can click on **Select Open Loop** and select the path directly from the canvas afterward.

- Finally, click on the green check mark in **SelectionManager** to confirm the open-loop selection. The open-loop selection will then show in the **Path Selection** field, as highlighted in *Figure 14.27*:

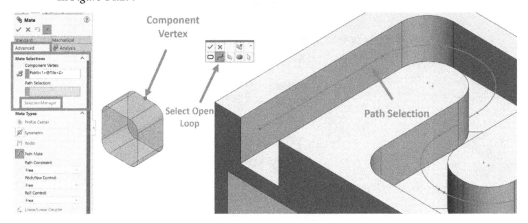

Figure 14.27 – The selection of the path and vertex

- **Path Constraint**: This allows us to determine the movement of the part (tile) along the path. In this exercise, we can choose the **Free** option, which will allow us to simulate the movement of the tile along the path by dragging it. Other options include **Distance Along Path** and **Percent Along Path**. These will fix the tile in a certain position along the path, which we can determine by distance or percentage.

- **Pitch/Yaw Control**: This has to do with the orientation of the object – the tile, in this case. We can pick the **Follow Path** option; this will allow us to constrain one axis of the moving part to be tangent to the selected path throughout the movement. Note that the selection of the **X**, **Y**, and **Z** axes are all in relation to the selected path, not in relation to the part or assembly. Another option we can select for this control is **Free**, which will not impose any type of limitation. A good practice is to experiment with different options before confirming what fits your needs.

- **Roll Control**: This also has to do with the orientation of the moving object. We can pick the **Up Vector** option; this will constrain one axis of the moving part to align with a vector of our choosing. With this option, we can set one side of the tile to be facing one direction throughout its movement. In our example, we can pick the edge highlighted in *Figure 14.28*. Another option we can select for this control is **Free**, which will not impose any limitation related to rolling along the path. Similarly, when setting up **Pitch/Yaw Control**, a good practice is to experiment with the different options to see their effect. You might need to mix and match between the different options in **Roll Control** and **Pitch/Yaw Control** to see which combination of options fits your requirements. Keep experimenting until you get the orientation shown in *Figure 14.28*.

 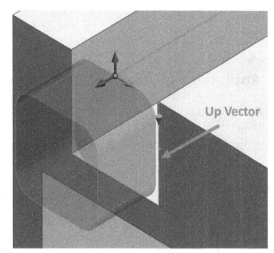

Figure 14.28 – Vectors allow us to define the orientation of the part while following the path

4. Click on the green check mark to confirm the mate. Once you confirm the mate, try moving the tile; you will notice that it will follow the path without any rotation around the path.

This concludes our application of the path mate. Next, we will cover the linear/linear coupler mate.

Defining the linear/linear coupler

The **linear/linear coupler** advanced mate allows us to create a linear transitional relationship between two different parts. Here, if one part moves, that other part will move as well according to a specific ratio. *Figure 14.29* highlights the effect of the linear/linear coupler mate. Note that from *Figure 14.29*, the drawer and drawer support are coupled together in such a way that the drawer and the drawer support will follow each other's linear movements at a 1:2 ratio where the drawer will move twice as much as the drawer support. This ratio can be adjusted to our needs.

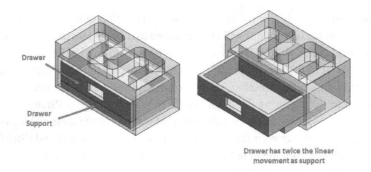

Figure 14.29 – The impact of the linear/linear coupler mate

Now that we know what the linear/linear coupler mate does, we can start applying it in our assembly.

Applying the linear/linear coupler

In this section, we will learn how to use the linear/linear coupler mate. We will do that by setting a transitional relationship between the drawer and the drawer support, as highlighted in *Figure 14.30*. In this exercise, we are continuing our work on the assembly we started with the path mate:

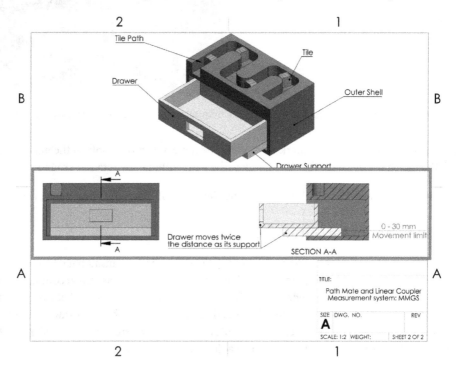

Figure 14.30 – The linear/linear coupler command we will apply in this exercise

To apply this mate, we are required to have two different components that will move linearly in relation to each other. We can follow these steps to set it up:

1. Select the **Linear/Linear Coupler** mate by selecting the **Mate** command, the **Advanced** tab option, and then the **Linear/Linear Coupler** mate, as highlighted in *Figure 14.31*.

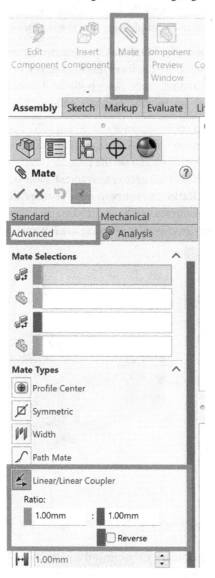

Figure 14.31 – The location of the linear/linear coupler mate

2. Once we select the mate, we will be prompted to set it up, as shown in *Figure 14.32*. Here is a brief description of the different settings as well as what to select in this exercise:

- **Entity to Mate**: In these two fields, we can select faces or edges upon which the relationship will be built. Note that this selection will also indicate the direction in which the movement will happen. For example, note the face selections we have in *Figure 14.32*; this indicates that the movement we want to couple/pair is in the normal direction to these surfaces.

- **Reference Component for Mate Entity**: Here, we can select a part to act as a reference for the moving component. We can leave this field empty, in which case, the movement will use the assembly origin as a reference. Since both the drawer and the drawer support movement are not referenced to any moving part, we can leave it blank.

- **Ratio**: In this, we can set the distance movement ratio. For example, in a ratio of 1:2, one part will move twice as much compared to others in the same linear direction of movement. As indicated in *Figure 14.32*, we can set the ratio to 1 : 2, where 1 refers to the drawer support movement and 2 refers to the drawer movement. Note that, in the SOLIDWORKS interface, the selection of the **Ratio** and entity to mate settings are color-coded for easier reference. By default, both movements will be in the same direction. However, we can also have them moving in different directions by selecting the **Reverse** option.

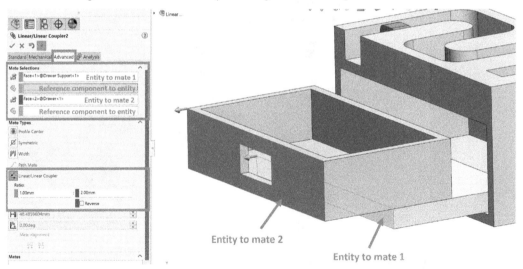

Figure 14.32 – The selection and PropertyManager of our linear/linear coupler mate

3. Click on the green check mark to confirm the mate. After that, try moving any of the mated parts; you will notice that the drawer will always move twice as much as the drawer support.

This covers the basic application of the linear/linear coupler mate. However, in practice, this mate is often coupled with other mates to simulate a more realistic application. Let's do that next.

Fine-tuning the linear/linear coupler mate

If we look back at the resulting assembly we have after applying the linear/linear coupler mate, we will notice that the part does not have a starting point from the backend of the outer shell. Also, the linear movements are not restrained for the two parts. To set a starting point and limit the linear movement, we will utilize the coincident and distance range mates respectively.

Before creating those adjustments, we need to suppress the linear/linear coupler mate we applied earlier. To do this, click on the mate in the design tree and select the suppress icon, as highlighted in *Figure 14.33*. We will un-suppress the mate after adjusting the other settings:

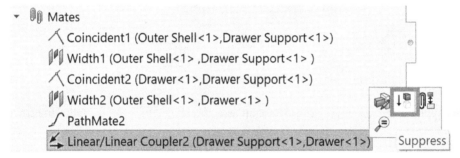

Figure 14.33 – Suppressing the linear/linear coupler allows us to adjust the starting position

Now, we will start by setting our starting point using the coincident standard mate. Here, we will create a coincident mate between the back of the drawer and the drawer support to coincide with the back of the outer shell, as indicated in the cross-section shown in *Figure 14.34*.

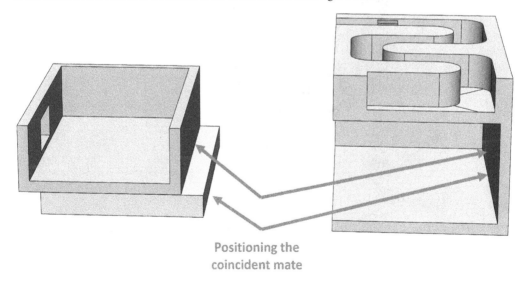

Positioning the
coincident mate

Figure 14.34 – We set up the initial positions of the parts

Note that we are applying this mate for positioning purposes only. To limit the mate to the positioning effect, we can select the **Use for positioning only** option, as highlighted in *Figure 14.35*. We can find this option at the bottom of the PropertyManager when setting up the mate. Checking this option will not list the mate in the design tree as an active mate; rather, it will only bring the parts to the position as if the mate were applied.

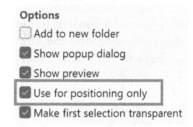

Figure 14.35 – We can set up mates for positioning purposes only

After applying the two positioning mates, our assembly should look like the following cross-section:

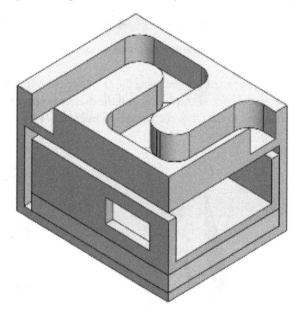

Figure 14.36 – The starting positioning of the drawing assembly

Now, we can use the distance range advanced mate to limit the movement of the drawer support from 0.00mm to 30.00mm. Make sure that you set the initial distance to 0.00mm to keep the positioning. After that, we can un-suppress the linear/linear coupler mate we suppressed earlier. Then, you can start testing the functionality of the mate by moving the drawer or the drawer support. An alternative approach to this exercise is to set the positioning and the movement limitation first, and then apply the linear/linear coupler mate.

This concludes our coverage of the linear/linear coupler advanced mate. We covered what it is, how to set it up, and how to use different mates to support it. This mate can help to simulate different common sliding mechanisms, such as the ones in common drawers, making it an important tool in our mechanical design toolkit.

Summary

In this chapter, we learned about all of the advanced mates available in SOLIDWORKS assemblies. These include the profile center, symmetric, width, distance range and angle range, path mate, and linear/linear coupler mates. For each of the mates, we learned what the mate does and how to use it in the SOLIDWORKS assemblies environment. These advanced mates allow us to create more flexible assemblies that are closer to a variety of dynamic design intents, which is the objective of this chapter. At this point, we are well acquainted with both standard and advanced mates. With those, we are able to simulate different types of interactions as well as many different types of dynamics.

In the next chapter, we will continue working with assemblies to cover configurations and design tables, which will enable us to generate different assembly versions in a single SOLIDWORKS assembly file. This is in addition to different assembly tools, such as collision and interference detection and assembly features.

Questions

1. What does the profile center mate do?

2. What does the symmetric mate do, and what are the elements needed to apply it?

3. What does the path mate do, and what are the different constraints we can use with the mate?

4. What does the linear/linear coupler mate do, and what is a common application for this mate?

5.   Using the part files provided for this question, generate the assembly highlighted in the following figure. Note that both sticks move symmetrically on the path in the base:

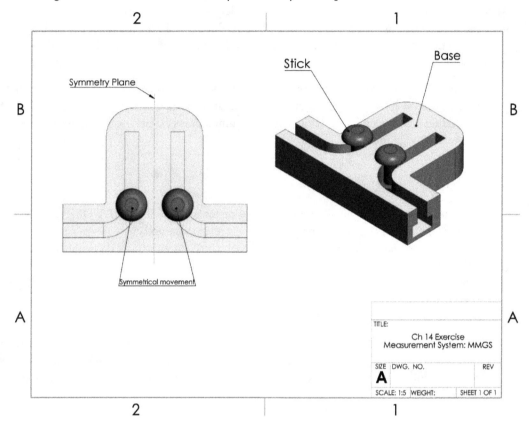

Figure 14.37 – The drawing for Question 5

6. Using the part files provided for this question, generate the assembly highlighted in the following figure:

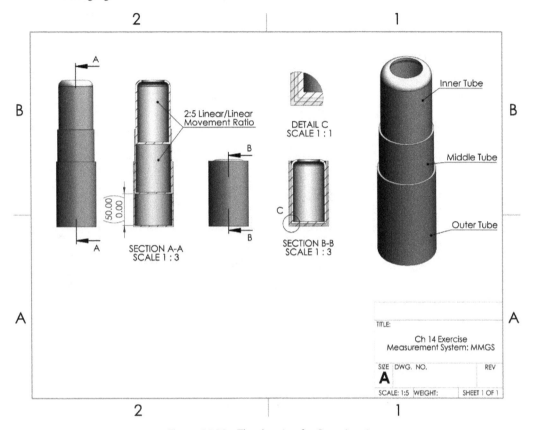

Figure 14.38 – The drawing for Question 6

> **Important note**
> The answers to the preceding questions can be found at the end of this book.

Get This Book's PDF Version and Exclusive Extras

UNLOCK NOW

Scan the QR code (or go to packtpub.com/unlock). Search for this book by name, confirm the edition, and then follow the steps on the page.

Note: Keep your invoice handy. Purchases made directly from Packt don't require an invoice.

Advanced SOLIDWORKS Assembly Competencies

Working with assemblies is more than just putting parts together and linking them with mates. In this chapter, you will learn about additional assembly competencies that can help us build more technically feasible products. First, we will learn about the **Interference Detection** and **Collision Detection** tools, through which we can identify when parts are interfering or colliding with each other. Then, we will cover assembly features that are applied in the assembly file rather than the part file. Finally, we will learn about configurations and design tables within the assembly's context. These will enable us to generate and manage multiple variations of an assembly within one file.

The following topics will be covered in this chapter:

- Understanding and utilizing the Interference Detection and Collision Detection tools
- Understanding and applying assembly features
- Understanding and utilizing configurations and design tables for assemblies

By the end of this chapter, we will be able to put together more advanced assemblies that are verified for interferences and collisions and include variations and features outside the scope of parts. The overall goal of this chapter is to acquire additional skills to make our assemblies more flexible and able to cater to more specific applications.

Technical requirements

This chapter requires access to SOLIDWORKS and Microsoft Excel on the same computer. The project files for this chapter are available at the following GitHub repository: `https://github.com/PacktPublishing/Learn-SOLIDWORKS-2025-Third-Edition`

The CiA video for this chapter can be found at `https://packt.link/hzyxf`

Understanding and utilizing the Interference Detection and Collision Detection tools

By now, we know that most of the products we deal with in our everyday life consist of more than one part. This is the case regardless of whether we are looking at a simple pen or a complex engine. As SOLIDWORKS professionals, we will require more tools than just mates to help us evaluate our assembly. In this section, we will cover the Interference Detection and Collision Detection tools. We will cover what they are and how to use them in SOLIDWORKS assemblies. Throughout this section, we will be working with the assembly we linked with this chapter.

Interference detection

Within an assembly, we might end up having different parts interfering with each other, meaning there might be one part inside another. This could happen because of a deficiency in our design or because it was intentional. Regardless, SOLIDWORKS assembly provides us with the Interference Detection tool to determine how much interference is taking place in our assembly. This tool lets us know whether or not any interference is taking place; it also allows us to determine the location of the interference, the parts involved, and how much volume is interfering.

To learn how to use the tool, we will apply it to the assembly shown in *Figure 15.1*. You can download the parts and the assembly file from this chapter. Open up the assembly file after downloading it. Our task is to find out whether some parts are interfering with each other:

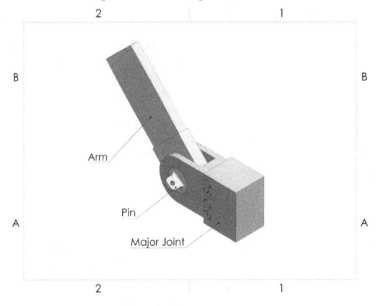

Figure 15.1 – The assembly we will apply the Interference Detection tool on

To apply the Interference Detection tool, we can follow these steps:

1. Select the **Evaluate** tab, then select the **Interference Detection** command, as shown in *Figure 15.2*:

Figure 15.2 – The location of the Interference Detection command

2. In **PropertyManager**, the assembly will be listed under **Selected Components**, as shown in *Figure 15.3*. Click on **Calculate**. Then, the interference detection result will be shown in the space under **Results**:

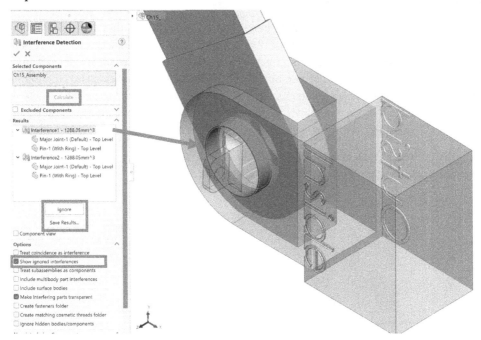

Figure 15.3 – The interference detection calculation and the result

Notice that the result will show how many interference incidents are taking place in the assembly alongside the interference volume. In this example, we have two interference incidents, each with an interfering volume of **1288.05 mm^3**. If you expand the interference incident menu, you will see a list of the parts involved in that particular interference. Also, if you select the interference incident, it will be highlighted in the canvas as a visual indication of where the interference is taking place.

> **Note**
>
> To see all the interfaces, make sure the **Show ignored interferences** option is checked under **Options**. This option is highlighted in *Figure 15.3*. Otherwise, SOLIDWORKS can ignore repeated interferences involving the same parts, as is the case in our example.

Once we know where the interference is taking place, we can choose to address, ignore, or save results to an external file, as indicated in *Figure 15.3*. Generally, we can address those interferences by adjusting the design of the parts or adjusting how the different parts are joined together. A common perception in practice is that interferences are always undesired when modeling products. However, that is not always the case. Let's address this practical point.

Interferences in practice

In practice, interferences are not negative in an absolute sense. Rather, they could also be the desired outcome, depending on the application. For example, in industrial machines such as pumps and compressors, ball bearings can have an interference fit with shafts to ensure both shaft and bearing will rotate together. Maintenance personnel will then use methods such as heating to install ball bearings with their interfering shafts. As SOLIDWORKS users, we need to be able to determine the interference instance in our assembly. It is also essential for us to understand that, in some instances, interference is desirable for fitting applications.

In the **Interference Detection** tool's **PropertyManager**, we have different options that we can utilize depending on our needs. The options are highlighted in *Figure 15.4*:

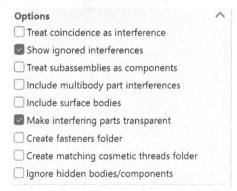

Figure 15.4 – More options for the interference detection calculation

Let's take a look at what each of those options does, as follows:

- **Treat coincidence as interference**: We can find this option if we expand the **Options** menu in **PropertyManager**. This will list all instances with two parts touching each other in the interference result—that is, parts with a transitional fit.

- **Show ignored interferences**: When using the **Ignore** command with interference, it will get hidden from the results in **PropertyManager**; checking this option will have it shown as a grayed-out listing. Also, this will display all repeated interferences in our results.

- **Treat subassemblies as components**: This will have the software treating subassemblies as a single component, meaning that interfering parts within the subassembly will not be included in the interference detection study.

- **Include multibody part interferences**: Having this checked will ask the software to check whether different bodies within one part are interfering and, if so, to report those interferences.

- **Make interfering parts transparent**: This will change the display of the interfering parts to be transparent, making it easier to visually study the interference.

- **Create fasteners folder**: This will create a folder under the results only for fasteners involved in the assembly.

- **Create matching cosmetic threads folder**: This will create a separate folder for components with matching cosmetic threads. Unmatching threads will not be listed in the same folder.

- **Ignore hidden bodies/components**: This will exclude hidden parts and bodies from the interference detection study.

This concludes our exploration of the Interference Detection tool. We learned what it does and how to use it. Next, we will learn about the Collision Detection tool.

Collision detection

The Collision Detection tool allows us to get notified when two parts collide with each other. This is commonly used with dynamic assemblies. As we are moving different parts within the assembly, the tool can notify us when the collision happens and highlight which parts are colliding.

To illustrate how to use the tool, we will apply it to the same assembly we used with the Interference Detection tool earlier. To start, make sure the assembly file attached to this chapter is open. To use the tool, we can follow these steps:

1. Select the **Move Component** command in the **Assembly** tab, as highlighted in *Figure 15.5*:

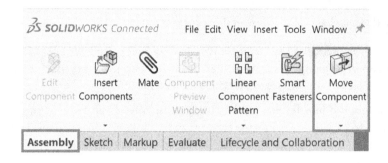

Figure 15.5 – The location of the Move Component command

2. In **PropertyManager**, select the **Collision Detection** option. Also, check the **Stop at collision** and **Dragged part only** options and the **Highlight faces** and **Sound** advanced options, as shown in *Figure 15.6*:

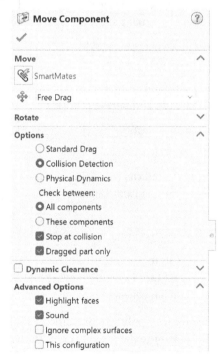

Figure 15.6 – Collision Detection is within the Move Component command

3. Start moving the arm; you will notice that the movement will stop when the *arm* hits the *major joint*. The software will also give us a sound indicator, notifying us that the collision took place. Also, the colliding faces will be highlighted to help us to identify where the collision is happening.

Once we have the parts at the colliding position, we can use the **Smart Dimension** command to get the exact collision angle or distance measurements. For instance, in this exercise, we can measure the collision angle with the following steps:

1. Drag the arm clockwise until it collides with the major joint, as shown in *Figure 15.7*.

2. Use the **Smart Dimension** command in the **Layout** tab to measure the angle between the arm and the major joint, as shown in *Figure 15.7*.

This way, we will be able to identify the collision angle between the arm and the major joint as **90** degrees, as highlighted in *Figure 15.7*. Those numerical measurements can then help us to make different design decisions:

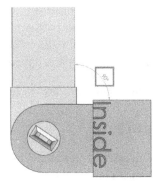

Figure 15.7 – Smart Dimension can be used to generate measurements relating to the collision

> **Note**
> Interfering parts will be flagged as being already colliding when using the Collision Detection tool. In the preceding example, if we keep the **Dragged part only** option checked, SOLIDWORKS will give us the message that the model is in a colliding position because the *pin* is in an interference relationship with the *major joint*, as explored earlier in this section.

This concludes our coverage of the Collision Detection tool. We learned what it is, what it does, and how to use it. We also learned about the Interference Detection tool in an earlier section titled *Interference detection*. Next, we will learn about assembly features, which refer to applying features in the assembly's context.

Understanding and applying assembly features

Assembly features refer to applying features such as extruded cuts and fillets within the assembly file rather than the part file. In this, the features will be recorded and applied on the assembly file and listed in the assembly file design tree without impacting the design of the original part. In this section, we will learn about assembly features and how to apply them.

Understanding assembly features

We can look at assembly features as standard features that are stored in the assembly file rather than the part file. This makes them unique in delivering a particular design intent. To apply assembly features, we need to be working in the assembly file. Compared to the features that we can apply as we are modeling parts, assembly features are limited and only include subtractive features. These include extruded cuts, swept cuts, revolved cuts, fillets, and holes. By default, assembly features do not propagate to the parts; however, we can make them propagate if needed. We use assembly features for two main reasons, outlined as follows:

- To deliver the message that the material subtraction takes place after manufacturing the parts themselves and during the assembly process.

- To apply a subtractive feature across multiple parts, which will make some design intents easier to accomplish, such as having a hole perfectly aligned and going through multiple parts. In this case, we can use the **Propagate feature to parts** option.

Next, we will apply the **Extruded Cut** assembly feature to our assembly.

Applying assembly features

Here, we will apply the **Extruded Cut** assembly feature to the assembly we generated earlier. We will use the feature to drill a hole in the major joint and the pin, as shown in *Figure 15.8*:

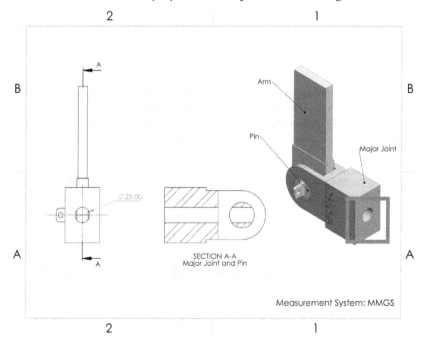

Figure 15.8 – The assembly we will generate in this exercise

To apply the **Extruded Cut** assembly feature, follow these steps:

1. Select the **Assembly** tab, then expand the **Assembly Features** command and select **Extruded Cut**, as shown in *Figure 15.9*:

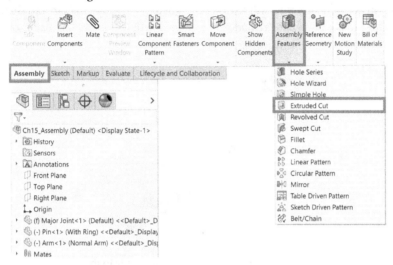

Figure 15.9 – The location of the different assembly features

2. Apply the features as usual by selecting the sketch face and drawing the circle sketch, then exit the sketch to set up the extruded cut.

3. Set the end condition to **Through All**; you will notice that the cut preview will go through all of the parts. However, we only require it to apply to the major joint and the pin. We can set that up on **Feature Scope**, as shown in *Figure 15.10*:

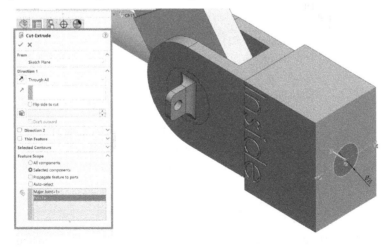

Figure 15.10 – Feature Scope allows us to apply the assembly feature to specific parts

4. Click on the *green checkmark* to apply the extruded cut.

We can view the cut on the two parts using the cross-section view, as shown in *Figure 15.11*. Note that the hole only goes through the major joint and pin. One important aspect to note is that the holes on the two parts (major joint and pin) will always be aligned. If we rotate the pin and then rebuild the assembly, the hole in the pin will be remade to be aligned with the one in the major joint:

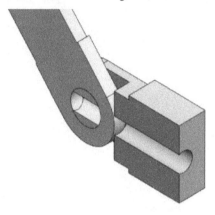

Figure 15.11 – The resulting cut did not apply to the arm

Note that, after applying the extruded cut, it will then be listed in the assembly's design tree and not under any particular part, as highlighted in the following design tree:

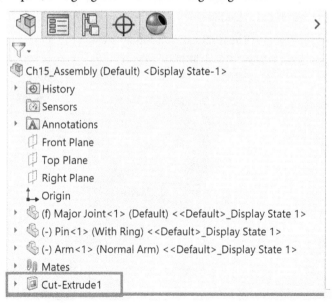

Figure 15.12 – The assembly feature is listed in the assembly design tree

Also, if we open the part file for the major joint or pin, we will notice that the new extruded cut is not there. If we want the extruded cut to reflect on the part, we can check the **Propagate feature to parts** option under **Feature Scope** when setting up the extruded cut. You can see that option under **Feature Scope** in *Figure 15.10*.

This concludes this section on assembly features. We learned what they are, why we should use them, and how to apply them. Next, we will use the same model to learn about configurations and design tables in the assembly's context.

Understanding and utilizing configurations and design tables for assemblies

With configurations and design tables, we will be able to create different versions of an assembly in the same assembly file. This is very similar to what we learned previously when creating configurations for a part. Within an assembly, our configurations can vary based on the mates applied and the existing parts within the assembly. We can also link them to configurations within the individual parts themselves. In this section, we will generate different versions of the assembly file we worked on in the previous section. We will start by generating a configuration *touchpoint* by manually adding a configuration.

Using manual configurations

We will manually create the configuration shown in *Figure 15.13*. As illustrated in the drawing, this configuration includes an additional mate compared to the default setup, which fixes the *arm* at a 90-degree angle to the major joint:

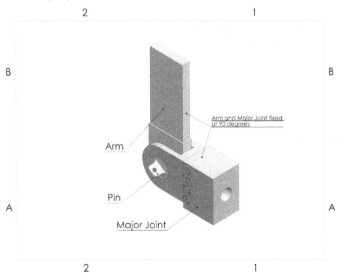

Figure 15.13 – The configuration we will generate in this exercise

The procedure for adding a configuration in assemblies is the same as in parts. The difference here is that we will be dealing with different elements, such as mates. To generate the indicated configuration, we can follow these steps:

1. Go to **ConfigurationManager**, right-click, then select **Add Configuration...**, as shown in *Figure 15.14*. Assign the name TouchPoint to the configuration. Then, make sure we are operating within this new configuration:

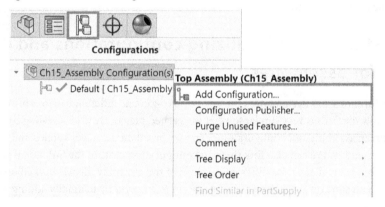

Figure 15.14 – The location of the Add Configuration… command

2. Apply the standard mate angle to set the *arm* at 90 degrees with the *major joint*. This will fix the *arm* at the position shown in the initial drawing in *Figure 15.13*.

This will apply the mate to the new **TouchPoint** configuration. As usual, go back to the **Default** configuration to double-check how the two configurations are different. You will notice that in the **Default** configuration, the *arm* can rotate freely, while in the **TouchPoint** configuration, the *arm* is fixed at 90 degrees. Also, you will notice that the additional angle mate we added will be shown as suppressed in the **Default** configuration.

> **Note**
>
> The **Default** configuration will have the angle mate suppressed if the **Suppress new features and mates** option is checked in its properties. You can find more information about this option in the next section, *Configuration advanced options*.

Configuration advanced options

When starting a new configuration, three notable advanced options are important for our workflow. These are highlighted in *Figure 15.15*:

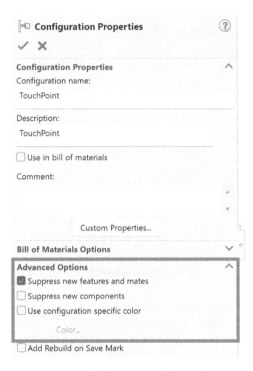

Figure 15.15 – Configuration advanced options

Let's explore the impact of each option, as follows:

- **Suppress new features and mates**: Having this option checked means that any new features and mates applied to other configurations would be suppressed in this configuration.

- **Suppress new components**: This option means that any other components that get added to other configurations will be suppressed in this particular configuration.

- **Use configuration specific color**: This allows us to apply a specific color appearance to the configuration for easier identification.

> **Note**
> The first two suppress advanced options relate to how the configuration at hand interacts with changes in other configurations, not to how other configurations interact with the configuration at hand.

After adding the configuration, we can access and adjust the advanced configuration by accessing **Properties...** in **ConfigurationManager**, as highlighted in *Figure 15.16*:

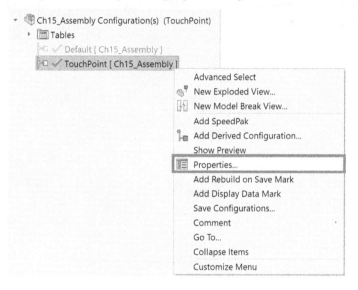

Figure 15.16 – We can access the advanced options through properties

This concludes how to add a new configuration within an assembly. Next, we will generate different configurations using design tables.

Design tables

Using design tables, we will generate the configurations highlighted in *Figure 15.17* for our assembly. Note that all new configurations are based on the **Default** one, not the **TouchPoint** one. Also, the difference between each configuration is the shape of the *arm* and the *pin*. Both the *arm* and *pin* parts already have two-part configurations. For the *arm*, we have the **Normal Arm** and **Fork Arm** configurations, while for the *pin*, we have **With Ring** and **Plain**, as illustrated in *Figure 15.17*:

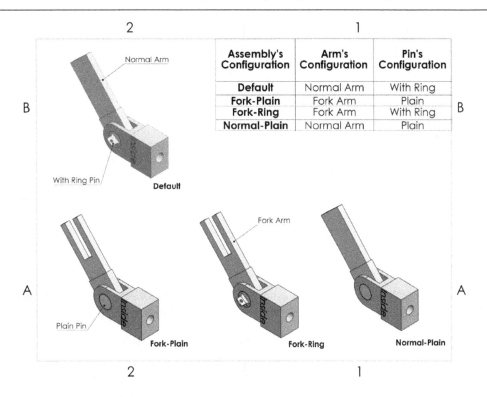

Assembly's Configuration	Arm's Configuration	Pin's Configuration
Default	Normal Arm	With Ring
Fork-Plain	Fork Arm	Plain
Fork-Ring	Fork Arm	With Ring
Normal-Plain	Normal Arm	Plain

Figure 15.17 – The design table we will generate in this exercise

We can add a design table to an assembly in the same way we add it to a part. However, we can control multiple parts within the assembly. To generate the four configurations shown in the preceding diagram, follow these steps:

1. Insert a new design table by going to **Insert**, **Tables**, then **Excel Design Table**. Set the table to **Auto-create**, then click on the green checkmark.

> **Note**
> Once at the design table, we will notice the two configurations we have are already listed. These are **Default** and **TouchPoint**. This is in addition to all of the differences between them, including the angle mate we generated earlier when working with configurations.

2. To recall the parts configurations, we can use the $configuration@partName<instance> title format. Call the configurations for two parts, the *arm* and the *pin*. We can find both the part name and instance in the assembly's design tree. The design table will look as in *Figure 15.18*:

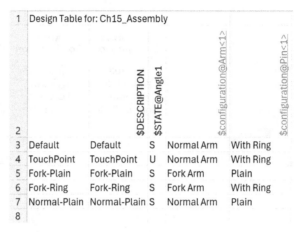

1	Design Table for: Ch15_Assembly				
		$DESCRIPTION	$STATE@Angle1	$configuration@Arm<1>	$configuration@Pin<1>
2					
3	Default	Default	S	Normal Arm	With Ring
4	TouchPoint	TouchPoint	U	Normal Arm	With Ring
5	Fork-Plain	Fork-Plain	S	Fork Arm	Plain
6	Fork-Ring	Fork-Ring	S	Fork Arm	With Ring
7	Normal-Plain	Normal-Plain	S	Normal Arm	Plain
8					

Figure 15.18 – The format of the design table

3. Apply the design table by clicking anywhere on the canvas. Then, double-check the different new configurations generated. The configurations should match the ones shown in *Figure 15.17*.

This concludes how to generate a design table for part configurations within an assembly. Note that the part configurations we used in this exercise were generated within part files. We can also manually adjust the part configuration within the assembly. To do this, we can simply click on the part listing in the design tree and change the configuration from the drop-down menu, as shown in *Figure 15.19*. We can also access the same drop-down menu by clicking on the part in the canvas:

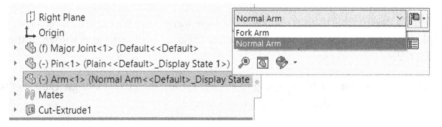

Figure 15.19 – We can adjust the active part configuration in the assembly from the design tree

In this section, we learned about configurations and design tables within an assembly context. This enabled us to generate multiple variations of an assembly within one assembly file.

Summary

In this chapter, we learned various skills relating to SOLIDWORKS assembly tools. We started with the Interference Detection and Collision Detection tools to determine whether two or more parts are interfering with each other in a static position or colliding with each other in a dynamic movement.

Next, we learned about assembly features, which are subtractive features that are applied and stored with the assembly file rather than the part file. Finally, we learned about configurations and design tables, which enabled us to generate multiple variations of an assembly within one assembly file. Overall, the skills we have learned in this chapter will enable us to better meet different design intents by evaluating, adjusting, and varying our assemblies' outcomes.

Next, we will work on a 3D modeling project to create a remote-controlled helicopter model. The project will provide you with a comprehensive practice for many of the topics covered in this book.

Questions

Answer the following questions to test your knowledge of this chapter:

1. What do the Interference Detection and Collision Detection tools do?
2. What is the difference between assembly features and the features we add within parts?
3. What is the difference between dealing with configurations and design tables in assemblies versus parts?
4. Download the parts and assembly linked to this exercise. Find the interference volume between **Helical Gear** and **Fixture** and between **Worm Gear** and **Ball Bearing**. All parts are as indicated in the following diagram:

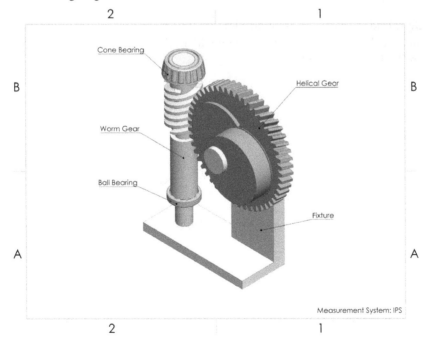

Figure 15.20 – Drawing for question 4

5. Using the assembly from the previous question, generate a design table introducing the following configurations:

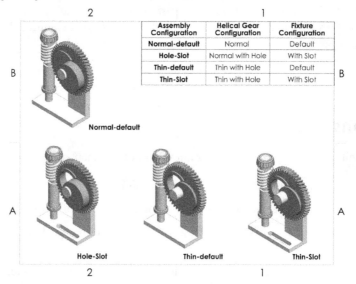

Figure 15.21 – Drawing for question 5

6. Download the parts and assembly files with this exercise. Use the assembly features to create the hole shown in the following diagram. Make the assembly feature propagate to the parts involved in their part files:

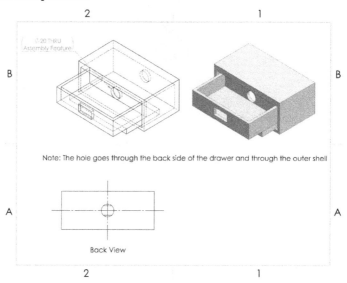

Figure 15.22 – Drawing for question 6

7. Using the assembly from the previous question, generate two additional configurations, one showing the drawer fully open and the other showing it fully closed, as shown in the following diagram. Make sure the default configuration remains flexible:

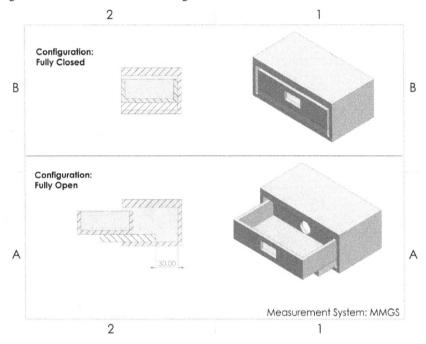

Figure 15.23 – Drawing for question 7

Project 2
3D Modeling an RC Helicopter Model

SOLIDWORKS is a 3D design tool. Just like all tools, the more you use it, the better you become at it. In this project chapter, you will be provided with project work that you can do to hone your skills. In this project, you will be 3D modeling and assembling a helicopter toy model from a set of engineering drawings.

This project chapter will cover the following topics:

- Understanding the project
- 3D modeling the individual parts
- Creating the assembly

By the end of this chapter, you will have more confidence in using the different SOLIDWORKS tools for practical projects.

Technical requirements

You will need to have access to SOLIDWORKS to complete the project.

Understanding the project

Understanding what the project entails is essential before starting the work. This will allow you to draw up a plan and manage your work expectations for completing the project. In this project, you will be 3D modeling a **remote control** (**RC**) helicopter toy, as shown in *Figure P2.1*:

Figure P2.1 – The RC helicopter you will 3D model in this project

The model consists of 16 parts, 12 of which are unique. *Figure P2.2* shows the bill of materials with the names of the parts, their quantity, and their position in the assembly:

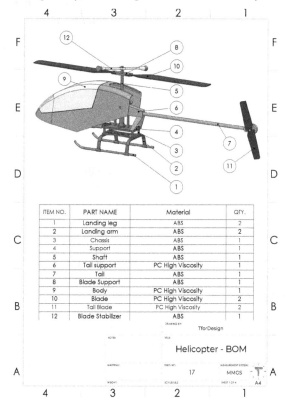

ITEM NO.	PART NAME	Material	QTY.
1	Landing leg	ABS	2
2	Landing arm	ABS	2
3	Chassis	ABS	1
4	Support	ABS	1
5	Shaft	ABS	1
6	Tail support	PC High Viscosity	1
7	Tail	ABS	1
8	Blade Support	ABS	1
9	Body	PC High Viscosity	1
10	Blade	PC High Viscosity	2
11	Tail Blade	PC High Viscosity	2
12	Blade Stabilizer	ABS	1

Figure P2.2 – The RC helicopter and its bill of materials

At this point, you already have an idea of the project's outcome and the complexity of the required parts and assembly. In the following section, we will provide you will the engineering drawings needed to replicate all the parts and assembly. Now that we have an idea about the project's final output, we can discuss how to tackle it in this project.

> **Note**
> The drawings and 3D models presented in this project are for practice rather than manufacturing purposes.

There are two ways in which you can tackle this project depending on your 3D modeling level. They are as follows:

- **Moderate level**: Take a look at the drawings and the provided sample procedure to complete the project.

- **Advanced level**: Only take a look at the drawings without using the sample procedure.

You can also follow your own way to 3D model the RC helicopter utilizing the provided drawings and sample procedures. Keep in mind that the sample 3D modeling procedures provided are meant as a sample guide. They are not meant to present an optimal procedure; rather, you can look at them as one possible procedure to generate the models and assemblies. To grow your own 3D modeling style, you can experiment with modeling the project using different modeling procedures.

> **Tip**
> You can treat the project as your own and customize the provided RC helicopter to end up with your unique design.

In this project, we will first explore the individual parts, and then move on to the assembly. So, let us get started with the parts. We will also provide you with hints that can assist you with your work.

3D modeling the individual parts

In this section, we will explore the different part drawings that represent the RC helicopter. The provided drawings have enough information for you to replicate all the parts to end up with an identical result to the one shown in *Figure P2.1*. Thus, one option for handling the project is to create an exact replica of the given drawings. However, you can also choose to customize and adjust different elements of the design to make it your own. Keep in mind that this is your project, so feel free to treat it as such.

Exploring the individual parts

The provided RC helicopter consists of 16 parts. However, 12 of these that you will need to 3D model are unique, as highlighted in *Figure P2.2*. The parts you will need to 3D model are as follows. In the text, we will mod

- Landing leg
- Landing arm
- Chassis
- Support
- Shaft
- Tail support
- Tail
- Blade support
- Body
- Blade
- Tail blade
- Blade stabilizer

> **Note**
>
> The names of the parts presented in the bill of materials might be different than the names used in the industry.

Your task is to use the presented drawings to 3D model the individual parts. As you are 3D modeling the different parts, keep in mind that there is no one correct way of 3D modeling any of the parts. However, we will provide you with some hints for one approach that can push you forward if you find yourself getting stuck. Also, feel free to customize your design using the given drawings as a base of inspiration. The order in which the drawings are presented in this chapter is arbitrary.

> **Note**
>
> All engineering drawings are presented using the third-angle projection.

Let us start exploring the drawings one after the other. The first drawing is for the *landing leg*, which is shown in *Figure P2.3*.

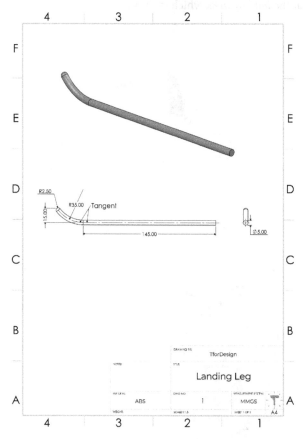

Figure P2.3 – Detailed drawing for landing leg

Here is a sample procedure for 3D modeling the landing leg:

1.   You can start with a swept boss to create the main shape, as shown in *Figure P2.4*:

Figure P2.4 – Swept boss to be used to start the landing leg

2. Apply fillets to get a rounded edge.

Next, we can look at the *landing arm*, which is shown in *Figure P2.5*:

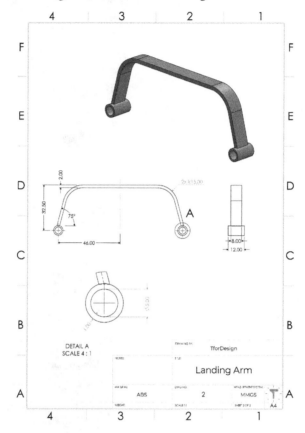

Figure P2.5 – Detailed drawing for the landing arm

Here is a sample procedure for 3D modeling the landing arm:

1. Create a sketch highlighting the entire shape of the arm. Then, use an extruded boss with selected contours to extrude the long-connected shape. This will result in the shape shown in *Figure P2.6*.

Figure P2.6 – A possible first step in creating the landing arm is using an extruded boss

2. Reuse the sketch from *step 1* to extrude boss the rings to get the final shape shown in *Figure P2.7*.

Figure P2.7 – One sketch can be used for more than one feature

After the landing arm, we can explore the *chassis*, which is shown in *Figure P2.8*:

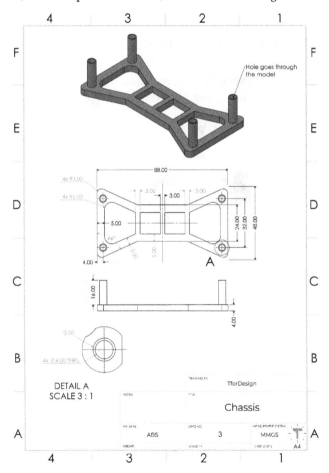

Figure P2.8 – Detailed drawing for the chassis

Here is a sample procedure for 3D modeling the chassis:

1.   Use an extruded boss to create the mainframe of the chassis, as shown in *Figure P2.9*.

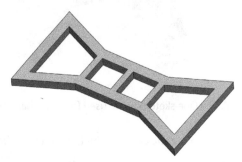

Figure P2.9 – Extruded boss can be used to create the mainframe of the chassis

2.   Use the fillet feature to round the corners. Then, use an extruded boss and cut to create the extensions and holes in the corners. You will get the final shape shown in *Figure P2.10*.

Figure P2.10 – The chassis can be finalized with fillets, an extruded boss, and an extruded cut

---

Tip

In general, it is a good practice to add fillets and chamfers at the end of the modeling process.

After the chassis, we can take a look at the *blades*. There are two blades in the RC helicopter, top blades and tail blades. We can utilize configurations or design tables to create both blades in one part file. This is because both blades have similar design features. We will first create the top blade and then generate the tail blade out of it. *Figure P2.11* shows the default (top) blade.

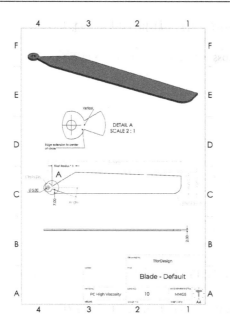

Figure P2.11 – Detailed drawing for the blade

Note that the drawing has many variables without numerical values. The numerical values for both the default and tail configurations are presented in *Figure P2.12*:

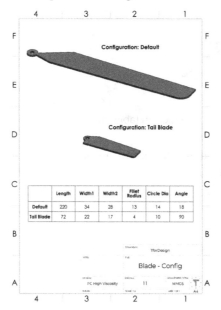

	Length	Width1	Width2	Fillet Radius	Circle Dia	Angle
Default	220	34	28	13	14	18
Tail Blade	72	22	17	4	10	90

Figure P2.12 – The two configurations for the blade

Here is a sample procedure for 3D modeling the two blades:

1. Use an extruded boss with the sketch shown in *Figure P2.13* to create the main blade.

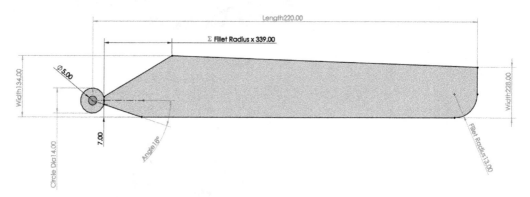

Figure P2.13 – Extruded boss can create the bulk of the blade

2. Introduce a design table, as shown in *Figure P2.14*, to create the tail blade. Use the variables named in *Figure P2.13*.

	A	B	C	D	E	F	G	H
1	Design Table for: Blade							
2		Length@Sketch1	Width1@Sketch1	Width2@Sketch1	Fillet Radius@Sketch1	Circle@Sketch1	Angle@Sketch1	
3	Default	220	34	28	13	14	18	
4	Tail Blade	72	22	17	4	10	90	
5								
6								

Figure P2.14 – A design table can be used to create the tail blade

3. After introducing the design table, double-check the generated configurations for the default top blade and the tail blade.

Next, we can look at the *shaft*, which will connect most of the helicopter parts. The shaft's specifications are highlighted in *Figure P2.15*.

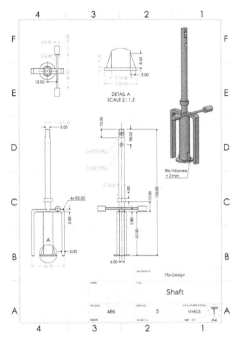

Figure P2.15 – Detailed drawing for the shaft

Here is a sample procedure for 3D modeling the shaft:

1. Use a revolved boss to create the long part of the shaft, as shown in *Figure P2.16*.

Figure P2.16 – A revolved boss can generate the long rod of the shaft

2. Use the extruded boss and fillet commands to create the two legs on the sides of the shaft. You can create one side and then mirror the other side to get the result shown in *Figure P2.17*.

Figure P2.17 – Extruded boss, fillets, and feature mirror can create the two leg-shaped figures

3. You can create the side-ways-looking antenna using a revolved boss. You might need to generate a new plane for that, as shown in *Figure P2.18*. You can get the exact shape using multiple applications of the extruded boss feature as well.

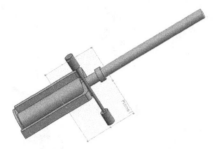

Figure P2.18 – A new plane was used as a base for a revolved boss

4. Using extruded cuts, you can create a straight surface on top of the shaft and the two holes indicated in *Figure P2.19*:

Figure P2.19 – The holes and straight surface can be made with an extruded cut

5. Using the rib feature, you can create the two ribs on the lower part of the shaft. You can apply the feature twice or apply it once and mirror the result to the other side. This can result in what is shown in *Figure P2.20*.

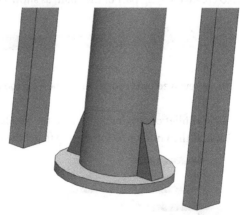

Figure P2.20 – The rib feature can be used to create the ribs

After the shaft, we can start working on the *blade support*, which will connect the blades to the shaft. The details of the blade support are shown in *Figure P2.21*:

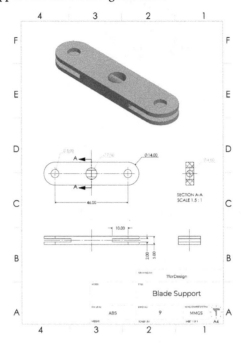

Figure P2.21 – Detailed drawing for the blade support

Here is a sample procedure for 3D modeling the blade support:

1. Use the extruded boss feature to create the following shape. When doing so, the end condition mid-plane might make it easier to do the cuts in *step 2*. The result will look as shown in *Figure P2.22*.

Figure P2.22 – One extruded boss application can generate most of the shape

2. Use an extruded cut to make the slot in the middle of the shape, getting the shape shown in *Figure P2.23*.

Figure P2.23 – The middle cut can be made with an extruded cut

3. Apply another extruded boss to create the cylindrical rod located in the middle hole. You can use the end condition **Up to next** to have the circular extrusion bounded by the surface to get the shape shown in *Figure P2.24*.

Figure P2 –24 – The 90 mm-long cylindrical part can be created with an extruded boss

Next, we can start looking at the *tail support*, which will connect the *tail* to the *support*. The specifications of the tail support can be seen in *Figure P2.25*.

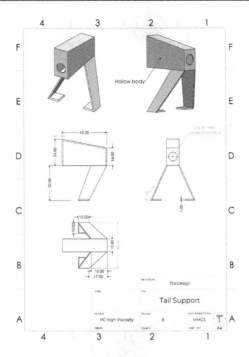

Figure P2.25 – Detailed drawing for the tail support

Here is a sample procedure for 3D modeling the tail support:

1. We can start by creating the bulky hollow part shown in the following figure. This can be made with the extruded boss, shell, and extruded cut features, in that order. Note that the bottom surface is removed using the shell feature. Also, note that the circular cut goes through both sides of the tail support. This should give you the shape highlighted in *Figure P2.26*.

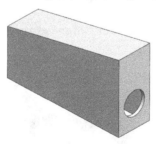

Figure P2.26 – The hollow part of the tail support can be made with the shell feature

2. The leg part of the support can be made with a lofted boss, as shown in *Figure P2.27*. Note that even though both ends are rectangular, they have different sizes, with the lower end not aligned with the side of the boxy area.

Figure P2.27 – Lofted boss can be used to create the long leg

3. Use an extruded boss to create the lower straight foot indicated in *Figure P2.28*. Note that the foot is drafted. You can toggle the draft within the extruded boss feature or apply the draft as a separate feature.

Figure P2.28 – A drafted extruded boss can be used to create the foot

4. Use the feature mirror to mirror the lofted and extruded boss to the other side to get the shape shown in *Figure P2.29*.

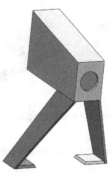

Figure P2.29 – The feature mirror command can quickly replicate features

Next, we can have a closer look at the *tail,* which will connect to the tail support on one end, and the tail blades on the other end. The specifications of the tail are highlighted in *Figure P2.30.*

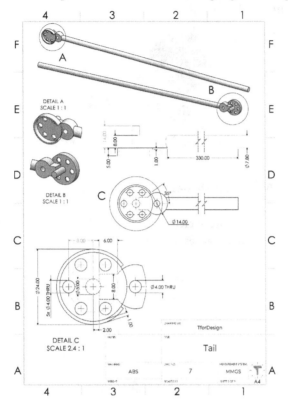

Figure P2.30 – Detailed drawing for the tail

Here is a sample procedure for 3D modeling the tail:

1. Use two extruded bosses to create the base of the tail, getting the result shown in *Figure P2.31.* To make modeling simpler, you can create one sketch and use it to apply more than one extruded boss feature.

Figure P2.31 – The same sketch can be used twice for different features

2. Use two other extruded boss features to create the disk-shaped part and its boundary. We can also use one sketch two times for two extruded boss applications. This will get you what is shown in *Figure P2.32*.

Figure P2.32 – Two extruded bosses with the same sketch can be used to create the disk

3. Use the extruded cut feature to create the five cuts shown in *Figure P2.33*. Note the cuts follow a circular pattern of six instances, with the one on the far right being skipped:

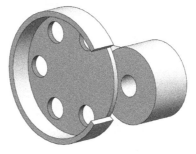

Figure P2.33 – The cuts on the disk follow a circular pattern.

4. Use an extruded boss to create the long part of the tail. As shown in *Figure P2.34*, you might need to create a new plane as a base for the extruded boss:

Figure P2.34 – An extruded boss with a new reference plane used to create the long rod

Next, we will take a look at the support shown in *Figure P2.34*. Note that this can be a multi-body part.

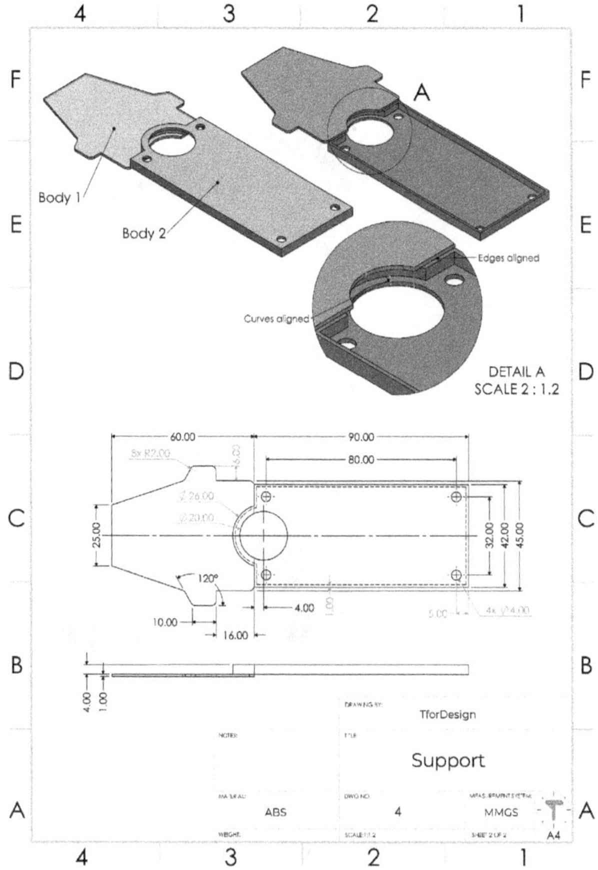

Figure P2.35 – Detailed drawing of the support

Here is a sample procedure for 3D modeling the support:

1. Use an extruded boss to create the flat part of the support, as shown in *Figure P2.36*. We can include the fillets in the sketch or apply them separately as a feature.

Figure P2.36 – An extruded boss can be used to create the shown body

2. Use another extruded boss to create the second body, as shown in *Figure P2.37*. Make sure that the bodies are not merged by unchecking the **merge result** option. Note the first body was made transparent for clarity only:

Figure P2.37 – An extruded boss can be used to create another body

3. Use the shell feature to shell the second body, as shown in *Figure P2.38*.

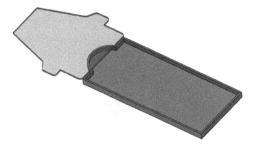

Figure P2.38 – A shell can be used to shell only one body

4. Use the extruded cut feature to create the final cuts. Note the larger cut applies to both bodies. Thus, we have to make sure both bodies are included in the scope of the cut. This should finish off the part, giving us the result shown in *Figure P2.39*.

Figure P2.39 – An extruded cut applied to more than one body at once

Now, we can work on the *blade stabilizer* shown in *Figure P2.40*.

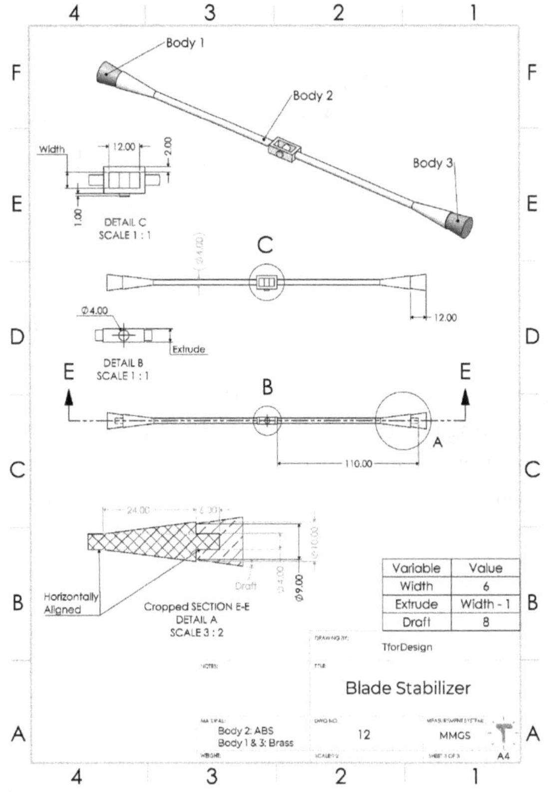

Figure P2.40 – Detailed drawing for the blade stabilizer

The blade stabilizer consists of three different bodies, two of which are identical. Also, part of the dimension of the indicated **Extrude** is related to the **Width** dimension. Here is a sample procedure for 3D modeling the blade stabilizer:

1. Use two extruded boss features to create the middle part of the blade stabilizer, as shown in *Figure P2.41*.

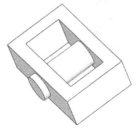

Figure P2.41 – The extruded boss can be used to create the middle part of the blade stabilizer

2. Use a revolved boss to create the rod-like part, as shown in *Figure P2.42*.

Figure P2.42 – A revolved boss can create the long rod with one application

3. Create the brass part toward the tip of the stabilizer. Note that the part is drafted by 8 degrees, as shown in the drawing in *Figure P2.40*. We can set the draft angle from the PropertyManager of the extrude boss feature. Since the tip is a different body, we can uncheck the **Merge result** option when applying the extruded boss. This will give the shape shown in *Figure P2.43*.

Figure P2.43 – A drafted extruded boss can be used to create the tip body

4. Create the hole on the tip part, as indicated in *Figure P2.44*.

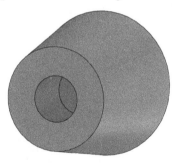

Figure P2.44 – The hole applied to a specific body

The location of the hole is covered by the main rod. To access the hidden location, you right-click on the body listed in the design tree, then select **Isolate**, as highlighted in *Figure P2.45*. Alternatively, we can change the transparency of the other body.

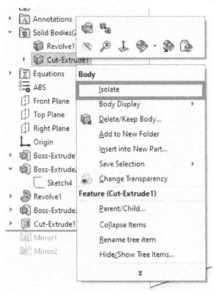

Figure P2.45 – We can isolate bodies to make them easier to work with

5. Mirror the revolved boss feature and the tip body to the other side to get the final result shown in *Figure P2.46*. To do this, we will have to apply the mirror feature twice: once to mirror the revolved boss feature and another time to mirror the body at the tip. This should finish off the part, giving the result shown in *Figure P2.46*.

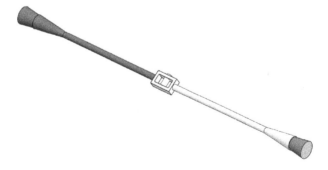

Figure P2.46 – Mirroring features and bodies can make modeling the part faster

After the blade stabilizer, we can move on to 3D modeling the last part of the RC helicopter, the *body*. A detailed drawing of the body is shown in *Figure P2.47*.

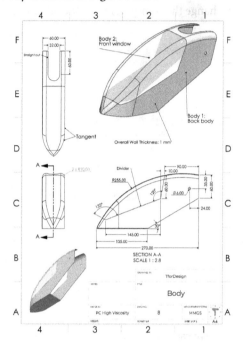

Figure P2.47 – Detailed drawing of the body

Note that the body is a multi-body part with two bodies. One is transparent while the other is not. We will 3D model it as one part. Then, we will split the part into two. Here is a sample procedure for 3D modeling this part:

1. Use an extruded boss to create the overall shape of the body. This will get you the shape shown in *Figure P2.48*.

Figure P2.48 – The bulk of the body can be made with an extruded boss

2. We can make the tip of the body using an extruded cut, as shown in *Figure P2.49*. The sketch used for the extruded cut is shown in the same figure. After that, we can use the fillet feature to round the long top edges to a 10 mm radius, as indicated in *Figure P2.47* and *Figure P2.49*.

Figure P2.49 – The tip of the body can be made with an extruded cut

3. Use the shell feature to hollow the shape to 1 mm thickness. Make sure to remove some of the faces to end up with the shape shown in *Figure P2.50*.

Figure P2.50 – The shell feature can both hollow the body and remove unwanted faces at the same time

4. Use two extruded cut features to generate the cuts, ending up with what is shown in *Figure P2.51*. Those are the cuts on the top and side of the body.

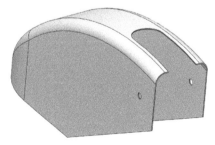

Figure P2.51 – Extruded cuts can cut the holes and top slot

At this point, we have created all the overall design features for the body. We are left with splitting the body into two to end up with the front window and the back solid body part. Next, we will explore how to split our body.

Splitting a body into two

To split a body, we will first create a sketch that highlights where the split is happening. Then, we will use the **Split** command to split the body. Let us do this by following these steps:

1.   Create the cut sketch shown in *Figure P2.52*.

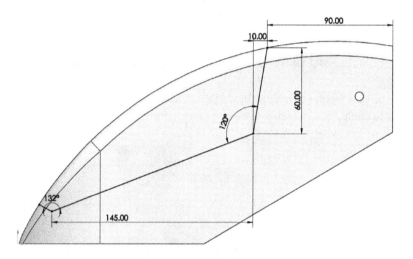

Figure P2.52 – A sketch can be used to split bodies

2. Select the **Split** command by going to **Insert | Molds | Split…**, as shown in *Figure P2.53*.

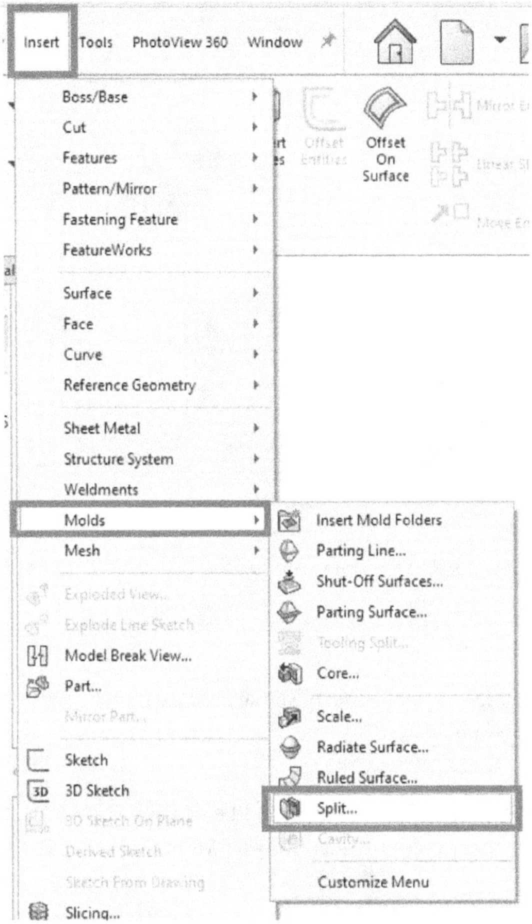

Figure P2.53 – The location of the Split command

3. Under the **Trim Tools** field, select the cut sketch we created in *step 1*, then click on **Cut Part**. This will create two resulting bodies under that title. Check the two bodies, then click on the green checkmark. The sequence is highlighted in *Figure P2.54*.

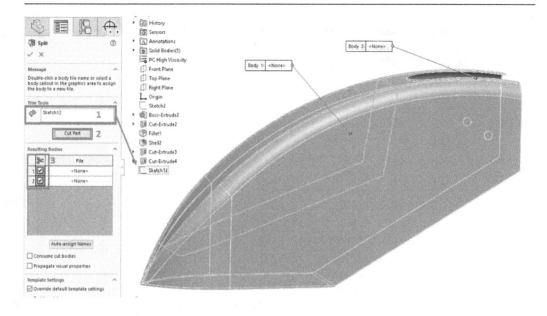

Figure P2.54 – The settings used for the Split command

> **Note**
>
> Giving names to the files in the PropertyManager shown in *Figure P2.54* will save the named body in a separate part file.

At this point, the body part should be complete, like the one shown in *Figure P2.55*. We can assign different materials and appearances to each body as we see fit.

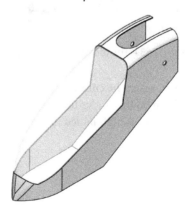

Figure P2.55 – The final body part

At this point, we are done with 3D modeling all the unique parts required for our RC helicopter model. Next, we will work on assembling the parts.

Creating the assembly

Now that we have all the parts 3D modeled, we can start exploring the assembly and joining all the parts together. We will do that in this section. *Figure P2.56* shows the fully assembled RC helicopter model:

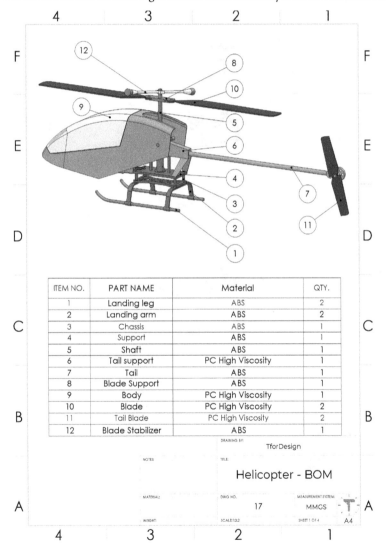

ITEM NO.	PART NAME	Material	QTY.
1	Landing leg	ABS	2
2	Landing arm	ABS	2
3	Chassis	ABS	1
4	Support	ABS	1
5	Shaft	ABS	1
6	Tail support	PC High Viscosity	1
7	Tail	ABS	1
8	Blade Support	ABS	1
9	Body	PC High Viscosity	1
10	Blade	PC High Viscosity	2
11	Tail Blade	PC High Viscosity	2
12	Blade Stabilizer	ABS	1

DRAWING BY: TforDesign

NOTES: TITLE:

Helicopter - BOM

MATERIAL: DWG NO.   MEASUREMENT SYSTEM:
17   MMGS
WEIGHT: SCALE1:3:2   SHEET 1 OF 4   A4

Figure P2.56 – The fully assembled RC helicopter model with all the parts

Let us explore more drawings that showcase the mates between the different parts within the assembly. *Figure P2.57* highlights the connections between the shaft and the support. It also shows the relationship between the tail support and the support parts. Note that it also highlights an additional hole created in the context of the assembly.

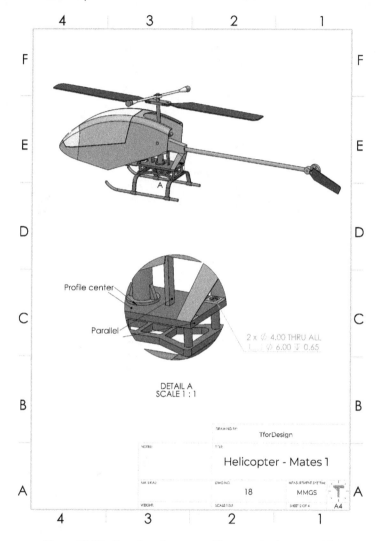

Figure P2.57 – Drawing shows specific mates in the assembly

Figure P2.58 shows an exploded view of the RC helicopter as well as selected mates. It also highlights an additional cut in the landing arm that was made in the assembly context.

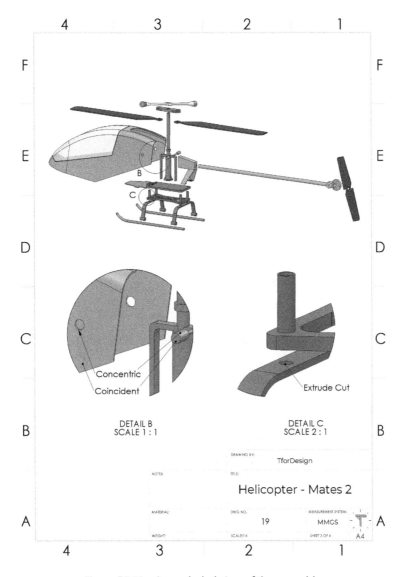

Figure P2.58 – An exploded view of the assembly

Figure P2.59 shows how the body part is angled with the chassis.

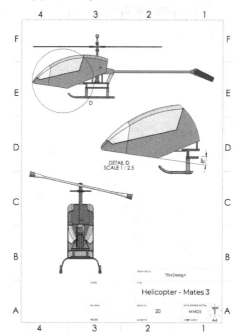

Figure P2.59 – The body has an 8-degree angle with the chassis

The assembly figures explored contain all the main information needed to build the assembly. Note that not all the mates are specifically mentioned. Many of the mates can be concluded from the drawings of the overall assemblies. There are two different methods we can use to create the assembly:

- Adding all the parts in one assembly file and mating them
- Creating multiple subassemblies of fewer parts and then joining them together to form a larger assembly

In this chapter, we will follow the second approach. Keep in mind that there is no one correct approach to generating an assembly such as this. Thus, treat the information presented here as hints and food for thought. Feel free to adapt or experiment with different approaches.

In the approach we are adapting, we will first build four smaller subassemblies, then join them to form the larger assembly. Each of the subassemblies we are building will consist of four or fewer unique parts. Building smaller subassemblies and then joining them to a larger one can help make them more manageable and easier to work with.

> **Note**
>
> A subassembly is an assembly file that is inserted into another assembly file.

The first subassembly consists of the *landing leg* and the *landing arm*, as shown in *Figure P2.60*. Overall, this subassembly consists of four parts, two of which are unique. The figure also highlights the major mates used to build the subassembly.

> **Tip**
>
> You can experiment with using the linear pattern to build the shown subassembly.

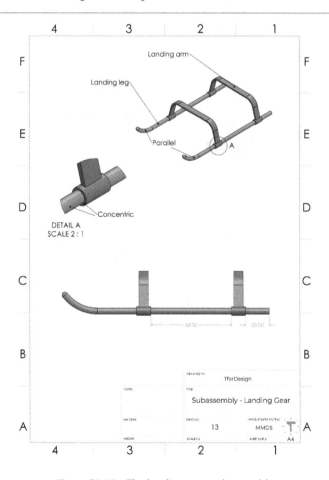

Figure P2.60 – The landing gear subassembly

The next subassembly consists of the *support* and the *chassis*, as shown in t *Figure P2.61*.

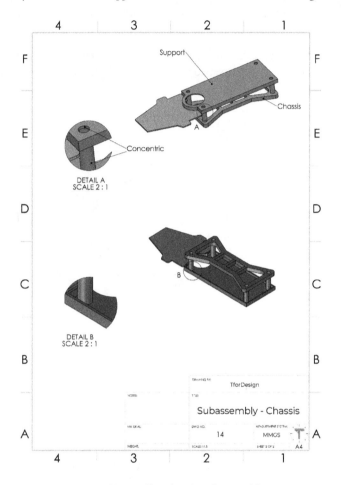

Figure P2.61 – The chassis subassembly

The next subassembly consists of the *shaft*, *blade stabilizer*, *blade support*, and *blades*, as shown in *Figure P2.62*. We can set the shaft as the fixed part for this subassembly as it is not a dynamic part in the final assembly. The figure also highlights selected mates that we can apply in the assembly.

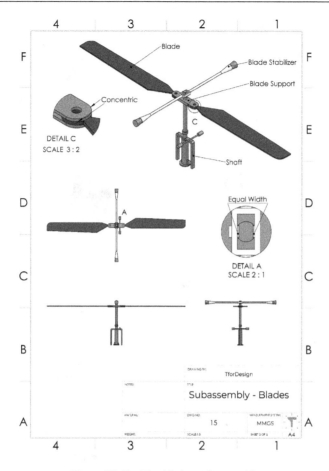

Figure P2.62 – The blades subassembly

The last subassembly consists of the *tail support*, *tail*, and *tail blade*, as shown in *Figure P2.63*. Overall, this subassembly consists of four parts, three of which are unique. The figure also highlights the major mates used to build the subassembly. We can pick the tail support as the fixed part for this subassembly as it is a non-moving part in the final assembly.

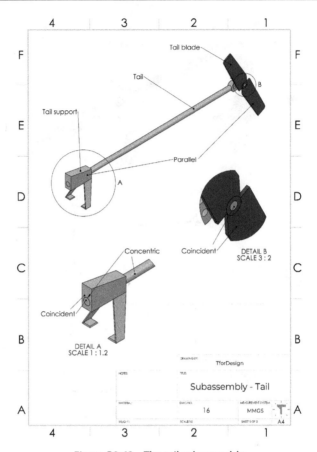

Figure P2.63 – The tail subassembly

At this point, we can create a new subassembly that will join the four subassemblies together with the body. The following figure highlights the major mates connecting the different subassemblies with the main RC helicopter body. You can revisit *Figures P2.56* to *P2.59* explored earlier for more information about how the different parts in the assembly interact with each other.

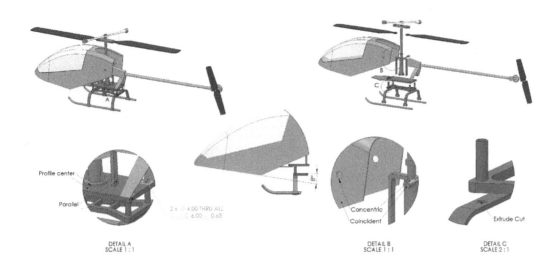

Figure P2.64 – The body connecting with the other subassemblies

By completing the assembly, you have completed the project work to 3D model an RC helicopter. As additional activities, you can use different evaluation tools, such as interference detection, to find undesirable interferences and adjust the design as applicable. You can also further customize the RC helicopter model to make your own.

Summary

In this project chapter, you worked on 3D modeling an RC helicopter toy model. To do that, you had to interpret engineering drawings, 3D model different parts, and then join them in an assembly. The skills you used to work on this project include many advanced skills often used by professional users, such as working with multibody parts, building configurations, and using design tables, advanced mates, and assembly features. In the process, you have also constructed a project that can be included in your personal portfolio to highlight your 3D modeling skills.

In the next chapter, we will explore some of the basic cloud services that are offered with your SOLIDWORKS license.

16

Introduction to SOLIDWORKS Cloud Services

SOLIDWORKS Cloud Services offers an array of data and collaboration tools on the cloud, enhancing design sharing, management, and **Computer-Aided Design (CAD)** file storage. The service provides seamless integration with 3DEXPERIENCE, increasing productivity and flexibility in design tasks. In this chapter, we will introduce the interconnected features of SOLIDWORKS Cloud Services and the 3DEXPERIENCE platform.

We will explain the fundamentals of Cloud Services, emphasizing collaboration, storage management, and the link between SOLIDWORKS and 3DEXPERIENCE. The various roles, applications, and cloud storage in the 3DEXPERIENCE platform will also be introduced.

We will learn how to utilize Cloud Services to boost productivity, streamline project management, and enable seamless team collaboration. The following topics will be covered in this chapter:

- Introducing SOLIDWORKS Cloud Services
- Getting started with SOLIDWORKS Cloud Services
- Saving and sharing your 3D designs
- Marking up your designs
- How to edit and save your work after feedback

By the end of the chapter, you will have gained familiarity with Cloud Services and will have learned how to use the different features of the cloud. You will be well equipped to save, share, and markup your designs smoothly. Through these features, you will learn to increase your productivity and enhance your SOLIDWORKS experience.

Technical requirements

This chapter will require that you have access to the SOLIDWORKS software. The project files for this chapter are available at the following GitHub repository: `https://github.com/PacktPublishing/Learn-SOLIDWORKS-2025-Third-Edition`

The CiA video for this chapter can be found at `https://packt.link/y8H0L`

Introducing SOLIDWORKS Cloud Services

Cloud Services makes working in SOLIDWORKS an elevated experience by allowing seamless file sharing, efficient markup tools, and secure storage. To understand how we can use the services in Cloud Services to improve our designing journey, let us first understand the basics of Cloud Services and 3DEXPERIENCE.

Cloud Services

SOLIDWORKS Cloud Services is a collection of data and collaboration tools on the cloud. It connects SOLIDWORKS to the 3DEXPERIENCE platform, providing access to an introductory set of tools on the 3DEXPERIENCE platform. Cloud Services allows for enhanced collaboration, easier storing, and efficient managing and control of 3D designs.

From July 2023, all new licenses of SOLIDWORKS come with an upgraded subscription that comes with Cloud Services. Existing users can upgrade to enjoy Cloud Services for a minimal fee. However, users have the flexibility to continue to renew their usual subscription if they prefer. In this chapter, we will only address the main Cloud Services capabilities included in a standard SOLIDWORKS license, which are file sharing, markup and collaboration, and life cycle management.

> **Note**
> Throughout the chapter, we will use the term *cloud service* to refer only to the included services with a standard SOLIDWORKS license. Otherwise, the cloud services available in the 3DEXPERIENCE platform is much larger.

Cloud Services offers several tools to elevate user's experience in SOLIDWORKS. It includes advanced tools for design sharing and markup, secure storage and revisioning of designs, and effective design management and control.

Cloud Services and the 3DEXPERIENCE platform

Cloud Services provides a glimpse into the world of the 3DEXPERIENCE platform. The 3DEXPERIENCE platform is a comprehensive business tool that connects to SOLIDWORKS through the integration of various software applications, data, and services to support the different stages of product development, manufacturing, and collaboration.

You can think of the 3DEXPERIENCE platform as an operating system designed for design and manufacturing. That operating system has different applications, that are given to individuals within an organization based on their roles.

There are different roles in the platform based on the user's responsibilities and tasks. Each role provides access to tailored applications and functionalities. For instance, the platform administrator role manages users and can perform administrative tasks such as setting communities, whereas a user with the role of 3D Creator will not be able to assign roles; instead, that user will be able to create 3D models with design applications.

The 3DEXPERIENCE platform is also equipped with various applications. These apps provide users with a comprehensive and integrated product development experience. For instance, **3DPlay** is a tool used to visualize the 3D models and conduct design reviews. When you share your designs, your guests will view them in 3DPlay. Other applications such as CATIA, SIMULIA, and DELMIA are also available on the platform.

Aside from role allocation and the various applications, the 3DEXPERIENCE platform also offers an integrated cloud storage and collaboration tool called **3DDrive**. 3DDrive provides users with cloud storage to store and manage their design data. It also allows users to easily share files stored in the 3DDrive with team members to facilitate seamless collaboration.

3DEXPERIENCE SOLIDWORKS allows SOLIDWORKS users to increase their productivity and simplify their project management with tools for collaboration, project management, and data storage. It bridges the gap between the various teams involved in a project.

The version of Cloud Services that comes with a standard SOLIDWORKS license is a simpler version of the 3DEXPERIENCE platform, providing enhanced collaboration, storage, and data management. However, its tools are limited, for instance, it does not include browser-based CAD. Regardless, Cloud Services can be optimized to increase productivity, save time and money, and ensure comfortable design and collaboration experiences. In this chapter, we will address the main cloud features you will likely have access to. Those include the following:

- Secure cloud storage
- File sharing and collaboration
- Mark-up and annotation features

In this chapter, we will learn how to utilize these features to maximize the benefits they offer for our design projects. In the following section, we will learn how to integrate the cloud services into our SOLIDWORKS.

Getting started with SOLIDWORKS Cloud Services

In this section, we will activate Cloud Services. This includes setting up the services and leveraging the 3DEXPERIENCE platform.

Setting up Cloud Services

Setting up Cloud Services is a straightforward process. It takes less than five minutes to activate Cloud Services on the 3DEXPERIENCE platform, which is the interface for managing Cloud Services. You will also be able to access some features directly through the SOLIDWORKS desktop interface after installing them. To install Cloud Services, follow these steps:

1. Go to the 3DEXPERIENCE platform at `https://my.3dexperience.3ds.com/welcome` and log in with your account. The website is shown in *Figure 16.1*.

Get the Full Power of the **3DEXPERIENCE**® platform

3DEXPERIENCE on the Cloud provides all organizations with a holistic real-time vision of their business activity and ecosystem, connecting people, ideas, data and solutions in a single collaborative and interactive environment available at all times. The easy-to-use interface helps everyone involved in innovation projects interact to imagine, design, simulate and deliver differentiated customer experiences. Break free of IT constrains to scale and ramp up faster than ever with the **3DEXPERIENCE** on the Cloud.

Figure 16.1 – The welcome page of the 3DEXPERIENCE platform

2. Once you are on the platform, click on **BROWSE PORTFOLIO** or the compass in the top-left corner of the page, and then select your role, as shown in *Figure 16.2*. Look for the **3DEXPERIENCE SOLIDWORKS** role. You can also directly look for the **SOLIDWORKS Connected** app.

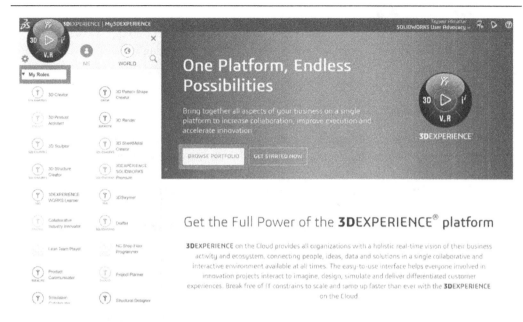

Figure 16.2 – Selecting your role

3. Click **Install** for SOLIDWORKS Connected. This will prompt you to download the 3DEXPERIENCE launcher. The Install option with then be replaced by the Open option shown in *Figure 16.3*.

4. Once installed, click on **Open**, as shown in *Figure 16.3*.

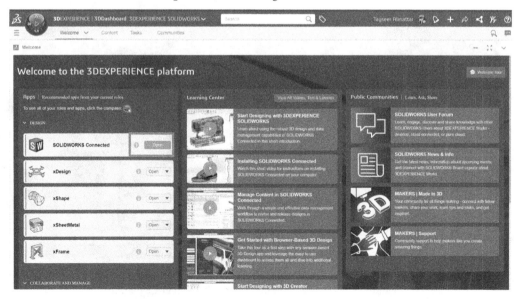

Figure 16.3 – The user interface of the 3DEXPERIENCE platform

Now that we have Cloud Services connected to our SOLIDWORKS desktop app, we can start utilizing some of the unique features it provides. We will start by addressing saving and sharing your 3D models through the cloud.

Saving and sharing your 3D models

Cloud Services provides an integrated platform to store, share, and access your files. It can act as a secure hub to store your designs to be easily accessed from anywhere with an internet connection. It also allows you to share your designs with external reviewers without requiring them to install any external applications. Let us start by going through how we can store our 3D models in the cloud.

Storing your designs with bookmarks and storage spaces

Having a safe place to store your designs is a crucial need for every designer. Cloud Services provides safe storage for your designs. Before learning how to save designs on Cloud Services, let us learn more about a special feature of Cloud Services – **bookmarks**.

Bookmarks are like folders; they provide a space to organize your files and make them easy to find.

To create a bookmark, follow these steps:

1. Navigate to the **Content** tab in the 3DEXPERIENCE platform, as shown in *Figure 16.2*.

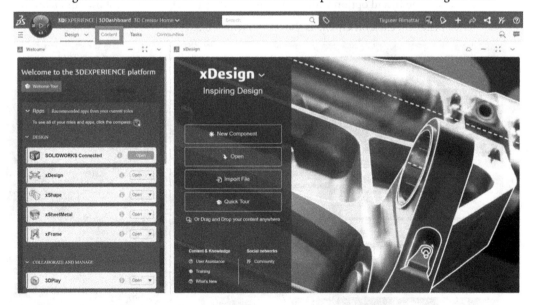

Figure 16.4 – The Content tab on the 3DEXPERIENCE platform

2. On the menu to the left, right-click on **Bookmark** and select **New Bookmark**, as shown in *Figure 16.5*.

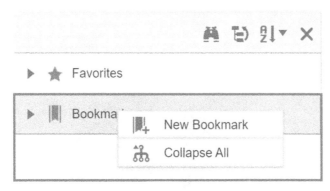

Figure 16.5 – New Bookmark

3. Type in your title and click **Create**, as shown in *Figure 16.6*. In our case, we will use the title `Design Collaboration`.

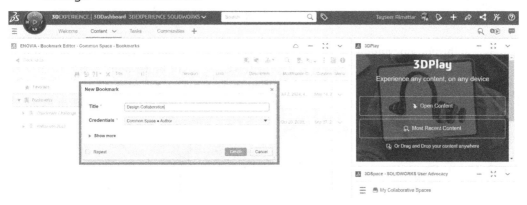

Figure 16.6 – Creating a bookmark

Now, let us see how you can store a file in the bookmarks.

To store a file, follow these steps:

1. Open your SOLIDWORKS file. You can download the 3D model we are using in the project download link at the beginning of the chapter or use your own model.

2. Navigate to the **Lifecycle and Collaboration** tab and click on **Save with Options**, as shown in *Figure 16.7*.

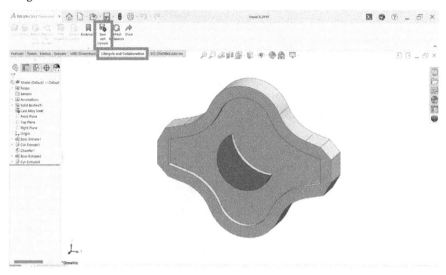

Figure 16.7 – Storing the designs with the Save with Options command

3. In the pop-up window, you will be asked to select a location to store your design. You have the option of storing it in your 3DEXPERIENCE space or under bookmarks. To store your work, click on the **Select Bookmark** button, and choose your designated bookmark. We will select the **Design Collaboration** bookmark, which we created earlier, as shown in *Figure 16.8*.

Figure 16.8 – Selecting a bookmark

4. Click **Apply** to store your design in the bookmark.

5. Then, click **Save**.

Showing the hidden Lifecycle and Collaboration tab

The **Lifecycle and Collaboration** tab might not be shown in your interface as it is in *Figure 16.7*. If that is the case, it might be hidden. To reveal it, you can follow these steps:

1. Right-click on the CommandManager, as indicated in *Figure 16.9*.

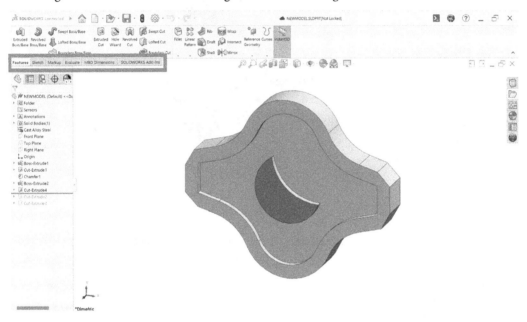

Figure 16.9 – Right-click on the CommandManager to adjust the displayed tabs

2. Select **Tabs** and then select **Lifecycle and Collaboration**, as shown in *Figure 16.10*.

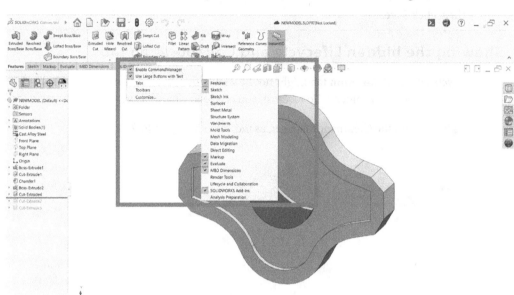

Figure 16.10 – Showing the Lifecycle and Collaboration tab

Now that the **Lifecycle and Collaboration** tab is displayed, you will be able to use all the functions it includes.

At this point, we have successfully stored our 3D model in the cloud; let us now address how to locate it and open it from the 3DDrive.

Opening your designs from the 3DDrive

Having stored our files in Cloud Services, we will now learn how to access them. To open files stored in Cloud Services, follow these steps:

1. Click on the *open* icon in SOLIDWORKS, as shown in *Figure 16.11*.

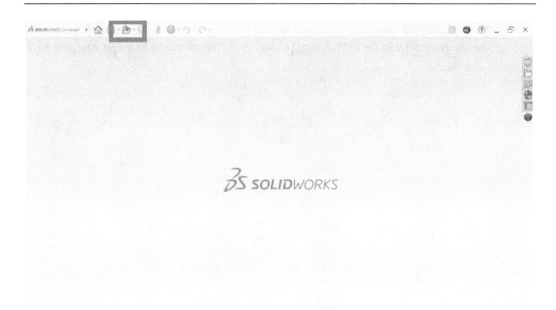

Figure 16.11 – Opening a file

2. Then, click on the **Open from 3DEXPERIENCE** button in the bottom-left corner, as shown in *Figure 16.12*.

Figure 16.12 – Opening from 3DEXPERIENCE

3. Now, locate your file under the title you have saved, as shown in *Figure 16.13*. You can also use the quick filter in the bottom-right corner or the search toggle to access your files easily.

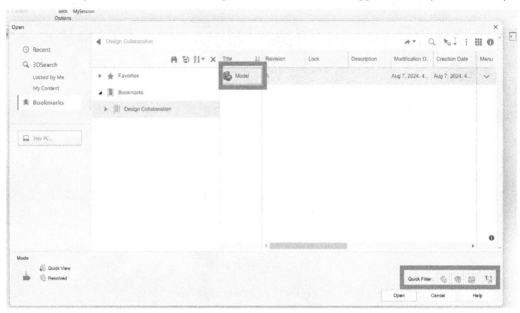

Figure 16.13 – Locating your file in the cloud directly from the desktop interface

After locating your file, simply click on **Open** to access it. You will now have your file successfully opened from the 3DDrive. You can edit your file as you normally would on SOLIDWORKS and save any updates directly on the cloud as well. The next section will talk about how we can share our files in the 3DEXPERIENCE platform.

Sharing your designs

One of the most exciting features of Cloud Services is perhaps the ability to share your designs with anyone and everyone. With the share feature in Cloud Services, you can seamlessly share your designs with reviewers without requiring them to have access to any special applications.

To share your designs, follow these steps:

1. Open the desired file and navigate to the **Lifecycle and Collaboration** tab, as shown in *Figure 16.14*.

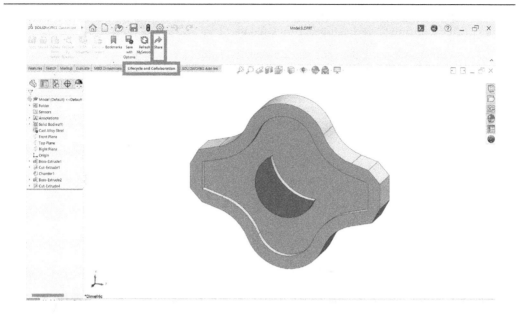

Figure 16.14 – Sharing your work

2.   Click on **Share** and select your desired file type.

3.   A window will pop up, as shown in *Figure 16.15*. You can edit the file name, file type, and the location of your file. This shares your file to the 3DDrive.

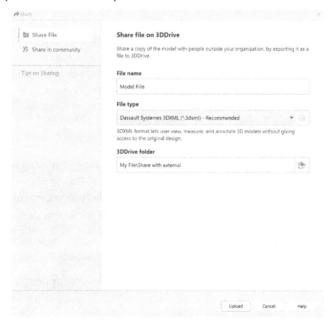

Figure 16.15 – Sharing your file to 3DDrive

4. After confirming the details, click on **Upload**.

5. Another window will then pop up, as shown in *Figure 16.16*, where you will be asked to control the access of your file. Turn on the toggles according to your preferences.

Figure 16.16 – Sharing files externally

6. Include a message if desired.

7. Click **Share**.

When you specify the email address of the recipients, the **Share** button will be activated. Once you click share, an email will be sent to the recipients notifying them that you have shared your design with them. If you are sharing your designs through the link, then simply click on **Copy external link** to copy the link. You need not click on the **Share** button.

Before moving to address marking and reviewing designs, let us elaborate more on the sharing options highlighted in *Figure 16.16*:

*   **Activate external sharing link**: This enables people to view your designs through a link. When you turn on this toggle, you can click on **Copy external link** to copy the link. Then, you can simply send this link to anyone, and they will be able to view your designs. Note that this link can be shared amongst people, and they will be able to view your designs.

*   **Enable guest comments**: This allows guests to comment on your designs. This is a particularly useful feature when you are seeking feedback on your designs.

*   **Restrict access to specific users**: This limits the access of your design. It only allows certain accounts to view your designs. Enter the email address accounts with whom you want to share your designs and the access will be restricted to them.

It is a good practice to include details about your designs in the **Add a message** box. You can let your reviewer know which file you are sharing, the reason you are sharing, or what they should look out for in the designs.

Sharing the files through the 3DEXPERIENCE platform is a quick and convenient way to collaborate with team members. Now that we have learned how to share our files, in the next section, we will learn how to comment on the designs if you are on the receiving end.

Marking up on designs

After sharing a file or having been shared a file, we may be given feedback or be asked to give feedback. In this section, we will learn how to use the markup feature with Cloud Services.

How to access the shared file

To give feedback, we need to access the file first. When you share your designs with your reviewer through their email account, they will receive a message, as shown in *Figure 16.17*. This email notification will include your account and the file that you are sharing.

Figure 16.17 – Email notification of file sharing

By clicking on the file name, the reviewer will be directed to the 3DEXPERIENCE platform. To view the design, if the reviewer does not have an account, they must first create a 3DEXPERIENCE account by simply filling in the required information.

The same process applied for sharing designs through a link – a 3DEXPERIENCE account must first be created.

How to annotate the design

When a reviewer is signed in, they can view the entire 3D model in the 3DPlay app. They can rotate it from all views, explode the assembly, and mark it up in the 3DPlay interface shown in *Figure 16.18*.

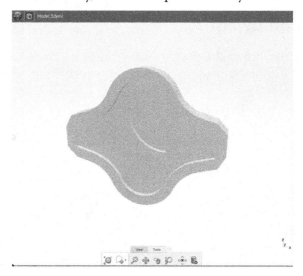

Figure 16.18 – Viewing the design in the 3DPlay app

Below the design, there are a handful of features that the guest can use. In the **View** bar, as shown in *Figure 16.19*, they can adjust the viewpoint and view modes of the model.

Figure 16.19 – The View bar

Figure 16.20 shows the **Tools** bar, which provides an array of tools to analyze the design. Tools such as **Measure**, **Clipping**, and **Annotation** can be utilized.

Figure 16.20 – The Tools bar

Let us look at these tools in detail starting with **Share as Comment**.

Share as Comment

This feature allows you to make comments on the design. Simply click on the **Share as Comment** icon and a window will pop up where you can make comments, as shown in *Figure 16.21*.

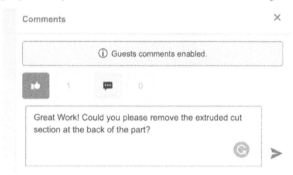

Figure 16.21 – Commenting on the design

When you make a comment, the person who sent you the file will get a notification of it. This lets them know that you have seen and reviewed their designs. Let us now explore the Measure tool.

Measure

Measurement is an important part of 3D modeling. The dimensions need to be correct and accurate to produce a worthy design. The **Measure** tool allows you to analyze and evaluate the dimensions of the model. You can choose from the various measurement tools provided, including point, center point, canonic properties, and parts.

To measure, follow these steps:

1. Go to the **Tools** bar and select the **Measure** tool, as shown in *Figure 16.22*.

2. Choose the desired measuring tool, as shown in *Figure 16.22*.

Figure 16.22 – Measurement tools

3.   Select the entity to be measured.

4.   The dimensions will be as shown in *Figure 16.23*.

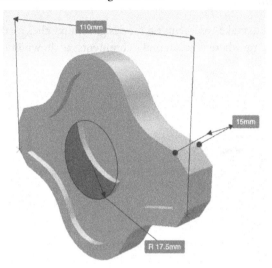

Figure 16.23 – Dimensions from the Measure tool shown on the model

By identifying the dimensions within the models, we can gain a clearer understanding of the design. To delve deeper into our analysis, we will now explore how to examine the model from various cross-sections to visualize its internal components via the **Clipping** tool.

Activate Clipping

To analyze the different cross-sections of the model, the **Clipping** feature can be used. This feature allows you to clip the model from different angles and sections, giving you a look at the model's internal structure.

To make a cross-section across a plane, follow these steps:

1.   Click on the **Edit Clipping** icon under the **Tools** tab, as shown in *Figure 16.24*.

Figure 16.24 – The Clipping feature command

2. Click on the plane icon, as highlighted in *Figure 16.25*.

Figure 16.25 – Plane icon to generate a cross-section

3. Right-click on the plane and from the drop-down menu, select the desired axis, as shown in *Figure 16.26*.

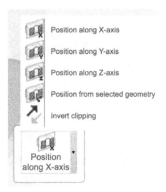

Figure 16.26 – Making a cross-section

4. Adjust the depth of the cut with the toggle. The toggle is located above the model and is shown in *Figure 16.27*.

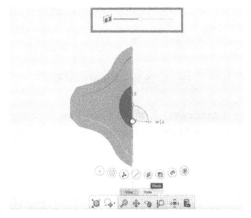

Figure 16.27 – Toggle to control the depth of the cut

To activate or deactivate the clipping, simply click on the **Activate Clipping** button. This allows you to view the internal structure of the model. This is helpful when there might be hidden features embedded in the design; for instance, there could be certain extruded cut features internally in the structure. Now, let us explore how to leave annotations on the design itself.

Annotation command and text annotation command

Other than making comments, you can also annotate the drawing to draw focus to a specific part. The annotation command allows you to draw and write handwritten comments on the design. The text annotation command is used for typed comments. The effect of both commands can be seen in *Figure 16.28*.

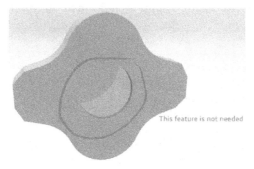

Figure 16.28 – Annotating the design

To annotate, follow the following steps:

1. Choose **Annotation** Commands or **Annotation Key** Command, as shown in *Figure 16.29*.
2. Draw or type your comments as prompted.

Figure 16.29 – Annotation tools

Those annotations made by the reviewer will make it easier for the designer to modify their designs. By directly annotating on the design, you can provide specific feedback, highlight areas for improvement, and communicate suggestions clearly.

The markup feature fosters communication and the flow of ideas between team members. This promotes efficient collaboration and clearer communication of design intent. Now, we will learn how we can respond to the feedback and how to save our designs after refining them.

How to edit and save your work after feedback

After receiving feedback from our reviewer, we need to access them so we can refine and edit our designs. We also need to save the updated designs to the 3DEXPERIENCE platform. Let us learn how to do each of them in the following section.

Viewing the comments

When your designs are marked up and commented on by an external reviewer or entity, you will receive notifications in the **3DEXPERIENCE** tab, as shown in *Figure 16.30*.

Figure 16.30 – Receiving notifications

You are also able to interact with the comments. From *Figure 16.31*, you can comment or like the comment.

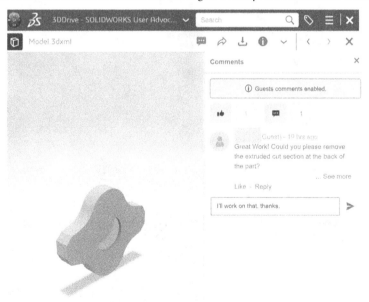

Figure 16.31 – Interacting with the comments

After reviewers leave a comment on your design, be sure to acknowledge their comments. It lets them know you value their feedback and have taken their comments into consideration.

Editing and saving your work

After receiving feedback, the next step is to fine-tune or edit your designs. To edit your work, you need to lock the file first to save your edits to the 3DEXPERIENCE platform. Locking prevents conflicts with concurrent edits by other team members and lets them know that the file part is being edited to avoid simultaneous edits by multiple team members.

To lock your files, follow these steps:

1. Go to the **3DEXPERIENCE** tab in the SOLIDWORKS desktop app, as shown in *Figure 16.32*:

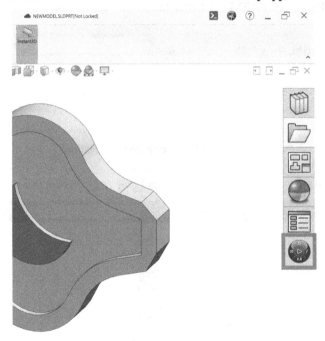

Figure 16.32 – The 3DEXPERIENCE tab in the SOLIDWORKS interface

2. Right-click on the unlocked lock icon, as shown in *Figure 16.33*.

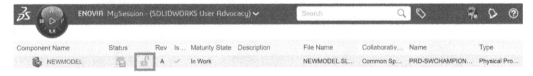

Figure 16.33 – Unlocked lock

3. Select **Lock** from the drop-down menu, as shown in *Figure 16.34*.

Figure 16.34 – Locking the Work

4. A green key icon will appear, indicating your file has been locked, as shown in *Figure 16.35*.

Figure 16.35 – Locked key indicating the design is already locked

Now that you have locked your work, you can modify your designs according to the feedback you received.

To edit and save your work, follow these steps:

1. Edit your design as you normally would on SOLIDWORKS.

2. To reflect the changes in the 3DEXPERIENCE platform, go to the **3DEXPERIENCE** tab task pane in the SOLIDWORKS desktop app.

3. Right-click on the icon under **Status**. The icon should have an orange save key, as shown in *Figure 16.36*. This indicates that the modifications to the design have not yet been updated.

Figure 16.36 – The status icon indicating unsaved work

4. Choose **Save** after right-clicking on the icon, as shown in *Figure 16.37*.

Figure 16.37 – Drop-down menu showing the save option

5. The icon will now have a green checkmark, as shown in *Figure 16.38*, indicating that the design on the 3DEXPERIENCE platform has been updated.

Figure 16.38 – The status icon indicating the file has been updated

Your design on 3DEXPERIENCE will now be updated. However, any link or email account you shared with the previous account will not reflect the changes. You would need to resend the file part.

After editing your design, your design may need to go for further review, or it might be ready to be released. To indicate this, you will need to change the maturity level of your file, which will be discussed in the following section.

Maturity level

The **maturity level** of the file indicates the status of the work. It lets people know whether the work is in progress or if it is completed. In Cloud Services, by default, all designs are set to an **In Work** status.

To change the maturity level of your design, you can follow the following steps:

1. Go to the **3DEXPERIENCE** tab in the SOLIDWORKS desktop app, as shown in *Figure 16.39*.

Figure 16.39 – The 3DEXPERIENCE tab

2. Right-click on the file name.

3. Choose **Maturity** from the drop-down menu, which is shown in *Figure 16.40*.

Figure 16.40 – Changing the Maturity level

4.   Select the desired maturity level.

There are various maturity levels, including **Private**, **In Work**, **Frozen**, **Released**, and **Obsolete**. Let us look at the different maturity levels in detail:

- **Private**: This level restricts access to your file. This can be helpful if you are still in the initial stage of development and would like to finalize certain affairs before sharing your designs.

- **In Work**: This is the stage most designs are at the longest. It shows that the design is in progress. It indicates to your team members they can review and comment on your designs.

- **Freeze**: This level indicates that final approval is pending. The project is at its final stage and is waiting for approval from the upper management. When a file is frozen, you will not be able to modify it.

- **Released**: This level indicates that the design is finalized and completed.

- **Obsolete**: As the name suggests, this level indicates that the design is outdated or no longer being used/developed. Files at the obsolete level are mostly stored for reference.

At different stages of your design, you can change the maturity level of your design to allow your colleagues to know the progress of the work.

Summary

In this chapter, we provided an overview of SOLIDWORKS Cloud Services and the 3DEXPERIENCE platform, highlighting their advantages in design sharing, management, and collaboration. We discussed how Cloud Services can improve user experience through features such as file sharing, markup tools, and secure storage, while also delving into elements such as role assignment, applications, and integrated cloud storage tools within the 3DEXPERIENCE platform.

We explored how to effectively utilize the 3DEXPERIENCE platform for storing, accessing, sharing, and annotating designs, focusing on tools for precise dimension measurement, internal component visualization, and feedback annotation. Additionally, we addressed key features for seamless collaboration, file management, and progress tracking, and outlined the process of editing and saving designs post-feedback, including best practices such as locking files before modifications and updating designs on the 3DEXPERIENCE platform.

Understanding how to leverage the capabilities of Cloud Services can elevate your design processes and promote team collaboration effectively, positioning you to enhance productivity and streamline workflows.

This brings our journey together to a close; you have transformed from a novice to a proficient SOLIDWORKS user, equipped with the skills and knowledge to tackle complex design challenges. You have gone a long way already, and we are all proud of the journey you have undertaken. Remember, the world of design is ever-evolving, and your continuous learning and curiosity will keep you at the forefront of innovation.

Questions

Answer the following questions to test your knowledge of this chapter. The following questions will reinforce the main topics we learned here:

1. What are the key features of SOLIDWORKS Cloud Services?

2. How does SOLIDWORKS Cloud Services differ from the broader array of cloud services available on the 3DEXPERIENCE platform?

3. What are some of the key applications available on the 3DEXPERIENCE platform, and how do they enhance the product development experience?

4. What is the purpose of the integrated cloud storage tool called 3DDrive on the 3DEXPERIENCE platform?

5. What are the key options for sharing files on the 3DEXPERIENCE platform?

6. How can reviewers access shared designs on the 3DEXPERIENCE platform?

7. What is the significance of locking files before editing designs on the 3DEXPERIENCE platform?

8. What does changing the maturity level of a file signify in the cloud service of the 3DEXPERIENCE platform?

Important note

The answers to the preceding questions can be found at the end of this book.

Get This Book's PDF Version and Exclusive Extras

UNLOCK NOW

Scan the QR code (or go to `packtpub.com/unlock`). Search for this book by name, confirm the edition, and then follow the steps on the page.

Note: Keep your invoice handy. Purchases made directly from Packt don't require an invoice.

17
Unlock Your Exclusive Benefits

Your copy of this book includes the following exclusive benefit:

- ☁ Next-gen Packt Reader
- 📄 DRM-free PDF/ePub downloads

Follow the guide below to unlock them. The process takes only a few minutes and needs to be completed once.

Unlock this Book's Free Benefits in 3 Easy Steps

Step 1

Keep your purchase invoice ready for *Step 3*. If you have a physical copy, scan it using your phone and save it as a PDF, JPG, or PNG.

For more help on finding your invoice, visit `https://www.packtpub.com/unlock-benefits/help`.

> **Note**
> If you bought this book directly from Packt, no invoice is required. After *Step 2*, you can access your exclusive content right away.

Step 2

Scan the QR code or go to `packtpub.com/unlock`.

On the page that opens (similar to *Figure 17.1* on desktop), search for this book by name and select the correct edition.

<packt>   Q Search...   Subscription 🛒⁰ 👤

Explore Products   Best Sellers   New Releases   Books   Videos   Audiobooks   Learning Hub   Newsletter Hub   Free Learning

Discover and unlock your book's exclusive benefits

Bought a Packt book? Your purchase may come with free bonus benefits designed to maximise your learning. Discover and unlock them here

Discover Benefits ———— Sign Up/In ———— Upload Invoice

Need Help?

✦ 1. Discover your book's exclusive benefits   ∧

Q Search by title or ISBN

CONTINUE TO STEP 2

⚇ 2. Login or sign up for free   ∨

☁ 3. Upload your invoice and unlock   ∨

Figure 17.1: Packt unlock landing page on desktop

Step 3

After selecting your book, sign in to your Packt account or create one for free. Then upload your invoice (PDF, PNG, or JPG, up to 10 MB). Follow the on-screen instructions to finish the process.

Need help?

If you get stuck and need help, visit
`https://www.packtpub.com/unlock-benefits/help`
for a detailed FAQ on how to find your invoices and more. This QR code will take you to the help page.

Note

If you are still facing issues, reach out to `customercare@packt.com`.

Assessments

Answers to questions

Here, we have answered afll of the questions asked at the end of each chapter. You can use these questions to review what you have learned throughout this book.

Chapter 1 – Introduction to SOLIDWORKS

1. SOLIDWORKS is **three-dimensional (3D)** design software. It is **Computer-Aided Design (CAD)** software that runs on Windows computer systems. This software was initially launched in 1995 and has grown to be one of the most common pieces of software used globally for engineering design.

2. Some of these industries are as follows:

   - Aerospace
   - Consumer products
   - Construction
   - High-tech electronics
   - Medicine
   - Oil and gas
   - Packaging machinery
   - Engineering services
   - Furniture design
   - Energy
   - Automobiles

3. The 3DEXPERIENCE platform is a cloud-based suite of integrated software applications developed by Dassault Systèmes, the parent company of SOLIDWORKS. It functions as an extensive digital workspace or "operating system" tailored for product design and development, offering tools for CAD modeling, simulation, project management, and data analytics. It complements SOLIDWORKS by providing a centralized platform for global team collaboration, data management, and seamless integration of various design and manufacturing processes to enhance productivity and innovation.

4. Two benefits of using the 3DEXPERIENCE platform for a team working on a product design project are as follows:

   - **Global Connection**: The platform facilitates real-time collaboration across team members, regardless of their geographic locations, by providing a shared digital environment that is accessible from anywhere, enhancing communication and reducing project timelines.

   - **Efficient Project Management**: With tools that offer project monitoring through dashboards and reporting features, the platform allows for effective tracking of project progress, document management, and decision-making, thus increasing the overall efficiency and organization of product development cycles.

5. In parametric modeling, the model is created based on relations and a set of logical arrangements that are set by the designer or draftsman. In the SOLIDWORKS software environment, they are represented by dimensions, geometric relations, and features that link different parts of a model to each other. Each of these logical features can be called a **parameter**.

6. Some of the advantages of parametric modeling are as follows:

   - Ease of capturing engineering-driven design intents

   - Ease of modifying and adjusting models throughout the design and production cycles

   - Ease of creating families of parts that are similar in parameters

   - Ease of communicating the design with manufacturing establishments for manufacturing

7. Parametric modeling is based on geometric logical elements, making it better for engineering applications. Direct modeling is more abstract, making it better for artistic applications.

8. SOLIDWORKS certifications are provided by SOLIDWORKS in certain topics after passing a skills-based exam. They are a good way to show employers or clients that you have mastery over a certain aspect of the software that would be needed for a specific project.

9. The certifications can be classified under four categories/levels: associate, professional, professional advanced, and expert.

Chapter 2 – Interface and Navigation

1. The three types of files we can use with SOLIDWORKS are part files, assembly files, and drawing files.

2. The difference is in the content of the files. A part file generally contains a model of a single part. An assembly file can link different parts together in one file. A drawing file has commands to generate 2D engineering drawings out of a part file or an assembly file.

3. The canvas contains a visual presentation of the 3D model we are creating.

4. The design tree shows all of the commands that were used to create the model shown in the canvas. It also lists the default planes and other information related to the model we are creating.

5. The **International System of Units (SI)**: This is also commonly known as the **metric system**. Currently, it is used in most countries in the world. Note that **SI** stands for the French **Système International**. Another common system is the Imperial system, which is mostly used in the United States.

6. To open a SOLIDWORKS file, you can click on **File** in the top-left corner and then select **New...**. After that, select **Part** and click **OK**.

7. To switch the measurement system to MKS, we can click on the unit system shortcut in the bottom-right corner of the interface and select **MKS (meter, kilogram, second)**, as highlighted in the following screenshot:

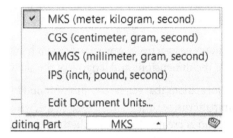

Figure A.1 – The shortcut menu to switch units found in the bottom-right corner of the interface

Chapter 3 – SOLIDWORKS 2D Sketching Basics

1. SOLIDWORKS sketches are the first step we take when creating a 3D model. A sketch provides a guide for how to implement and apply 3D features.

2. The two common stages we follow when sketching are outlining and defining.

3. First, make sure your measurement system is set to MMGS. To sketch the shape, we can start by drawing the circle and setting its radius to 100 mm. We can use the origin point as the center of the circle. Then, we can draw a right-angle triangle inside with one vertex being coincident with the origin and the other two with the perimeter of the square.

4. First, make sure your measurement system is set to IPS. We can start this sketch by sketching a center-rectangle and having the center of the origin. Note that it is a square, so we can also set the two perpendicular sides equal to each other. The two have circles that can be created with the arc command. We can use **Centerpoint Arc** and set the center of the arc to have a **Coincident** relation with the center of the square side and top lines. The isosceles triangle can be created with two different lines.

5. First, make sure your measurement system is set to CGS. We can start by using the **Straight Slot** command to generate the outer shape. We can then define the slot by setting its radius and length as 3 cm and 10 cm. The middle diamond can be created with four different lines. The top and bottom vertices are midpoints to the top and bottom lines of the slot, while the side vertices coincide with the center points of the side arcs.

6. First, make sure to set your measurement units to MMGS. We can then start the sketch by sketching a rectangle with 100 mm and 50 mm sides. Then, we can set up the corners using the fillets and chamfers commands. The fillet is simple, with a radius of 20 mm. For the chamfers, one is an angle-distance chamfer while the other is a distance-distance chamfer.

7. Under-defined, fully defined, and over-defined sketches are the different statuses that SOLIDWORKS uses to categorize our sketches. *Under-defined sketches* have parts that are loose or lack proper definition. *Fully defined sketches* have all parts of the sketch fully fixed. *Over-defined sketches* are those with more relations and dimensions than needed to have all of the sketch elements fully fixed.

Chapter 4 – Special Sketching Commands

1. Mirroring a sketch allows us to reflect/copy specific sketch entities around a specific reflection line. This will include the dimensions and lengths of the sketch.

2. Patterns are repeated formations that can be commonly found in consumer products, architecture, fabrics, and so on. In patterns, we often have a base shape, which is sometimes called a **base cell** or **patterned entity**, and is created from scratch; then the basic shape gets repeated multiple times. This is then repeated multiple times to form a bigger piece. In SOLIDWORKS, we can make linear patterns and circular patterns.

3. Trimming in SOLIDWORKS allows us to easily remove unwanted sketch entities or unwanted parts of sketch entities.

4. First, make sure your measurement system is set to MMGS. We can sketch half of the sketch using three lines to form the triangle. We can then use the **Circle** command to end up with the following sketch:

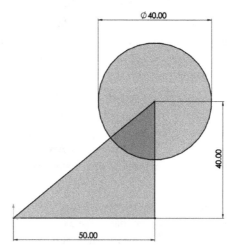

Figure A.2 – The original sketch before trimming and mirroring

After that, we can trim the unneeded part of the circle and then mirror the resulting shape.

5. First, make sure your measurement system is set to IPS. To create this sketch, we can start by sketching and fully defining one circle. Then, we can use the **Linear Pattern** command to make one row of three circles. After fully defining our linear pattern, we can use a circular pattern with six instances around 360 degrees to conclude and get the final shape.

6. First, make sure your measurement system is set to IPS. We can start by sketching and fully defining two ellipses, as shown in the following diagram:

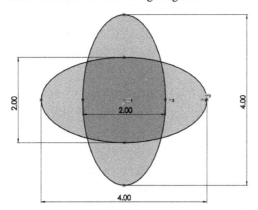

Figure A.3 – The base sketch before trimming

After that, we can use the power trim tool to trim the unwanted parts and end up with the final sketch required.

7. First, make sure your measurement system is set to MMGS. We can start the sketch by sketching only the teardrops. We can do it by sketching a circle, then have two lines tangent to the circle and intersecting each other at their endpoints. Make sure the intersection is vertical to the center of the circle. Our sketch will be as follows:

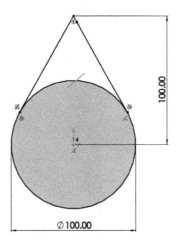

Figure A.4 – The original shape before the trim and circular pattern

We can then use the power trim tool to remove the unwanted part of the circle. Finally, we can use circular patterns in the other four instances.

Chapter 5 – Basic Primary One-Sketch Features

Note that when doing the practical questions and creating models, there is often no one right way of making a model. Hence, you might create the model in a different way than described here:

1. Features are our way into 3D in SOLIDWORKS. We usually use features directly after sketching to go from 2D sketches to 3D models. Features are commonly used after sketches. They are also mostly built based on sketches.

2. The extruded boss and cut features enable us to add or remove materials by extending the sketch perpendicular to the sketch plane.

3. Fillets and chamfers enable us to remove sharp edges or corners from our models and replace them with fillets or chamfers.

4. The revolved cut and revolved boss features enable us to add or remove materials by revolving a sketch around a revolution axis of our choosing.

5. First, make sure your measurement system is set to IPS. To create the shape, we can use the extruded boss, extruded cut, and fillet features. The two steps can be created with two extruded bosses. Then, the hole can be made with an extruded cut. The fillet feature can be applied to round the top edges with a 0.2-inch radius.

6. First, make sure your measurement system is set to MMGS. For this, we can use the revolved boss feature to create the sphere. We can start with a sketch of the letter D. Then, select the straight edge as the axis of revolution, as shown in the following diagram:

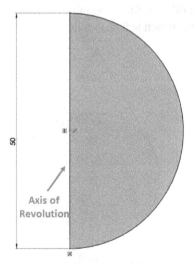

Figure A.5 – A sphere is a revolved letter D

7. First, make sure your measurement system is set to MMGS. Then, we can start by creating a profile in the shape of an ellipse, as shown in the following diagram:

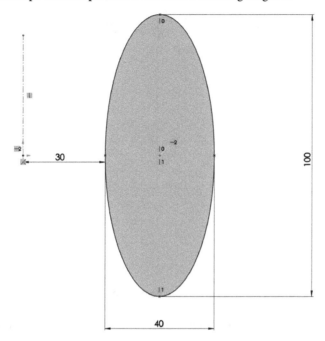

Figure A.6 – The exterior shape can be made with a revolved ellipse

We can then apply the revolved boss feature by 270 degrees to create the general shape. The circular hole shares the same center with the ellipse, which we can add in a different sketch and use the revolved cut feature. An alternative way is to sketch both the ellipse and the hole together in one sketch, and then apply the revolved boss feature.

8. First, make sure your measurement system is set to IPS. Then, we can create a rectangular base layer. After that, we can draw a circle and apply linear patterns to generate a total of 15 circles. Then, we can use extruded boss to extrude the pattern generated, as in the following diagram:

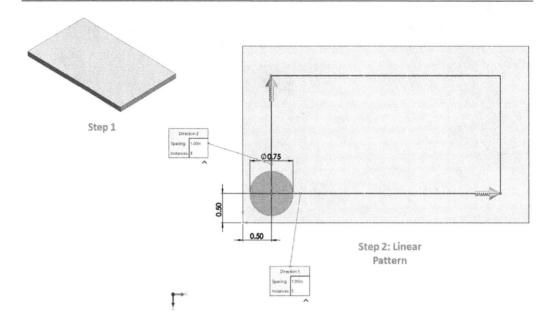

Step 1

Step 2: Linear
Pattern

Figure A.7 – A linear pattern can be used to create the repeated circles

The following screenshot highlights the PropertyManager for the linear pattern used in this exercise:

Figure A.8 – The PropertyManager for the linear pattern shown in Figure A.7

Chapter 6 – Basic Secondary Multi-Sketch Features

Note that when doing the practical questions and creating models, there is often no one right way of making a model. Hence, you might create the model in a different way than described here:

1. The eight common ways explored to define a plane are as follows:

    - Three points

    - One point and a line

    - Two parallel lines

    - Two intersecting lines

    - Other planes

    - A plane and a line

    - A plane and a point

    - A plane and a curve

2. Defining new planes makes it easier for us to apply features based on sketch planes that do not exist, which provides us with more flexibility in modeling.

3. Swept boss is a feature that allows us to add materials by sweeping a profile along a path. Swept cut allows us to remove materials in the same way.

4. Lofted boss is a feature that allows us to add materials by defining multiple cross-sections of the final shape we are modeling. Lofted cut allows us to remove materials in the same way.

5. In this exercise, we don't need to create a profile to create this using swept boss. We can start with the sketch shown here. We can also set the swept boss feature as shown in the following diagram:

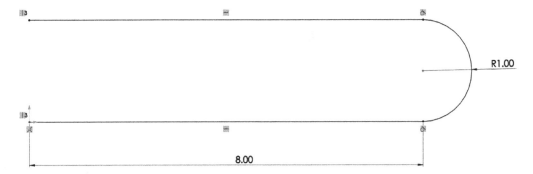

Figure A.9 – For a circular profile, just sketching the path is enough

After the sketch, we can set up the feature's PropertyManager as in the following screenshot:

Figure A.10 – Circular Profile can be selected directly from the swept boss PropertyManager

6. For this exercise, it is best to use lofted boss features to create the model. We can create the sections on three different planes with two guiding curves. The following diagram shows all three sections, the planes, and the guide curves:

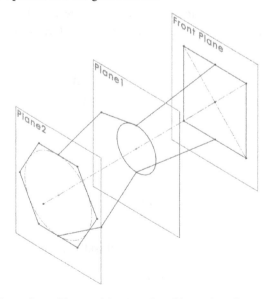

Figure A.11 – The combination of profiles and guide curves

The following diagram shows the sketches of the guide curves from the side view:

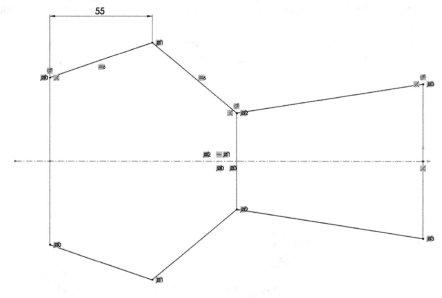

Figure A.12 – The side view of the guide curves

The last diagram shows the PropertyManager of the lofted boss with one guide curve selected:

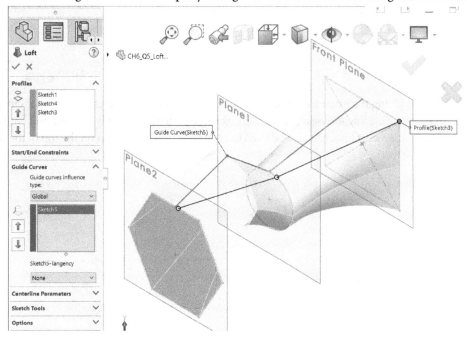

Figure A.13 – The PropertyManager with one guide curve selected

7. For this model, we will use reference geometries and the sweep cut feature. We can first start by creating the base rectangular prism, then define a new plane to create the cylinder. Lastly, we will create the cylindrical cut using the swept cut feature with a cylindrical profile. The following figure is the new reference plane used to build the model. The new plane can be used to generate the extruded cylinder with an extruded boss feature:

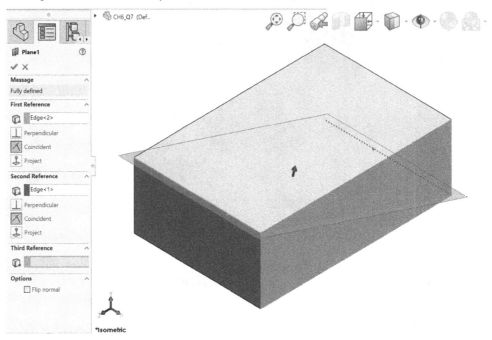

Figure A.14 – The new plane used to generate the tilted extruded boss

After applying the extruded boss, a swept cut can be used to generate the circular cut shown in the drawing. The following figure highlights the path used in applying the swept cut feature. Note that a swept boss can also be used to create the tilted extrusion on the top of the model.

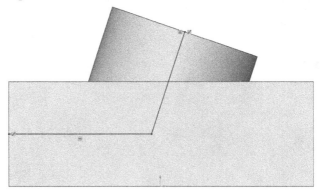

Figure A.15 – The path used to generate the cut going through the model

Chapter 7 – Materials and Mass Properties

1. A coordinate system helps us to orient ourselves in terms of directions and distances. This is by identifying your location in relation to a point of origin and the X, Y, and Z coordinates.

2. To define a new coordinate system, we need to define the origin point and the direction of the different X, Y, and Z axes. This can be done using vertices and edges or numerical values.

3. From mass properties, we can find the mass of the part, the volume, and the center of mass, as well as other information related to the design and the material of the part.

4. To create the model, we can use the swept feature, where our profile is a 40 x 40 square. The following diagram shows the path:

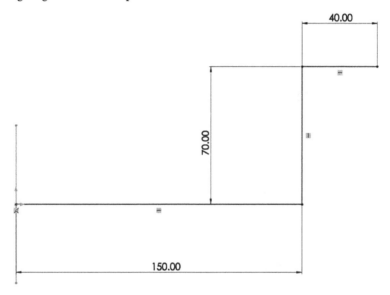

Figure A.16 – The path used with the swept boss feature to generate the model

After that, we can use the **Coordinate System | Reference Geometry** command.

5. For this exercise, we can first assign the material, **Aluminum Alloy: 1060 Alloy**, and then view the mass properties. The **Mass** value of the part is **1123.20 grams**. The **Center of mass** values in relation to **Coordinate System 1** (in millimeters) are **X = 20.00, Y = 69.81**, and **Z = -80.19**.

6. To create the model, we can use the feature loft boss by defining the two end hexagons, as shown in the following diagram. We can then apply the lofted boss feature to create the shape:

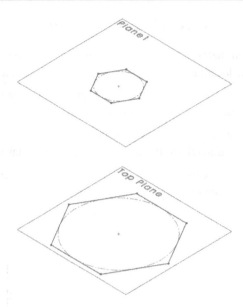

Figure A.17 – Two hexagons with different sizes can be used for the lofted boss feature

To create the new coordinate system, note that the *X* axis of the new coordinate system does not align with any edge in the model. To identify that axis, we can draw a construction line on the top surface using a sketch. After that, we can select that line as a reference in the reference coordinate system feature's PropertyManager. The following is a screenshot of the **Coordinate System** PropertyManager and the selection for the model. The origin can be the endpoint of the new construction line we created:

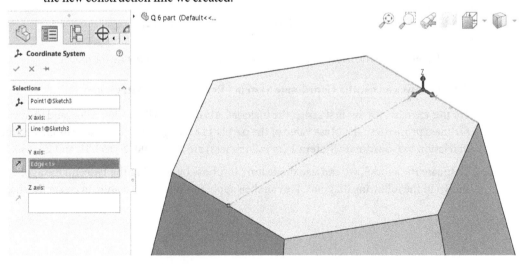

Figure A.18 – New construction lines and points can be used to aid the making of new coordinate systems

7. Note that the measurement system for the model is in MMGS. To find the mass in pounds and the center of mass in inches, we might need to change the units of measurement in the **Mass Properties** options. The mass of the part is 12.79 pounds. The **Center of mass** values in relation to **Coordinate System 1** (inches) are **X = 0.85**, **Y = 0**, and **Z = -3.75**.

Chapter 8 – Standard Assembly Mates

1. The SOLIDWORKS assembly environment allows us to combine more than one part to form an assembly. Most of the products we use in our everyday lives, such as phones, laptops, and pens, are made of more than one part put together.

2. Mates are what we use to link parts together. They allow us to set the type of interaction that would take place between different separate parts. Mates in SOLIDWORKS assemblies are very similar to relations in sketching. There are three types of mates: standard mates, advanced mates, and mechanical mates.

3. The standard mates are as follows:

   - Coincident

   - Parallel

   - Perpendicular

   - Tangent

   - Concentric

   - Lock

   - Distance

   - Angle

4. When making this assembly, we can insert the base part first to make it the fixed part. In terms of mates, we can utilize these mates: coincident, parallel, concentric.

5. Make sure to build on the assembly created in *Question 4* first to solve this question. Also, use coordinate reference geometries to generate the noted coordinate system in the drawing and assign the indicated material for each part as indicated in the table included. The numerical answers to the question are as follows:

   - **Mass**: 654.84 grams

   - **Center of mass (mm)**: X = 50.01, Y = 50.02, and Z = 19.46

6. In this assembly, we can insert the base box part first to make it a flexible base. **DETAIL A** in the drawing shows where the cover hinges on the box. Note that there are two separators so that, in the assembly, we can insert the same part twice to have more than one. An easy way to copy parts is to hold *Ctrl*, then click and drag a part; SOLIDWORKS will automatically create a copy of it. To put the assembly together, we can use the mates, coincident, distance, and angle. We can also use other mates and end up with the same result.

7. Make sure to build on the assembly created from the previous question to solve this. When evaluating mass properties, make sure to change the units for length and mass to inches and pounds. The numerical answers to the questions areas are as follows:

   - **Mass (pounds)**: 0.35 pounds
   - **Center of mass (inches)**: X = 1.18, Y = 4.13, Z = 1.42

Chapter 9 – Introduction to Engineering Drawing

1. Engineering drawings are used to communicate designs to other entities. When we produce a design for a specific product, we are often required to present an engineering drawing with it for communication.

2. This indicates a break in the object in the drawing. This is often used to fit relatively long objects within a drawing such as long construction beams. Note that there are many different types of break lines. This one is called a zigzag cut.

3. In comparison, dimension lines are thinner and have lighter coloring than visible object lines. Also, they serve different purposes. Object lines highlight the outer shape of the object while dimension lines highlight a specific dimension.

4. From the name, hidden lines highlight the hidden outline of the object from the drawing's viewpoint. This includes any details that are behind our first sight. They are drawn with equal dashed lines like the one shown here:

— —

Figure A.19 – Hidden lines as seen in standard engineering drawings

5. Detail views allow us to see and note small details that are otherwise hard to notice. We can think of them as a magnifying lens that we can use to zoom in on a particular part of the drawing.

6. Crop views are similar to a cropped image. In this, we can crop a specific part of the drawing and show it as a standalone view.

7. The screenshot on the right shows a break-out section view. This allows us to see what is behind a specific surface. Break-out section views allow us to see details that are hidden from view.

Chapter 10 – Basic SOLIDWORKS Drawing Layout and Annotations

1. We can open a new SOLIDWORKS drawing file by going to **New** and then selecting **Drawing**. After we open a new drawing file, the first thing we will have to decide is the drawing sheet format.

2. SOLIDWORKS provides five different display styles that we can use for our drawing views. Those are defined in SOLIDWORKS drawings as **Wireframe**, **Hidden Lines Visible**, **Hidden Lines Removed**, **Shaded With Edges**, and **Shaded**. Whenever we create a drawing, we have to decide which of the display styles will fit our application best.

3. There is no specific standard for scale to use when generating drawing views. This choice goes back to the team responsible for the drawing. However, a good practice is to take advantage of the drawing sheet space as much as possible.

4. The information block shows different information about the drawing, model, the team involved, and the company. That information includes the drawing scale, part material, weight, team members involved in making the model, a general statement from the mother company, and so on.

5. To create the following drawing, we follow the same steps explained in this chapter with the following notes:

    - The parent view is the **right view** instead of the **front view**. Hence, we will have to use the View Palette to add the right view first, and then project the back and top views from it.

    - The **left view** was inserted separately from the View Palette.

    - The scale 1:3 is not listed in the scale options. Hence, we will have to create and define the view by selecting the **User Defined** option under the scale.

    - Note the display is different for the different views as follows:

        • **Right, top, and left view**: Hidden-lines-removed display

        • **Back view**: Hidden-lines-visible display

        • **Isometric view**: Shaded-with-edges display

6. To generate the drawing, we can follow the same basic steps explained in this chapter. The drawing shown only uses the basic front and side views.

7. There is no correct way to create the model; you may use any of the skills we have considered so far to create the model. To double-check you have created the model correctly, change the material to **Alloy Steel** and check your mass is **436.73 grams**. If your mass is different, this indicates that your model is a bit different than the one we have created. One way to double-check your model is to create the drawing and double-check whether all of the dimensions match the drawing we have provided. The drawing is basic with orthographic and isometric views.

Chapter 11 – Bills of Materials

1. A **Bill Of Material** (**BOM**) is usually a table that includes specific information about the assembly or sub-assembly shown in the drawing.

2. The type of information listed in BOMs is open for the generator. The information listed often refers to the practice of the firm producing the bill, or else it follows a client's requirement. Common information found in a BOM would include names of parts, part numbers, the quantity of each part, costs, storage location, vendor, building material, and any other information related to the assembly at hand.

3. Linked information refers to information input within native SOLIDWORKS files. It also refers to assembly features such as the quantity of each part existing within an assembly. That linked information can be input into BOMs automatically. We can also input any information we desire into our bill manually. That information is unlinked.

4. To solve this question, you can follow the following points in order:

    • Open a new drawings file and input an isometric drawing view into the sheet.

    • Use the **Tables** command to insert a bill of materials.

5. To solve this question, you can follow the following points in order:

    • Highlight the **PART NUMBER** column from the default bill of materials, then select the column property.

    • Set the column type to **CUSTOM PROPERTY**, then select **PartNo** from the **Property name** drop-down menu.

    • Follow the same steps to change **DESCRIPTION** to **Vendor**.

    • Add a new column with the title Cost. Then, manually input the values shown in the provided table.

6. To solve the question, you can follow the following points in order

    • Add a new row at the bottom of the table.

    • In the last cell in the **Cost** column, use the **Total** equation function to sum all the costs.

7. To add balloon callouts, select the **Auto Balloon** command from the **Annotation** command category.

Chapter 12 – Advanced SOLIDWORKS Mechanical Core Features

1. Drafts are commonly applied to parts that are made with injection molding. A draft is a slight tilt between two different surfaces at different levels. The draft feature enables us to easily create this tilted surface. The **Shell** command enables us to make a shell of an existing model. It makes it easy to create objects such as cans and containers in a simple way. Ribs are often welded support structures that are added to help link different components/parts together. The rib feature enables us to create this support.

2. The Hole Wizard in SOLIDWORKS enables us to create holes in our model that match international standards for holes. This includes the drilling and threading of the holes as well. We usually use the Hole Wizard because it makes it easy and convenient to make those holes by selecting the hole standard and type and placing the hole directly on the part.

3. This model is simple to create when utilizing the draft and shell features. One way to create it is to boss extrude the overall cylinder, then apply an extruded cut at the bottom to form what you can use to apply the draft. For the draft, note that the inside surface is the natural surface. After that, you can apply the shell by setting it to have the top surface removed. Also, you can use the **Multi-thickness Settings** to have the lower surface of the cup thicker than the walls. To double-check your model, assign the **ABS plastic** material to the part, and then the mass should be about **39.78** grams. If the mass you get is more than 0.5% different, then you should double-check your model.

4. This model is simple to create when utilizing the rib feature and the Hole Wizard. The base model can be created with two extruded boss features, one to create the base and the other to create the pillar-shaped structure in the middle. After that, we can create the holes using the Hole Wizard, by selecting the specifications shown in the drawing. Note that we would need to apply the Hole Wizard command twice since we have two different types of holes. We can then end by creating the four ribs. Note that the four ribs are identical. Hence, we can create one and apply the circular pattern feature to create the other three. Also, we can use mirroring to mirror opposing ribs. Either method will save us time compared to creating the four ribs individually. To double-check your model, assign the **AISI 1020 Steel** material to the part, and then the mass should be about **1.32** pounds. If the mass you get is more than 0.5% different, then you should double-check your model.

5. This model is relatively complicated and includes a higher number of dimensions and features to create. One way to generate the model is to use the list of features highlighted in the following design tree:

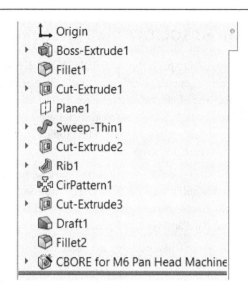

Figure A.20 – An example of a list of features used to generate the model

However, keep in mind that there is no one way to generate a model, and there is no specific order of isolated commands that we must use. Hence, take the design tree shown as a source of inspiration rather than following it to the letter. To double-check your model, assign the **Aluminum 1060** material to the part, and then the mass should be about **514.66** grams. If the mass you get is more than 0.5% different, then you should double-check your model.

6. Multi-body parts are models that were made within a SOLIDWORKS part file and that contain more than one separate body. Hence, they are called multi-body parts. Multi-body parts have advantages in working with certain applications that involve static interactions such as frames. This is due to the following advantages:

    - The frame and other static elements will all be contained in one file, making them easier to access and modify.

    - There's a faster work process as we will not need to use more than one SOLIDWORKS file. Also, we will not need to create mates to ensure the different parts fit together as is the case with assemblies.

7. This model can be created with a basic application of the extruded base feature. Each of the indicated parts can be created with one extruded boss. Toward the end, you can apply the extruded cut feature to create the slot shown to go through bodies 4 and 2. To save time, you can use the feature scope function to specify where the cut should go through.

    To double-check your model, assign the **Gray Cast Iron** material to the part; then the mass should be about **139.39** grams. If the mass you get is more than 0.5% different, then you should double-check your model.

8. Pattern B represents the best application of a circular pattern.

Chapter 13 – Equations, Configurations, and Design Tables

1. Equations in SOLIDWORKS are the same as equations in mathematics. In this, we get to define variables and introduce dimensions that are related to those values. SOLIDWORKS lists all of the variables and equations in the equations manager, which we can access to easily modify our variables and equations. Equations functions allow us to build a more connected 3D model that is easy to modify.

2. Configurations for a specific part are variations of the model of that part.

3. Design tables are tables that highlight the values and statuses of features. They enable us to create multiple configurations for a part by stating the variations on a *design table*, allowing us to create multiple new configurations at once.

4. To create this model, we can start by defining our **Global Variables** (**Y** and **X**) using the equations manager with the values **2.5** and **0.5** respectively. Then, model the part as normal by inputting the equations instead of numerical values. Make sure to start each equation with the = sign and have the variable around quotation marks. So, to enter the equation 2*X+0.5 as a dimension, we can enter [=2*"X"+0.5]. To double-check your model, assign the **AISI 304** material to the model and your mass should be **6.65** lbs. Note that the measurement system for this part is IPS.

5. We can use a manual configuration application to create this simple model. First, we can create the default configuration as normal. After that, we can add the new configurations one by one from the ConfigurationManager. For configuration C, select the **This Configuration** option when modifying the thickness from 10 mm to 5 mm.

6. We will use a design table to complete this exercise. We can start creating configuration **A** as the default configuration (you can rename the configuration from Default to A on the ConfigurationManager). To make the exercise easier, we can give names to the dimensions' width, length, and thickness as we are inputting the dimensions. Start a new auto-create design table and input the values shown on the table at the bottom of the drawing. Remember to double-check all of the configurations to ensure they are correct.

7. This exercise uses many elementary modeling techniques in addition to design tables and equations. To complete this exercise, first, make sure to set your measurement system to **MMGS**. Then, start by setting up the global variable and creating the default configuration. This is by applying the equations shown in the drawing as well. After creating the default configuration, you can assign the material to **AISI 1020 Steel** and check the mass is **3734.82** grams. If that is the case, then you have most likely got the initial model correct.

    We can start a new design table and input the table shown in the drawing. Note that we can also include equations and variables in the design table. However, when calling a variable, we have to quote the name with quotation marks. To double-check whether your configurations were set correctly, compare them with the masses given in the table listed in the second sheet of the drawing.

Chapter 14 – SOLIDWORKS Assemblies and Advanced Mates

1. The profile center advanced mate command helps us to center two surfaces in relation to each other in an assembly.

2. The mate symmetric allows us to set two different surfaces, edges, or points to have a symmetric dynamic relationship to each other around a symmetry plane. To apply the mate, we will need to have a symmetry plane as well as the mated elements such as surfaces, edges, and points.

3. The advanced path mate allows us to restrict the movement of a specific part to follow a designated path. We can restrain the movement with the path itself, in addition to the pitch/yaw control and roll control.

4. The advanced linear/linear coupler mate allows us to create a linear transitional relation between two different parts. In this, if one part moves, the other part will move as well according to a specific ratio. A common application for this mate is on drawers.

5. For this question, we can use the path mate and symmetric mates. When setting the path mate, notice that the sticks do not have any vertex for us to select. In this case, we can add a new point to act as the path's vertex. One way to do that is by editing the sketch within the part **Stick** and adding a new point, as shown in the following diagram:

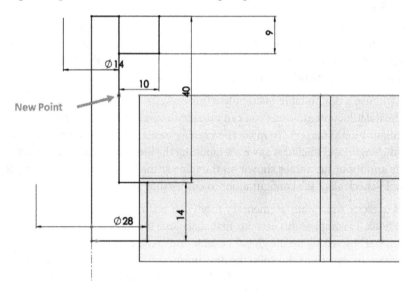

Figure A.21 – New geometries can be used to aid the applications of specific mates such as a path mate

A hint before placing the new point is to position the stick within the slot. We can use the mate width to create the position.

Once we have both parts in the assembly with the path mate set up, we can apply the mate symmetric to have both parts move symmetrically through the slot. First, we will need to introduce the symmetry plane. Setting the symmetric mate, we can choose the central vertical line in the stick sketch as an entity to mate. A common mistake will be to select one of the horizontal surfaces of the stick; this will not work, since the stick's movement is all horizontal as well.

A faster alternative to using two path mates and the symmetric mate is to use the **Mirror** command. After setting up one stick with the path, we can use the **Assembly** command **Mirror** to mirror the tick to the second half, which will copy all of the movement of the original stick.

6. In this exercise, we can use the linear/linear coupler and distance range mates to fulfill the movement requirements for the tubes. We can also use the concentric mate to ensure that all of the tubes are centered toward each other. To make the contracted position for tube assembly, we can use the standard coincident mate with the **Use for positioning only** setting.

Chapter 15 – Advanced SOLIDWORKS Assembly Competencies

1. **Interference Detection** identifies instances within the assembly where different parts are interfering with each other. This is as the assembly sits in a static position. **Collision Detection** identifies instances where two parts collide with each other as they are in a dynamic movement.

2. Assembly features are stored within the assembly file while features with parts are stored within the part file.

3. The implementation of new configurations and design tables are the same with parts and assemblies. However, the scope differs. Within assemblies, we can generate different assembly variations, while within parts, we can generate different parts variations.

4. This question is a direct application of the Interference Detection tool. However, when using the tool, you will notice there are many different interfering entities. Make sure to identify the required instances in the question. They should be as follows:

  - Between **Helical Gear** and **Fixture**: 0.22 in^3

  - Between **Worm Gear** and **Ball Bearing**: 0.02^3

5. Generating these configurations can be done via the direct application of design tables. The design table for this exercise would look as in the following screenshot:

1	Design Table for: gear Ass	$configuration@Helical Gear<1>	$configuration@Fixture<1>
2			
3	Normal-default	Normal	Default
4	Hole-Slot	Normal with Hole	With Slot
5	Thin-default	Thin with Hole	Default
6	Thin-Slot	Thin with Hole	With Slot

Figure A.22 – The design table used to generate the configurations in the assembly

It is important to have the syntax correct when calling out part configuration. Recall that the syntax should be in this format: `$configuration@partName<instance>`.

6. To generate this hole, we can use the extruded cut assembly feature. You can have the cut starting at the back of the drawer and going through everything to pass through the back of the outer shell. When applying the feature, make sure to check the **Propagate feature to parts** option in the feature's PropertyManager. One way to ensure that you have captured the design intent for the cut is to move the drawer to be fully open and fully closed. Then, rebuild the model and make sure the hole still goes through the back of the drawer, through the backside of the outer shell.

7. In this exercise, we can use manual configurations. We can start by adding the new configurations, fully closed and fully open. In both configurations, we can use the standard distance mate, mating the backside of the drawer support with the inner side of the outer shell.

For the fully closed configuration, the distance can be set to 0 mm. For the fully open configuration, the distance can be set to 30 mm.

After that, double-check that all of the configurations are as required with the default configuration still being flexible in opening and closing.

Chapter 16 – Introduction to SOLIDWORKS Cloud Services

1. The key features include enhanced collaboration, efficient design sharing, secure storage, and effective design management tools.

2. SOLIDWORKS Cloud Services specifically refers to the services included with a standard SOLIDWORKS license, while the 3DEXPERIENCE platform offers a wider range of cloud services.

3. Applications such as 3DPlay, CATIA, SIMULIA, and DELMIA provide comprehensive and integrated product development experiences through visualizations, design reviews, and other functionalities.

4. 3DDrive provides users with cloud storage for managing design data and facilitates seamless collaboration by allowing easy file sharing among team members.

5. The options include activating external sharing links, enabling guest comments, and restricting access to specific users.

6. Reviewers can access shared designs by creating a 3DEXPERIENCE account and following the provided link or the instructions in the email notification.

7. Locking files ensures that modifications are saved correctly to the platform and prevents conflicts with concurrent edits by other users.

8. Changing the maturity level indicates the status of the work, such as whether it is in progress, finalized, awaiting approval, or completed, providing clarity on the design's stage to team members.

Index

www.packtpub.com

Subscribe to our online digital library for full access to over 7,000 books and videos, as well as industry leading tools to help you plan your personal development and advance your career. For more information, please visit our website.

Why subscribe?

- Spend less time learning and more time coding with practical eBooks and Videos from over 4,000 industry professionals

- Improve your learning with Skill Plans built especially for you

- Get a free eBook or video every month

- Fully searchable for easy access to vital information

- Copy and paste, print, and bookmark content

Did you know that Packt offers eBook versions of every book published, with PDF and ePub files available? You can upgrade to the eBook version at packtpub.com and as a print book customer, you are entitled to a discount on the eBook copy. Get in touch with us at customercare@packtpub.com for more details.

At www.packtpub.com, you can also read a collection of free technical articles, sign up for a range of free newsletters, and receive exclusive discounts and offers on Packt books and eBooks.

Other Books You May Enjoy

If you enjoyed this book, you may be interested in these other books by Packt:

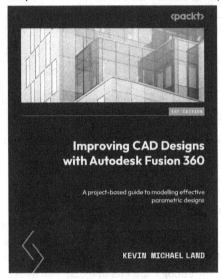

Improving CAD Designs with Autodesk Fusion 360

Kevin Michael Land

ISBN: 978-1-80056-449-7

- Gain proficiency in Fusion 360 user interface, navigation, and functionality
- Create and transform simple 2D sketches into 3D models
- Manipulate and control parametric 2D sketches using dimensions
- Become familiar with drafting on paper and taking measurements with calipers
- Create a bicycle assembly part with Fusion 360
- Use the form environment to create organic shapes
- Render a 3D model and understand how to apply materials and lighting
- Generate 2D assembly model drawings for documentation purposes

AutoCAD 2025 Best Practices, Tips, and Techniques

Jeanne Aarhus

ISBN: 978-1-83763-672-3

- Recognize hidden features in commonly used commands
- Explore innovative methods to make AutoCAD streamline your work for you
- Complete AutoCAD tasks in fewer steps using dynamic blocks and customizations
- Gain insider tips from AutoCAD veterans and gurus, saving you years of trial and error
- Use underutilized features of AutoCAD such as tables for data import and export
- Control how AutoCAD responds to your daily workflow and real-world environment

Packt is searching for authors like you

If you're interested in becoming an author for Packt, please visit `authors.packtpub.com` and apply today. We have worked with thousands of developers and tech professionals, just like you, to help them share their insight with the global tech community. You can make a general application, apply for a specific hot topic that we are recruiting an author for, or submit your own idea.

Share Your Thoughts

Now you've finished *Learn SOLIDWORKS 2025*, we'd love to hear your thoughts! Scan the QR code below to go straight to the Amazon review page for this book and share your feedback or leave a review on the site that you purchased it from.

`https://packt.link/r/1-835-46308-8`

Your review is important to us and the tech community and will help us make sure we're delivering excellent quality content.

www.ingramcontent.com/pod-product-compliance
Lightning Source LLC
Chambersburg PA
CBHW060634060326

40690CB00020B/4395